Plasma technologies for textile and apparel

Plasma technologies for textile and apparel

Edited by

S. K. Nema

and

P. B. Jhala

WOODHEAD PUBLISHING INDIA PVT LTD

New Delhi

Published by Woodhead Publishing India Pvt. Ltd.
Woodhead Publishing India Pvt. Ltd.,
303, Vardaan House, 7/28, Ansari Road,
Daryaganj, New Delhi - 110002, India
www.woodheadpublishingindia.com

First published 2015, Woodhead Publishing India Pvt. Ltd.
© Woodhead Publishing India Pvt. Ltd., 2015
Reprint 2017, 2018, 2019

Woodhead Publishing India Pvt. Ltd. ISBN: 978-93-80308-55-5
Woodhead Publishing India Pvt. Ltd. e-ISBN: 978-93-80308-96-8

Typeset by Mind Box Solutions, New Delhi
Digitally Printed and bound by Replika Press Pvt. Ltd.

Contents

Preface

Plasma, the fourth state of matter, is a hot ionized gas consisting of approximately equal numbers of positively charged ions and negatively charged electrons. Generally non-thermal or cold plasma is used for surface modification or sterilization of soft materials such as polymers, textiles, human skin, etc. The plasma consists of free electrons, radicals, ions, UV-radiation and different excited particles. These reactive species interact with material and modify both physical and chemical properties mainly at the surface typically up to 100 nm depth. Cold plasma interaction does not change the bulk properties of these materials. The cold plasma can be produced at low pressure as well as at atmospheric pressure and using low frequency, radio frequency and microwave power sources. Although, low pressure plasma (in vacuum) can be produced in a large volume easily using less than 600 V power source, however textile industry demands in-line treatment for large production. Therefore, plasmas at atmospheric pressure are very useful for textile treatment and surface modification. Atmospheric pressure plasmas such as corona discharge, dielectric barrier discharge, atmospheric pressure glow discharge and atmospheric pressure plasma jet are being used for in-line continuous treatment of textile materials and providing desired surface properties. As plasma is a green technology, it does not require large quantity of chemicals, water and energy for textile treatment. Therefore cold plasma technology is widely used for surface finishing applications to generate specific surface properties such as hydrophilicity or hydrophobicity, scouring of cotton; assists in removal of sizing agents, anti-bacterial properties by depositing nano-silver particles, shrink proofing; to improve dye-uptake properties, sterilize medical textiles, etc. In the world, few of the companies have developed plasma systems for in-line and batch processing of textiles, and these companies offer custom designed plasma machines. These companies include Dow Corning Plasma Systems (Ireland), Europlasma (Belgium), P2i (UK), AcXys (France), InspirOn Engineering Pvt. Ltd. (India), etc.

The main objective of this book is to popularize plasma-based technologies in textile industries and dissemination of knowledge gained over the years by

Indian institutes and organizations in the arena of plasma-based applications for textiles. The book describes basics of low temperature plasma production in vacuum as well as at atmospheric pressure and various applications of plasma in textile particularly in Indian context. It covers plasma diagnostics which includes electron temperature and electron energy distribution function using Langmuir probe. The book has 12 different chapters which covers the work carried out in India on plasma surface modification of textiles in different laboratories and research organizations in India over a period of time. This also includes the commercialization of Angora plasma technology, economic sustainability of plasma technologies in textile world, demand for plasma-based environment-friendly technologies from Indian industries and possible funding opportunities.

In Chapter 1, Ramkrishna Rane et al. describe the fundamentals of the plasma state and how it becomes a versatile industrial tool and how plasma interaction with textile surface actually modify the surface and how some of the plasma technologies are useful for coating applications on textiles. The fundamentals of plasma cover principle of gas breakdown, plasma discharges and inclusion of floating and biased electrodes in plasma and their effects in plasma. Authors described plasma interactions with textiles for surface modification and other plasma-based technologies that are used in textile sector.

In Chapter 2, S.K. Nema and P.B. Jhala present description of non-equilibrium (cold) plasmas and how they can be produced at atmospheric pressure. Further, they have discussed four types of non-thermal plasmas which work at atmospheric pressure and commonly used for textile surface modification. The second chapter discusses the reactions that take place when active species of cold, non-equilibrium plasma interact with the textile surface. These reactions tailor surface of textiles and provide desired properties in environment-friendly manner. Further, in this chapter they have covered different plasma and process parameters which govern gas phase and surface chemistry. This is followed by a review of the atmospheric pressure cold plasma technologies developed at FCIPT for some textile applications and the future growth direction of textile applications in India.

In Chapter 3, Shital S. Palaskar and A. N. Desai have written on plasma polymerization of HMDSO on different textile fibres. Plasma polymerization is the process of polymer synthesis in plasma environment. Plasma polymerization forms entirely new kind of material, which has unique properties such as excellent adhesion to substrate and strong resistance to most of chemicals. In this chapter, the studies on effect of plasma polymerization of hexamethyldisiloxane on natural and synthetic fibres are discussed.

Chapter 4, written by Ashwini K. Agrawal et al., is based on the theme of hydrophobic functionalization of textile substrates. Atmospheric pressure plasma has been used for surface modification of textiles by in-situ reaction of precursors. Effective surface functionalization can make the textile stain repellent/self-cleaning without affecting its bulk properties. Attempts are made to understand the mechanism of hydrophobic functionalization involving different non-fluoro precursor molecules along with helium. The influence of different plasma process parameters on the fragmentation of precursors and their reaction with substrate has also been discussed. The coated surface was characterized using X-ray photoelectron spectroscopy, FT-IR spectrometer and measuring contact angles.

In Chapter 5, Kartick K. Samanta et al. have discussed hydrophilic functionalization of polymeric/textile substrates. Cellulosic textile (cotton), ligno-cellulosic textile, protein fibres, and synthetic textile have been modified using either inert gaseous plasma or reactive plasma using various reactive precursors. Improvement in hydrophilic property is due to removal of impurities like, wax, fatty acid layer, residual size material, etc., formation of polar groups and increase in surface roughness/area. It has been found that formation of nano-sized channels may also help in improving wicking tendency of the substrate to both polar and non-polar liquids. The chapter also discusses the role of plasma parameters such as the discharge voltage, frequency, treatment time, and gas flow rate on fragmentation of precursors and functionalization of substrates.

In Chapter 6, N.V. Bhat and R.R. Deshmukh have written on processing of textiles to enhance their dyeing properties. The chapter reviews how the eco-friendliness of plasma technology scores over the existing wet processing used at present. Plasma processing can be used for surface modification, etching, functionalization, plasma coating (grafting) and desizing. The modified surfaces can be made either hydrophilic or hydrophobic depending on the conditions in the plasma reactor. It has been found that the dye uptake can be enhanced when fabrics are pre-treated with plasma on account of formation of hydroxyl, carboxyl and carbonyl groups on the surface.

Chapter 7 written by M.K. Bardhan and Jayant Udakhe is devoted to wool processing. Finer wool clothing is used for warmth, and this can be attributed to the fibre structure. However, woollen garments cannot be worn next to the skin due to the pricking and itching sensation caused by the wool fibres. To mitigate this problem, most of commercial treatments in practice use chemicals and are hazardous to the environment. New eco-friendly processes attempted include use of plasma, enzyme, UV/ozone and biopolymers alone or in combinations.

In Chapter 8, Shilin Sangappa and Arindam Basu have discussed on the composition of raw silk fibre. The silk fibre is surrounded by gum-like protein Sericin, which needs to be removed. Cold plasma treatment is a promising technology for partial degumming of silk without damaging the delicate fibroin fiber. In this chapter, low pressure (RF) plasma treatment of Mulberry (domesticated) and non-Mulberry (Vanya) degummed silk yarn has been presented. Effect of plasma treatment on silk surface morphology, physical properties, wicking properties, and shade depth has been discussed. It shows that there is a significant improvement in wicking and shade depth. However, there is a need to fine tune the plasma process parameters for different varieties of silk for industrial applications.

In Chapter 9, Dolly Gogoi et al. have discussed on RF plasma treatment to improve physical properties of Muga silk fibers without affecting their bulk properties. The chapter summarizes results of investigations on the effect of plasma parameters on surface chemistry and morphology, physical properties and thermal behavior of the plasma-treated fibers. The chapter further discusses the possibility of using plasma-treated Muga silk as resourceful decorative and clothing material.

In Chapter 10, G. Thilagavati and S. Periyasamy have discussed the principle of cold plasma treatment of textiles which is followed by treatments specific to various natural and synthetic textile materials for novel surface functionalities and improvements in surface-related applications. Advantages of cold plasma treatments have also been deliberated.

In Chapter 11, C. Balasubramanian has written on nano-titania application for imparting self-cleaning effect to cotton and silk fabrics. Various methods of generation of nanoparticles including plasma route are described. An extensive survey on efforts of various research groups for self-cleaning fabrics has been covered. A detailed report on the author's own work on nanotitania synthesis and its application to silk and cotton fabrics and the various tests and results obtained therein are provided. Finally a brief review of the study of potential hazards the nanoparticles of Titania can pose to humans is provided.

In Chapter 12, which is the last chapter, P.B. Jhala and S.K. Nema have made an attempt for a broad look at plasma textile technology with focus on status, techno-economics, limitations and potential. Though initiated in the 1960s, wide applicability limited by the low pressure requirement was overcome with the development of atmospheric pressure plasma sources. This has opened up a host of opportunities for the production of innovative finishes, branding and marketing. The chapter discusses the technology types, advantages and applications in textiles giving the current status in the world and in India. In the world, efforts have been made under various European

Union Framework Programs to develop innovative plasma solutions for various textile applications keeping in view the techno-economic aspects. In India too, FCIPT has commercialized Angora wool treatment technology for spinning of 100% Angora fiber in the cottage industry. The technology has also been successfully developed for technical textile in order to enhance the adhesion strength of coating on polyester substrate.

For Indian textile industry, plasma technology is an emerging high technology with wide application potential in textile finishing and nonwovens to keep an eye on newer developments in addressing specific quality as well as environmental related issues. Development of plasma technology for Angora wool and its successful implementation in the cottage industry has built the confidence in main stream industry to adopt the technology. Department of Science and Technology, Govt. of India, has been generous in supporting basic and applied research work in plasmas at various laboratories across the country. This book would be of interest to textile academicians and researchers in institutes as well as textile technologists and engineers in the industry.

P.B. Jhala
S.K. Nema

Foreword

The art, craft and the technology of textiles have witnessed dramatic changes over the thousands of years of its existence and evolution. Yet another fundamental transformation is happening to the textile industry with the intervention of cold plasma as a textile-manufacturing tool. This is a field where applications run far ahead of our understanding of their scientific basis.

That India, with its rich history of textile production and special weaves, should resonate to the winds of technological change is only natural. The post-independence revival of the Indian textile industry after its destruction during British colonization is already legendary. The fact that there are attempts to inject new vigour is symbolic of the dramatic growth of science and technology in India and the success of policies that promote the development and deployment of advanced science-based application in traditional industries.

It is appropriate that the very productive integration of plasma and textile technology which is happening in India is being celebrated with the publication of this book "Plasma Technologies for Textile Applications: Growth Potential in India", edited by Dr. S.K. Nema and Prof. P.B. Jhala, both pioneers in this field. The book covers a wide range of subjects relevant to the title. There are twelve chapters ranging from plasma sources to functionalization.

The first chapter describes the fundamentals of the plasma state and how it becomes a versatile industrial tool and how plasma interaction with textile surface actually modify the surface and how some of the plasma technologies are useful for coating applications on textiles. The second chapter discusses the reactions that take place when active species of cold, non-equilibrium plasma interact with the textile surface. This is followed by a review of the atmospheric pressure cold plasma technologies developed at FCIPT for some textile applications and the future growth direction of textile applications in India.

Plasma Polymerization of HMDSO on different textile fibres is the subject of another chapter. Plasma polymerization is the process of polymer synthesis in plasma environment. In this chapter, the studies on effect of plasma polymerization of hexamethyldisiloxane on natural and synthetic fibres are discussed. Hydrophobic functionalization of textile substrates is the theme of the next chapter. To understand the mechanism of hydrophobic functionalization involving different non-fluoro precursor molecules, the coated surface was characterized using X-ray photoelectron spectroscopy, FT-IR spectrometer and measuring contact angles.

Hydrophilic functionalization of polymeric/textile substrates is described in the next chapter. Cellulosic textile (cotton), ligno-cellulosic textile, protein fibres, and synthetic textile have been modified using either inert gaseous plasma or reactive plasma using various reactive precursors. Improvement in hydrophilic property is due to removal of impurities like, wax, fatty acid layer, residual size material etc., formation of polar groups and increase in surface roughness/area. It has been found that formation of nano-sized channels may also help in improving wicking tendency of the substrate to both polar and non-polar liquids.

The chapter on processing of textiles to enhance their dyeing properties reviews how the eco-friendliness of plasma technology scores over the existing wet processing used at present. Plasma processing can be used for surface modification, etching, functionalization, plasma coating (grafting) and desizing. The modified surfaces can be made either hydrophilic or hydrophobic depending on the conditions in the plasma reactor. It has been found that the dye uptake can be enhanced when fabrics are pre-treated with plasma on account of formation of hydroxyl, carboxyl and carbonyl groups on the surface.

The next chapter is devoted to wool processing. Finer wool clothing is used for warmth and this can be attributed to the fibre structure. However, woollen garments cannot be worn next to the skin due to discomfort caused by the wool fibres. Most of commercial treatments in practice use chemicals and are hazardous to the environment. New methods include use of plasma, enzyme, UV/ozone and biopolymers alone or in combinations.

Raw silk fibre is composed of a fibrous protein surrounded by gum-like protein Sericin, which needs to be removed. Cold plasma treatment is a promising technology for partial degumming of silk without damaging the delicate fibroin fiber. In this chapter, low pressure (RF) plasma treatment of Mulberry and non- Mulberry (Vanya) degummed silk yarn has been presented. Effect of plasma treatment on silk surface morphology, physical properties, wicking properties, and shade depth has been discussed. It shows that there is a significant improvement in wicking and shade depth. However there is a need to fine tune the plasma process parameters for different varieties of silk for industrial applications.

In the next chapter, RF plasma treatment is discussed, which improve physical properties of muga silk fibers without affecting their bulk properties.

The chapter summarizes results of investigations on the effect of plasma parameters on surface chemistry and morphology, physical properties and thermal behavior of the plasma-treated fibers. In another chapter, discussion of the principle of cold plasma treatment of textiles is followed by treatments specific to various natural and synthetic textile materials.

Nano-titania applicationfor imparting self-cleaning effect to cotton silk fabrics is the theme of another chapter. Various methods of generation of nanoparticles including plasma route are described. An extensive survey on the efforts of various research groups for self-cleaning fabrics has been included. A detailed report on the author's own work on nanotitania synthesis and its application to silk and cotton fabrics and the various tests and results obtained therein are provided. Finally a brief review of the study of potential hazards that the nanoparticles of titania can pose to humans is provided.

The last chapter attempts a broad look at Plasma Textile Technology with focus on status, techno-economics, limitations and potential. Though initiated in the 1960s, wide applicability limited by the low pressure requirement was overcome with the development of atmospheric pressure plasma sources. It has opened up a host of opportunities for the production of innovative finishes, branding and marketing. The chapter discusses the technology types, advantages and applications in textiles giving the current status in the world and in India. In the world efforts have been made under various European Union Framework Programs to develop innovative plasma solutions for various textile applications keeping in view the techno-economic aspects. In India too, FCIPT has commercialized Angora wool treatment technology for spinning of 100% Angora fiber in the cottage industry. The technology has also been developed for technical textile in order to enhance the adhesion strength of coating on polyester substrate.

Overall, this is a topical and informative book summarizing the status of this field and the work done in many Indian laboratories. I congratulate the authors for their efforts in making clear presentations of a technically complex field.

P. I. John
Consultant
Institute for Plasma Research
Gandhinagar – 382424

Contributors

Dr. A. N. Desai
The Bombay Textile Research
Association
Mumbai

Dr. Arindam Basu
Central Silk Technological Research
Institute
Bangalore

Dr. Arup Jyoti Choudhury
Department of Physics
Tezpur University
Assam

Dr. Arup Ratan Pal
Institute of Advanced study in
Science and Technology
Guwahati

Prof. Ashwini K. Agrawal
Department of Textile Technology
Indian Institute of Technology (IIT)
New Delhi

C. Balasubramanian
Facilitation Centre for Industrial
Plasma Technologies Division
Institute for Plasma Research
Gandhinagar

Dolly Gogoi
Institute of Advanced Study in
Science and Technology
Guwahati

Dr. G. Thilagavathi
Department of Textile Technology
PSG College of Technology
Coimbatore

J. S. Udakhe
Wool Research Association
Thane

Joyanti Chutia
Institute of Advanced study in
Science and Technology
Guwahati

Dr. Kartick K. Samanta
Central Institute for Research on
Cotton Technology
Mumbai

Dr. M. K. Bardhan
Wool Research Association
Thane

Prof. Manjeet Jassal
Department of Textile Technology
Indian Institute of Technology (IIT)
New Delhi

Dr. Mukesh Ranjan
Facilitation Centre for Industrial
Plasma Technologies (FCIPT)
Institute for Plasma Research
Gandhinagar

N. V. Bhat
Bombay Textile Research
Association
Mumbai

Prof. P. B. Jhala
Facilitation Centre for Industrial
Plasma Technologies (FCIPT)
Institute for Plasma Research
Gandhinagar

Prashanta Panda
Department of Textile Technology
Indian Institute of Technology (IIT)
New Delhi

R. R. Deshmukh
Institute of Chemical Technology
Mumbai

Ramakrishna Rane
Facilitation Centre for Industrial
Plasma Technologies (FCIPT)
Institute for Plasma Research
Gandhinagar

Dr. S. K. Nema
Facilitation Centre for Industrial
Plasma Technologies (FCIPT)
Institute for Plasma Research
Gandhinagar

Dr. S. Periyasamy
Department of Textile Technology
PSG College of Technology
Coimbatore

Shillin Sangappa
Central Silk Technological Research
Institute
Bangalore

Shital S. Palaskar
The Bombay Textile Research
Association
Mumbai

Dr. Subroto Mukherjee
Facilitation Centre for Industrial
Plasma Technologies (FCIPT)
Institute for Plasma Research
Gandhinagar

Basics of plasma and its industrial applications in textiles

Ramkrishna Rane, Mukesh Ranjan, and Subroto Mukherjee

Abstract: Plasma, which is considered to be the fourth state of matter, is a hot ionized gas consisting of approximately equal numbers of positively charged ions and negatively charged electrons. Based on the relative temperatures of the electrons, ions and neutrals, plasmas are classified as "thermal" or "non-thermal". It is the non-thermal plasma that is usually used for surface modification since it consists of free electrons, radicals, ions, UV-radiation and a lot of different excited particles. These reactive species are responsible for modifying both physical and chemical properties of the material. During plasma processing of textiles in particular, only minimal thermal degradation takes place as the species readily interacts with the surfaces, causing reactions that would otherwise occur only at elevated temperatures. Surface modification processes such as sputtering or etching of semiconductor, insulator or metal surfaces using plasma have been well established. However use of plasma in the field of textile is rather new.

In this chapter basic inherent properties of plasma have been introduced, which include principle of gas breakdown, plasma discharges and inclusion of floating and biased electrodes in plasma and their effects in plasma. In the later part, plasma interactions with textiles for surface modification and other plasma-based technologies used in textile sector are described.

1.1   Introduction

The fourth state of matter, popularly known as plasma, has very high energy content per unit volume compared to other states of matter. When solid is heated, it melts to become a liquid; and if more heat is supplied it becomes a gas (Fig. 1.1). On further heating, the molecules/atoms of the gas get ionized and start conducting electricity. Thus plasma is an ionized gas consisting of approximately equal numbers of positively charged ions and negatively charged electrons. It also contains neutrals that are not ionized. The charged particles are influenced by electric and magnetic fields, and hence these fields can be used to guide these charged particles to a material surface for modification (Roth, 1995). Classically, plasma is defined as "a quasi-neutral collection of charged and neutral particles which exhibits collective behavior" (Chen, 1983). Quasi-neutrality essentially indicates that despite the existence of localized charge concentrations, on average there are approximately equal numbers of positively charged ions and negatively charged electrons distributed in such a way so that their charges cancel. So for practical purposes, plasma can be considered as electrically neutral. The interaction between charged particles is governed by Coulomb's law, where the force of interaction between charged particles is long ranged. Thus the movement of charged particles is governed by fields originating at larger distances. Thus the plasma as a whole responds to fields exhibiting collective behavior (Chen, 1983). The individual behavior of the particle is not there and the entire plasma responds and adjusts itself to the fields. It can be understood as the difference between a person's individual response to a situation versus the response of a mob, where the mob can respond in a way which is different than the individual's response.

States of matter

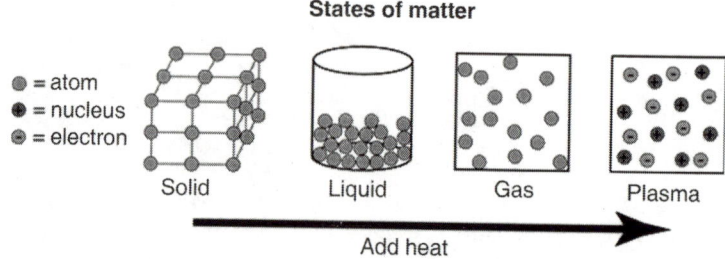

Figure 1.1 Representation of various states of matter

It is understood from investigations that 99% of the matter in the observable universe is in the plasma state. Sun and stars, interplanetary medium, the magnetospheres and ionospheres of the earth and other planets, as well as the ionospheres of comets lightning, etc., are examples of natural plasmas. Even in space around the earth, the plasma has densities dramatically lower than those achieved in laboratory plasma sources. In the laboratories, the plasma density is about 10 billion particles per cubic centimeter, whereas the density of the densest magneto-spheric plasma region is about 1000 particles per cubic centimeter.

The temperatures of plasmas are very high, ranging from several tens of thousands Kelvin. As the temperature is very high, it is convenient to express it in terms of the average kinetic energies of their constituent particles measured in "electron volts." An electron volt (eV) is the energy that an electron acquires as it is accelerated through a potential difference of 1 volt and is equivalent to 11,600 Kelvin. As plasma comprises of charged particles (ions and electrons), which are caused by ionization of the background, hence to make plasma the energy given to the gas must be higher than the ionization potential. For hydrogen, the ionization potential is 13.6 eV, i.e. the energy needed to eject one electron from the hydrogen atom. It is a big number if it has to be expressed in degree Kelvin.

The ions, electrons and neutrals may have different temperatures. Normally in plasmas, if the electron temperature is much higher than the ion and neutral temperature, the plasma is called as "non-thermal", as the plasma cannot be defined with one temperature. Usually these plasmas have low heat content/volume and do not heat the material surface with which it is interacting, to temperatures where it can modify the bulk properties of the material. This type of plasma is usually used for surface modification. Here the electrons can be at temperatures of few electron volts with ions being at temperature of 0.1 eV. There are plasmas where the electron, ion and neutral are at similar temperature. This plasma is called as "thermal" and has high heat content/volume. This type of plasma is used as heat source and can be used for bulk material disintegration. The temperature of the plasma species can be of the order of 1.0 eV. Figure 1.2 shows the entire plasma parameter space. Out of these the glow discharges and thermal processing plasmas are used for industrial plasma applications.

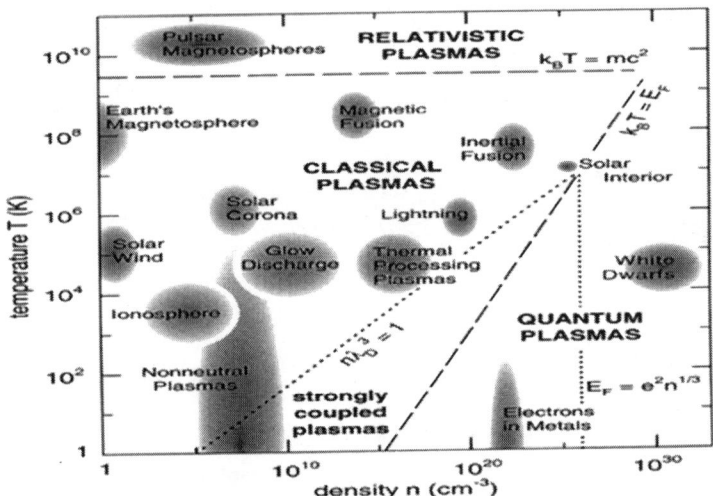

Figure 1.2 Plasma parameter space showing plasma with different density and temperature

1.2   Electric breakdown of gases

The electric breakdown of gases is a process of formation of conductive channel which occurs when sufficiently strong electric field is applied to it

(Raizer, 1991). As a result of the breakdown, different kinds of plasmas are generated. Mostly all the plasma starts with the electron avalanche. Electron avalanche is multiplication of some primary electrons in cascade ionisation. Let us consider at first the simplest breakdown in a plane gap of length 'd' between electrodes connected to a DC power supply (with voltage V), which provides the homogeneous electric field E = V/d. Background primary electrons due to cosmic ray, etc., can provide the very low initial current. Each primary electron drifts to anode ionising the gas (produce secondary electrons) and thus generates an avalanche. The avalanche develops both in time and in space, because the multiplication of electrons proceeds along with their drift from cathode to anode (Fig. 1.3). Ionisation is described by the Townsend ionisation coefficient 'α' showing electron production per unit length or the multiplication of electrons (Raizer, 1991)

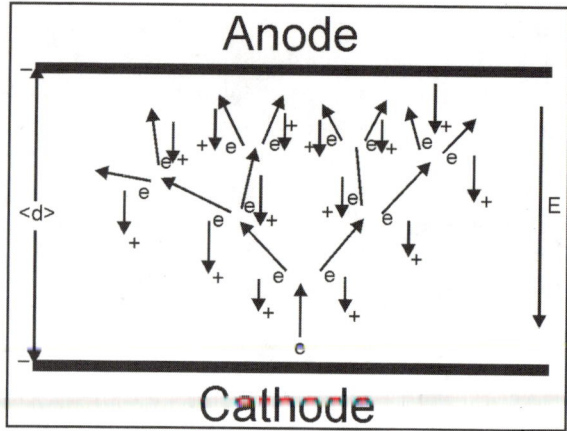

Figure 1.3 Electron Avalanche from Cathode to Anode

$dn_e/dx = n_e$ or $n_e(x) = n_{e0}\exp(\alpha x)$.

Where, n_{e0} is initial density per unit length along the electric field.

According to the definition of the Townsend coefficient α, each one primary electron generated near cathode produces $[\exp(\alpha d) - 1]$ positive ions in the gap. All the $[\exp(\alpha d) - 1]$ positive ions produced in the gap per one electron are moving back to the cathode and altogether knock out $[\gamma(\exp(\alpha d) - 1)]$ electrons from the cathode in the process of secondary electron emission. Where 'γ' is the secondary emission coefficient defined as probability of a secondary electron generation on the cathode by an ion impact. The secondary electron emission coefficient depends on cathode material, state of surface, type of gas and reduced electric field E/n_0 (defining the ions energy). The typical value of 'γ' in electric discharges is 0.01–0.1; affect of photons and meta-stable atoms and molecules (produced in avalanche) on the secondary electron emission is usually incorporated in the same. The current in the gap is non-self-sustained as long as $[(\exp(\alpha d) - 1)]$ is less than one, because positive ions generated by electron avalanche must produce at least one electron to start new avalanche. When electric field and Townsend coefficient become high enough, transition to self-sustained discharge or breakdown takes place.

1.3   Paschen's law for gas breakdown

Plasma formation happens by ionisation of the neutral gas. One of widely used ways of forming plasma is by accelerating electrons by an electric field and making the accelerated electrons collide with the neutral and ionising them. Electric field is produced by applying a potential difference between two electrodes separated by a distance. The law that relates the breakdown voltage (V) to the gas pressure (P) and separation distance (d) is described by Paschen's law (Raizer, 1991).

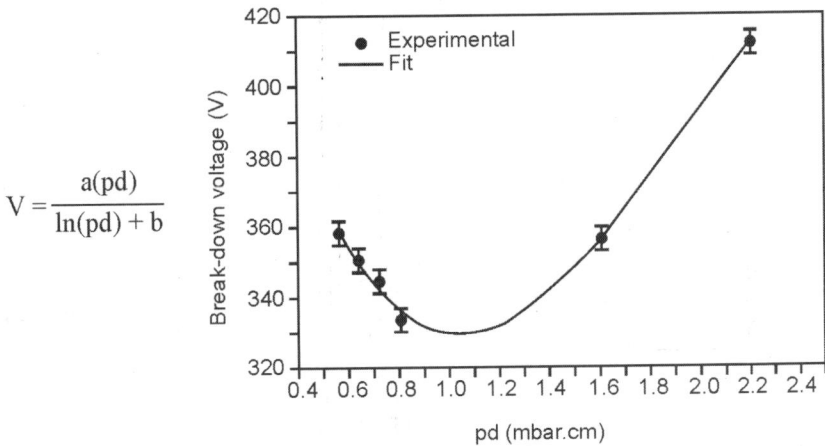

$$V = \frac{a(pd)}{\ln(pd) + b}$$

Figure 1.4 Equation of Paschen's law and graph V = break-down voltage (in volts), p = gas pressure, d = Electrode separation distance, a & b = constants, depend on gas composition (for air a = 43.6 × 106 V/atm-m and b = 12.8) and the graph of the above equation is called Paschen curve

Paschen's law states that the breakdown characteristic of a gap is function (generally non-linear) of the product of the gas pressure (P) and the gap length (d). Initially the breakdown voltage decreases as the Pd increases, and then increases gradually once again. The voltage at which the breakdown occurs is described by Paschen's law. The breakdown voltage increases at small values of Pd because there are insufficient neutral atoms to provide ionising collisions over the separation 'd'. Above the Paschen's minimum, the breakdown voltage raises because the electron collisions become frequent as the parameter Pd increases removing energy from the electrons. Hence stronger electric fields are required as one goes to higher values of Pd. As shown in Fig. 1.4 to the left of the Paschen's minimum, breakdown voltage increases with decreasing Pd. Here the gas is not very dense or the electrodes are very close; thus even if a large number of secondary electrons are emitted, there is a low probability that any electron will collide with neutral atoms during the journey from cathode to anode. As Pd decreases, collisions are less likely, and therefore the breakdown voltage is higher, thus the Paschen curve has a negative slope in this region.

1.4   Discharge current-voltage characteristics

Current-voltage characteristic of dc low pressure electrical discharge is shown in Fig. 1.5. If one takes low pressure discharge tube and raises voltage V, while measuring the current I flowing through the discharge, the discharge will follow highly non-linear voltage-current curve shown in Fig. 1.5.

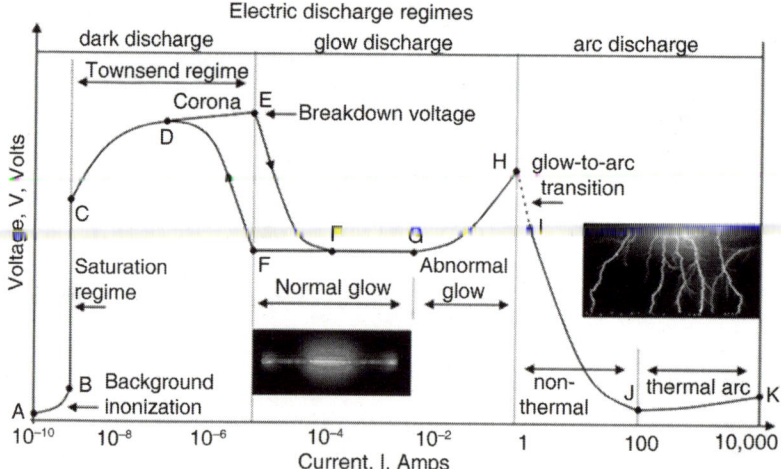

Figure 1.5 Universal I-V characteristic of DC low pressure electrical discharge

1.4.1   Dark discharge

Starting at very low voltage at left side, the region between A and B is background ionisation regime in which increasing voltage draws large fraction of electrons and ions created by cosmic rays and other forms of background radiations. In the saturation region between B and C, all the electrons and ions generated by radiations are removed from the discharge volume and electrons do not have enough energy to create new ionisations. In the regions from C through E, electrons gains sufficient energy from the electric field so that they ionise some of the neutral background gas. This results in exponentially increase of current as a function of voltage. In the region between D and E, due to the local electric field concentration on the surface of the electrode at sharp points, sharp edges corona discharge will occur. When the voltage is still increased, electric breakdown will occur at point E. The voltage at this point is called as breakdown voltage. The region between A and E on the current-voltage characteristic is called as dark discharge because except corona discharge and electrical breakdown point the discharge is invisible to the eye.

1.4.2   Glow discharge

Once the electric breakdown occurs discharge makes transition to glow discharge region. In this regime, the current is high enough. Amount of excitations of neutral background gas is high enough so that plasma is visible to the eye. In the normal glow region of voltage-current characteristics, voltage across the discharge is almost independent of the current. If the current is increased further, fraction of the cathode occupied by the plasma increases until the plasma occupies whole cathode surface. At point G, plasma covers whole cathode surface. At this point discharge enters in regime in which voltage increases with current. This region from G to H is called as abnormal glow region. At point H cathode current density increases and transition from glow to Arc occurs. Glow discharge plasma is used for large number of applications (Roth, 1995). It is used as a spectroscopic source for analytical chemistry (Bogaerts et al., 2002). Glow discharge plasma is used in micro-electronics industries and materials technology. In fact glow discharges are used for etching, modification, pre-treatment, cleaning of surfaces and for deposition of thin film on the surfaces. Some of the applications are described in later section. It is used in various types of light sources and in display panels also.

1.4.3   Low pressure glow discharge system

Different types of plasma systems based on electrode geometry, power supply, etc., are used for plasma processing (Lieberman and Lichtenberg, 1994). Some of them are mentioned in the following section. Simple glow discharge plasma system is direct current glow discharge which is explained above and is shown in Fig. 1.6(a). When a constant potential difference is applied in between cathode and anode, continuous current will flow through the discharge which gives direct current glow discharge. In this type of discharge, electrodes play important role in sustaining the plasma by secondary electron emission. The applied potential difference is not equally distributed between the cathode and anode (Raizer, 1991). Most of the voltage is dropped in first few millimetres in front of the cathode. This region has strong electric field near the cathode, and it is called as cathode dark space or 'sheath'. After this region, potential is nearly constant and slightly positive, this region is called as negative glow. This is the brightest region in between cathode and anode. When the distance between cathode and anode is relatively long, two more regions i.e. Faraday dark space and positive column are seen.

To sustain the DC, discharge electrodes have to be conducting; otherwise on insulating surface charge will pile up and glow discharge will eventually vanish. Discharge can also be sustained by applying alternating voltage in between electrodes covered with insulating layer; as in this case each electrode will act as a cathode and anode alternately, and charge accumulated in the first half of the cycle will remove from the surface in another half of the cycle. Most commonly and commercially used alternating voltage is radio frequency (RF) of 13.56 MHz. The glow discharge can be produced either by capacitive coupling of radiofrequency power (Fig. 1.6b) or by inductively coupling. In capacitive coupling of discharge, the power is coupled into the discharge by means of two electrodes and their sheath. In inductively coupled discharge, electrodes are not in direct contact with the discharge. The power to the discharge is coupled by induced electric field which is generated by changing magnetic field. In capacitively coupled RF discharge, ion flux and energy of ions that interact with the electrode cannot be controlled independently. If different frequencies are used, it can be controlled but it is not always practical. The ion flux in such type of discharge is low due to low or moderate plasma density. Hence, in order to overcome this problem, new type of low pressure, high density plasma systems like electron cyclotron resonance (ECR) plasma, Helicon plasma, etc., are generally used. In such cases power is coupled to the plasma through a dielectric window instead of direct connection to an electrode in the plasma. In ECR plasma source, resonance between microwave frequency (e.g. 2.45 GHz) and electron cyclotron frequency takes place and plasma is generated. As shown in Fig. 1.6(c), ECR plasma system consists of two parts: resonance region and process region. Low sheath voltages are achieved in such type of plasma systems. To control the ion energy, the electrode on which substrate is placed for surface modification can be driven by separate RF power supply. The plasma flows from resonance region to process region along the axial magnetic field lines. As heating of the electrons occurs through absorption of wave energy, this type of discharge is also called as wave-heated discharges. But another type of wave-heated discharge is Helicon discharge. A dielectric cylinder surrounded by magnetic coil is used as source chamber. An antenna connected to RF power source launches RF waves that propagate along the tube. The electrons are heated by absorbing the energy from RF waves. In comparison to ECR, smaller magnetic field can be used for wave propagation and absorption. Also instead of microwave, RF power source is used in case of Helicon discharges.

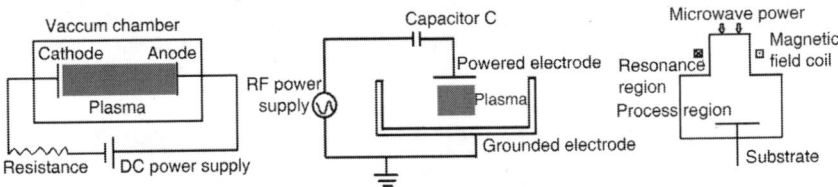

Figure 1.6 Schematic of diagram of (a) DC discharge, (b) capacitively coupled RF discharge, (c) microwave-coupled ECR discharge

1.4.4 Arc discharge

After transition in Arc regime (Fig. 1.5), the discharge settles down at some point between I and K depending upon internal resistance of DC power supply. The Arc regime from I to K is one in which discharge voltage decreases as the current increases. This regime is known as non-thermal Arc regime in which electron, ion and gas temperatures are unequal. The portion with positive slope between J and K is the thermal Arc regime in which all the species are at approximately equal temperatures. In this region plasma is close to thermodynamic equilibrium.

1.4.5 Dielectric barrier discharge (DBD)

Glow discharges can operate over a wide range of pressure regime. Low pressure glow discharge operates at typical pressure range of 10–100 Pascal. Atmospheric pressure glow discharge is also possible but it leads to arcing if typical conditions are not used. Hence according to classical Paschen law, if one goes for atmospheric pressure, inter-electrode distance has to be reduced. For most of the atmospheric pressure glow discharges, the inter electrode gap is of few millimeters. But for textile surface modification, such type of plasma is more suitable as continuous treatment is possible at atmospheric pressure. As mentioned above, stable atmospheric pressure glow discharge is obtained when other conditions like electrode geometry, carrier gas and frequency of the applied voltage are fulfilled (Kanazawa et al., 1988). At atmospheric pressure, dielectric barrier discharge (DBD) is used (Kogelschatz, 2003). Typically in DBD, one of the electrodes is covered with dielectric material. As the dielectric cannot pass DC current, the discharge operates in alternating voltages. The dielectric constant and thickness of dielectric layer determines the displacement current passing through dielectric. In most applications the dielectric limits the average current density in the gas space. Materials like glass, ceramic materials, and thin enamel or polymer layers are preferred as dielectric materials. For some applications additional protective or functional coatings are applied. Schematic of dielectric barrier discharge is shown in Fig. 1.7. Also the stability of the discharge is mostly dependent upon type of discharge gas. The helium gas gives stable glow discharge, while gases like nitrogen, oxygen, argon easily cause transition from glow discharge to filamentary discharge. However, it is possible to generate stable glow discharge using these gases by changing the electrode configuration (Okazaki et al., 1993).

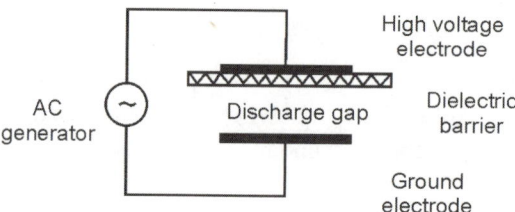

Figure 1.7 Schematic of Dielectric Barrier Discharge

At atmospheric pressure when the electric field in the discharge gap is high enough to cause breakdown, in most gases a large number of micro-discharges are observed (Fig. 1.8). Each micro-discharge has an almost cylindrical plasma channel, typically of about100 micron radius, and spreads into a larger surface discharge at the dielectric surface. Micro-discharge development occurs at a nano-second time scale. But as the applied alternating voltage cycle is of much longer duration (micro to millisecond range), this type of discharges are characterised by a large number of micro-discharges per unit cycle (Fig. 1.8). The power is consumed in large number of such short-lived micro-discharges. Hence experimentally it is difficult to measure the actual power consumed in the discharge. But based on the work of Manley, voltage-charge Lissajous figures are used to determine average power (Manley, 1943). This is done by putting a capacitance in series with the DBD experiment. The voltage across this measuring capacitor is proportional to the charge and close loop area of the applied voltage vs. charge, represents the consumed discharge energy.

Corresponding dissipated power can be calculated by multiplying discharge energy by operating frequency.

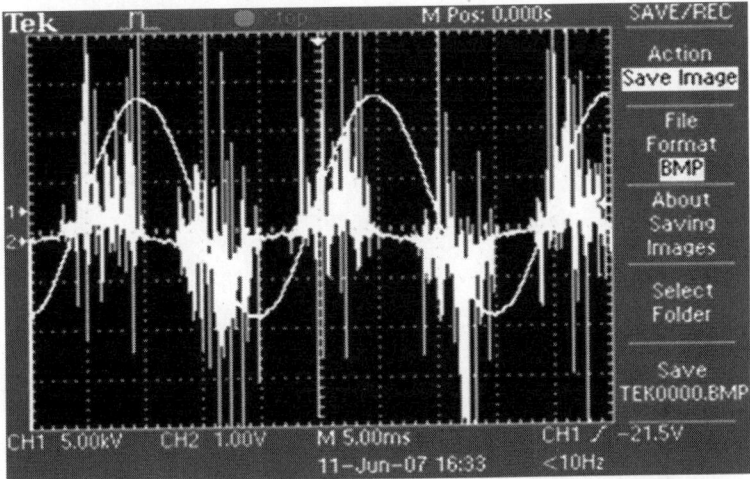

Figure 1.8 Voltage and current waveform showing micro-discharges

1.5   Plasma surface interactions

In industrial applications, low pressure or atmospheric pressure plasma is used for the physical or chemical modifications of the material surface. During plasma treatment highly energetic electrons in the plasma can activate number of reactions. Also the species in the plasma i.e. atoms/molecules, positive and negative ions, radicals, electrons and photons interact with the surface of the material to activate number of processes. Grafting, activation, film deposition, etching are the most common processes happen during plasma surface interaction. As these processes are very complex, they are not described in detail. The fundamental collisions in the plasma like ionisation, excitation, and recombination among the plasma species help for surface interaction processes.

Following examples are covered to demonstrate the use and application of plasma surface interaction in textile industry.

(i)   Surface modification by low energy plasma: Glow discharges, immersing electrode in the plasma etc.

(ii) Thin film growth on textiles with plasma methods: Physical Vapour Deposition (PVD)

(iii) Improving the wear resistance of textile machinery: Plasma nitriding

(iv) Plasma sterilization for cleaning the textile used in health industry: Glow discharge, pencil plasma torch

1.5.1   Surface modification by low energy plasma

Textile industry is conventionally using wet chemical methods for surface modifications; in this case, water and hazardous chemicals are utilized in huge quantities and residual water need to be disposed. This is an issue in terms of ecological impact on the environment and influence to human health. Due to this increasing environmental concerns and requirement for an environment-friendly process for textiles, new technologies are emerging. Plasma technologies can prove to be a suitable option as a green technology. Plasma-technology-based methods are environment friendly and modify surface properties without much modification to the actual surface and provide new and improved properties in environment-friendly manner. Bulk properties of textiles after the plasma treatment remain unmodified as compared to wet chemical processes, which penetrate deep into the fibers. Plasma produces surface reaction, the properties given to the material being limited to the surface layer of a few nanometers.

The modification of textile substrate using plasma generates various effects on the textile surfaces from the surface activation to a thin film deposition. At first, plasma reacts with the substrate surface and creates active species and new functional groups, which subsequently changes substrate activity. Plasma etching process changes the surface morphology of fibers and induces the nano- or micro-roughness on fibers surfaces. The nano-structured textile surfaces have a higher specific surface area, and lead to improved properties of the treated surface, i.e. increased surface activity, hydrophilic or hydrophobic properties, and increased absorption capacity towards different materials, i.e. nanoparticles and nano-composites.

The surface properties alterations obtained by a plasma treatment on fibre surface are very complicated. Particle-induced reactions usually take place in the upper ten nanometers of a surface. Short wavelength UV-radiation as it is emitted by low pressure plasmas initiates reactions in a thicker layer (about 100 nm) (Gorjanc et al., 2013; Shah and Shah, 2013; Chinta et al., 2012). The relation between the two and the extent of both can be controlled by the process gas and other process parameters. The outermost surface, only some atom layers, sometimes less than 1 nm, determines the interaction with other media. The chemical composition of this part of a fibre is responsible for good or bad adhesion with the laminates. In order to improve the adhesion properties, this part of a fibre can be modified by plasma. Therefore in plasma the process parameters play dominant role for the success of desired results after the treatment. Of course surface to be altered is also important, as different surfaces will have different effects after plasma interaction. Trace amounts of sizings, for example, can modify the reaction condition substantially and have to be taken into consideration for almost every process. In the following section, authors present the basic understanding of plasma interaction with an object immersed in plasma with and without a potential and later present some examples how it can affect the surface properties.

1.5.2   Floating object immerged in plasma

Irrespective of all kinds of plasmas used in textiles, basic physics of plasma interaction remains the same. When an object is immersed in plasma, there is no way for the charge to drain from the surface; we have a situation in which the wall is electrically floating. This gives rise to a 'floating sheath'. In this case, the object will charge up until the flux of electrons and ions match. The potential at which the flux matches is known as the floating potential of the plasma, which is typically few volts to few tens of volts in typical glow discharge as is given by:

$$\Phi_w = -\frac{kT_e}{e}\ln\left(\sqrt{\frac{2\pi m_e}{M_i}}\right)$$

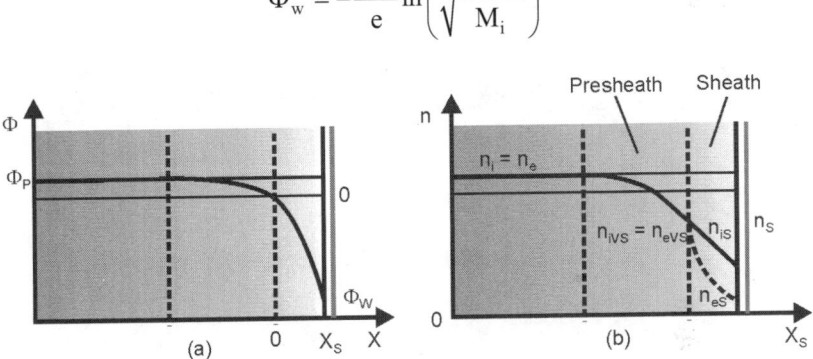

Figure 1.9 Floating object immerged in plasma (a) potential profile (b) electron/ion density profile

Here φ_w is floating potential, Te electron temperature, m_e and M_i are the mass of electron and ions, respectively. Figure 1.9 shows the potential distribution starting from plasma potential (φ_p) to floating potential (φ_w). It can be observed in between these two potentials there is an intermediate region in which the plasma potential started decreasing and then suddenly goes down. These regions are called pre-sheath and sheath regions, respectively. Normally plasma follows the charge neutrality, i.e. electron and ion densities are nearly the same but it is not the case in the sheath region (Lieberman and Lichtenberg, 1994; Raizer, 1991). In the sheath region ion density is much more than electron density. Therefore according to Bhom's sheath criteria ions with ion acoustic velocity enters in the sheath and hits the surface. Floating potential is usually not enough to modify any surface properties of hard materials like metal or semiconductors, but for textile even floating potential can do some modification to the surface.

As discussed above normally floating object immersed in plasma develops a very small floating potential, this potential may or may not be causing any surface modification, activation or cleaning. Therefore, normally objects are kept on a biased electrode so that incoming ions can gain some additional energy. There is a well-known ion matrix sheath theory applied in this case. The ion matrix sheath occurs when the potential on an object changes very rapidly. In this case, the electric field change is such that the electrons leave the immediate region while the ions remain fixed for a small instance. The heavy ions cannot respond as fast as the light electrons. Then there is an electric field that occurs through a uniform distribution of ions as shown in Fig. 1.10. Ion matrix sheath thickness is given by (Lieberman and Lichtenberg, 1994):

$$S_{wall} = \sqrt{-\Phi_{Wall} \frac{2\varepsilon}{en_0}}$$

$$= \lambda_{Debye} \sqrt{\frac{-2\Phi_{Wall}}{kT_e}}, \quad \lambda_{Debye} = \sqrt{\frac{\varepsilon kT_e}{en_0}}$$

n_0 is plasma density and T_e is electron temperature.

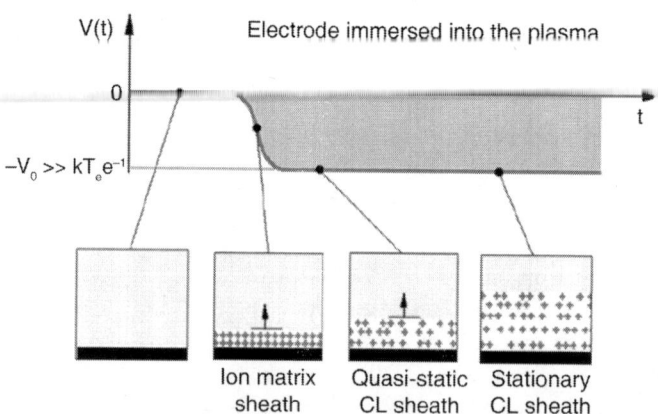

Figure 1.10 Various types of sheath formation on a biased substrate

It can be seen that ion matrix sheath thickness directly depends on the applied potential to the wall. Ion matrix sheath is not stable, very soon it starts expanding and its behavior is normally given by well-known Child's-

Langmuir Sheath. It confirms that flux remains constant but the sheath voltage increases with applied wall potential.

$$x_x = \frac{\sqrt{2}}{3} \exp(1/2)\lambda_D \left(\frac{2eV_0}{kT_e} \right)^{3/4}$$

Normally, textile fibres are non-conducting in nature; therefore, charge accumulation may occur on the surface. To avoid this situation, normally one should biased the surface with fluctuating DC like RF plasma (13.56 MHz) or pulsed DC (μs) pulses as discussed above. Such pulses will attract the ions to the surface during on time and will repel them in the off time of the pulse, but this will not hinder the process of surface cleaning or modification and will avoid surface charging.

1.5.3   Surface cleanliness by plasma

Plasma treatment is a surface treatment influencing only the top layer. In a cleaning process, inert (Ar, He) and oxygen plasmas are used. The plasma-cleaning process removes, via ablation, organic contaminates such as oils and other production releases on the surface of most industrial materials. These surface contaminants as polymers undergo abstraction of hydrogen with free radical formation and repetitive chain scissions, under the influence of ions, free radicals and electrons of the plasma, until molecular weight is sufficiently low to boil away in the vacuum. Activation plasma processes happen when a surface is treated with a gas, such as oxygen, ammonia or nitrous oxide and others that do not contain carbon. The primary result is the incorporation of different moieties of the process gas onto the surface of the material under treatment. Let us consider the surface of polyethylene, which normally consists solely of carbon and hydrogen: with a plasma treatment, the surface may be activated, anchoring on it functional groups such as hydroxyl, carbonyl, peroxy, carboxylic, amino and amines. Hydrogen abstraction produces free radicals in the plasma gases and functional groups on the polymeric chain. Almost any fibre or polymeric surface may be modified to provide chemical functionality and which enhances the adhesion characteristics significantly. This is a great improvement in the production of technical fabrics.

The SEM analysis strongly modified surface morphology of plasma-treated cotton as shown in Fig. 1.11, while smaller changes of the surface morphology of cotton samples treated with atmospheric air corona plasma and water vapor low-pressure plasma are noticed. SEM image of untreated cotton (Fig. 1.11a) shows a typical grooved surface morphology with macro-fibrils oriented predominantly in the direction of the fiber axis. The outlines of the macro-fibrils are still visible, and they are smooth and distinct due to

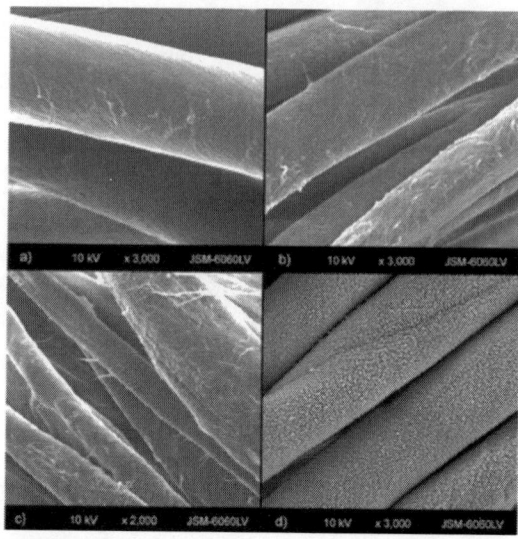

Figure 1.11 SEM images of (a) untreated cotton fibre, (b) corona plasma-treated cotton fibre, (c) water vapor low pressure plasma-treated cotton fibre, (d) CF4 plasma-treated cotton (Gorjanc et al., 2013)

the presence of an amorphous layer covering the fiber. The surface of corona plasma-treated cotton has striped, cleaned and more distinct macrofibrilar structure (Fig. 1.11b). The same effect can be noticeable on the surface of cotton fibers treated with water vapor low-pressure plasma (Fig. 1.11c). The plasma-treated fiber surface remains grooved and the macrofibrile structure has gained a much sharper outline. The individual macrofibrils (0.2–1.0 nm) and their transversal connections are visible in the primary cell wall. Between them, narrow voids thinner than 10 nm are noticeable. The surface morphology of CF_4 plasma-treated cotton (Fig. 1.11d) is very different comparing to those treated with water vapor low-pressure or corona plasma. CF_4 plasma-treated cotton has an extremely rough and nano-structured surface with dimensions of the grains roughly between 150 and 500 nm. Dissociation energies of plasma molecules are the reason for such rich etched surface. Both CF_4 and H_2O molecules get dissociated in plasma. The dissociation energy of CF_4 is 12.6 eV while the ionization energy of CF_4 is about 16 eV, which is much higher than the dissociation energy of water vapor or OH molecules, which is at about 5 and 4 eV, respectively. The electrons with energy of few electron volts are therefore likely to dissociate water and OH molecules rather than dissociating the CF_4 molecules (Gorjanc et al., 2010). The result is an extremely high dissociation fraction of water molecules and a moderate dissociation fraction of CF_4 molecules. Since the partial pressures of water and

CF_4 are comparable, it can be concluded that the density of O and OH radicals in our plasma is at least as high as the density of CF_x radicals if not more so. The textile sample exposed to plasma is therefore subjected to interact with CF_x, O, OH, F and H radicals. The density of H is probably as high as that of F so extensive recombination to HF is expected. CF_x radicals tend to graft onto the textile, but the grafting probability on cellulose is low compared to the interaction probability with O and OH radicals. O and OH radicals are extremely reactive and cause etching of cellulose. Fluorine atoms are efficient at abstracting hydrogen in the first step of the oxidation reaction. In addition, the presence of fluorine atoms in the plasma enhances the dissociation of the oxygen, further increasing the ashing rates. CF_4 is the gas most commonly added to oxygen to enhance the generation of atomic oxygen in plasma and to increase polymer etch rates. While the surface of plasma-treated cotton was changed, the mechanical properties of cotton fabrics did not alter after plasma modification.

1.5.4   Surface modification by physical vapour deposition (PVD)

Plasma techniques contribute not only to the surface modification by plasma interaction but also to various special kinds of thin film growth using physical vapor depositions methods. Now-a-days thin film with specific properties are grown on textile; such growth of solar cell on textile is used for making tents, silver coating on baby sleeping cloths, various decorative coatings, conductive fibers, growth of nanoparticles for antibacterial properties, etc. Here the PVD coatings are routinely used (Gries, 1993). PVD processes are atomistic deposition processes in which material is vaporized from a solid source in the form of atoms or molecules, transported in the form of a vapor through a vacuum or low-pressure gaseous (N_2, for example) environment to the substrate where it condenses. Typically, PVD processes are used to deposit films with thickness in the range of a few nanometers to thousands of nanometers; however they can also be used to form multilayer coatings, graded composition deposits and very thick deposits on various substrates.

The two most important technologies are evaporation and sputtering.

Evaporation

In evaporation the substrate is placed inside a vacuum chamber, in which a block (source) of the material to be deposited is also located. The source material is then heated to the point where it starts to boil and evaporate. The vacuum is required to allow the molecules to evaporate freely in the chamber, and they subsequently condense on all surfaces. This principle is the same for all evaporation technologies, only the method used to heat (evaporate) the source material differs. There are two popular evaporation technologies, which are e-beam evaporation and resistive evaporation each referring to the heating method. In e-beam evaporation, an electron beam is aimed at the source material causing local heating and evaporation. In resistive evaporation, a tungsten boat, containing the source material, is heated electrically with a high current to make the material evaporate. Many materials are restrictive in terms of what evaporation method can be used (i.e. aluminium is quite difficult to evaporate using resistive heating), which typically relates to the phase transition properties of that material. A schematic diagram of a typical system for e-beam evaporation is shown in Fig. 1.12. Electron emitted in thermionic emission are energized and guided on the material surface. Due to electron heating, material evaporates and can be deposited on the substrate.

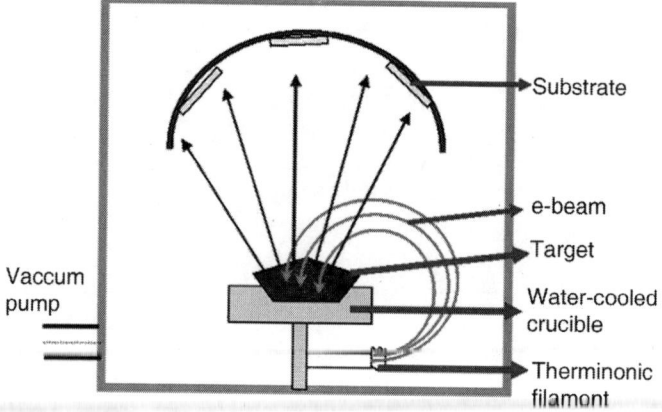

Figure 1.12 Schematic diagram of typical E-beam evaporation setup

Magnetron sputtering

Sputtering is a technology in which the material is released from the source at much lower temperature than evaporation. The substrate is placed in a vacuum chamber with the source material, named a target, and an inert gas (such as argon) is introduced at low pressure. In magnetron sputtering a normal glow discharge is formed with using target material as cathode. Permanent magnets are used to confine the electron in front of the target surface to make instance ionization and sustain the plasma at low pressure of typically 10^{-3} mbar. The ions are accelerated towards the surface of the target, causing atoms of the source material to break off from the target in vapour form and condense on all surfaces including the substrate (Lieberman and Lichtenberg, 1994). As for evaporation, the basic principle of sputtering is the same for all sputtering technologies. The differences typically relate to the manner in which the ion bombardment of the target is realized. A schematic diagram of a typical sputtering system is shown in Fig. 1.13. Depending on the requirements, multi-magnetron system, cylindrical magnetron, rectangular magnetrons are available in the industries for the large area coatings.

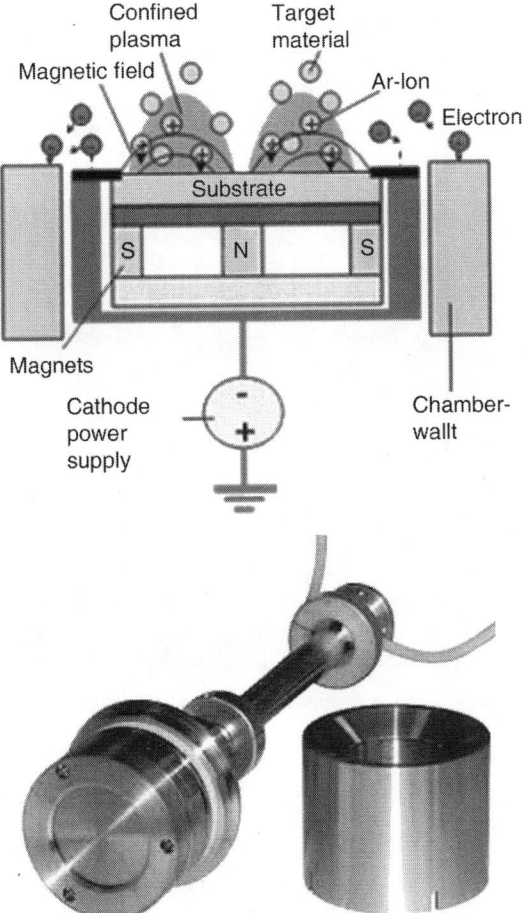

Figure 1.13 Schematic diagram of magnetron sputtering system

Nearly all metallic materials can be deposited on textile substrate by sputtering. Metallization is a metal coating process that adds value to and improves the functions of textile materials. Textile materials modified with different metals have attracted a great deal of attention owing to their potential applications in technology and design fields. The functionalization of textile materials using a sputter coating of metal can significantly modify surface properties of materials. The development of modified materials with improved properties will open up new possibilities for applications of these materials. Metals are considered to be safe for humans, as demonstrated by the widespread and prolonged use by women of copper intrauterine devices (IUDs). Copper/silver surface kills over 99.9% of bacteria (*Escherichia coli,* Enterobacteraerogenes, MRSA, *Pseudomonas aeruginosa, Staphylococcus aureus*) for 24 hours. An example of Cu film grown on textile surface is given below. Optical microscope picture (Fig. 1.14) (Vihodceva et al., 2011) evince that copper coating is not a flat film on cotton textile surface and copper particles are deposited on fibers without changing the textile surface structure.

Figure 1.14 Cu film grown on textile surface (Vihodceva, 2011)

Another example is of nano-silver film, which is an ideal functional material, as the substrate of textile materials that can be used to develop UV-shielding materials, fiber solar cells, medical antibacterial material, etc.; preparation of silver film at present is mainly done by Chemical Vapour Deposition (CVD), sputtering, electro less method and electroplating method. From Fig. 1.15(a), it can be seen that silver particles uniformly covered the surface of substrates, the size of particles were regular, a small number of particles occurred reunion in Fig. 1.15(b), the gap of particles were smaller, regular distribution in Fig. 1.15(c), it illuminated that the compactness of silver films was good (Lingling et al., 2011). Sheet resistance of films decrease and electrical conductivity gradually increase with the increasing of sputtering time (Lingling et al., 2011). Nano-silver particles deposited on the surface of polyester fibers gradually increased with the extending of sputtering time, and the rate of film growth was faster, the density and uniformity on the surface of films improved and the conductivity of films gradually increased.

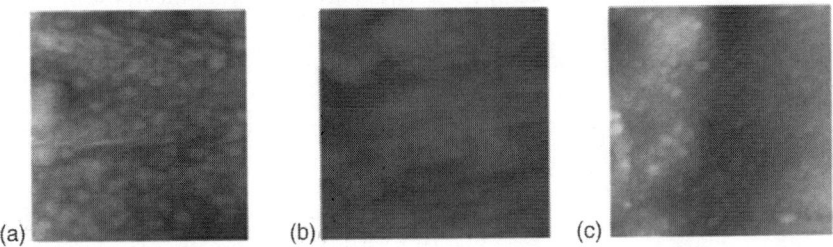

Figure 1.15 AFM images of Ag thin films under different sputtering power (a) 80 W; (b) 120 W; (c)160 W (Scanning scale is 5000 × 5000 nm) (Lingling et al., 2011)

1.5.5   Plasma nitriding

Plasma nitriding is a subsurface modification technique, where the hardness of steel components is enhanced by diffusing nitrogen to large depths inside the component surface in the presence of plasma. Figure 1.16 shows a typical plasma nitriding reactor used at Facilitation Centre for Industrial Plasma Technologies (FCIPT) for nitriding industrial components. A large number of components have been plasma nitrided at FCIPT and other places, and a data of life enhancement of various steels is seen after doing plasma nitriding. For most steels after plasma nitriding hardness, wear resistance and corrosion resistance increases (Basu et al., 2008).

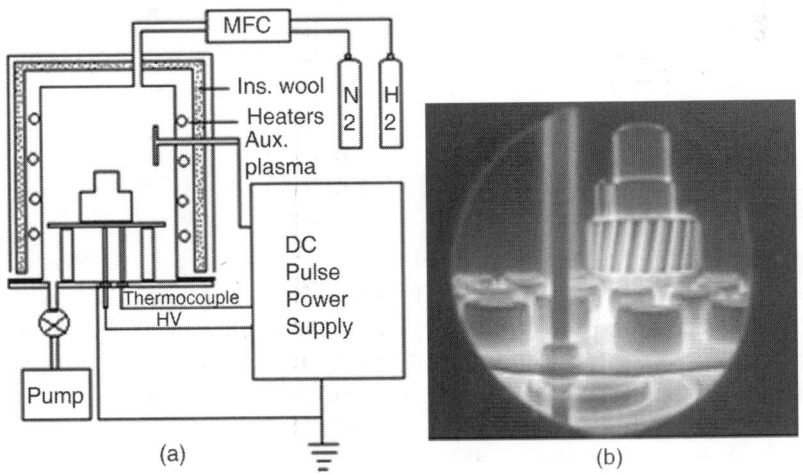

Figure 1.16 (a) Typical plasma nitriding reactor; (b) glow discharge plasma for nitriding of steel components

In textile machines, minute fiber particles can increase abrasive wear by several orders of magnitude. The profitable operation of modern, high productive textile machines depends, to a large extent, on how well the key components that make contact with the fibers and threads resist wear and tear. At many stages in the processing line, the surface finish has a direct influence on the quality of the final product. Hence, in order to improve the quality of the final product and the life of the key components, the surface layers of these components must be smooth and wear resistant. Any technique that would modify these surfaces should ensure:

- uniform thread quality guaranteed by the components smooth surface that remains unchanged throughout its service life
- increase in processing speed met by an improved wear resistance

Plasma nitriding is one such technique that improves the surface hardness without impairing the surface finish, making it an ideal method to apply for such components like the weft clamps of looms, yarn guides, grippers and cutting edges (Farkaz et al., 1980). In plasma nitriding process, the components are negatively biased with respect to the vacuum chamber which acts as an anode. The nitrogen–hydrogen gas mixture introduced into the chamber gets ionized when certain DC voltage is applied across the electrodes at an optimum background pressure. These ions accelerate towards the cathode and bombard the surface with high kinetic energy thereby increasing the temperature of the components and involving diffusion of nitrogen to higher depths with time. As a result the top few layers are nitrided and have a maximum hardness on the surface. Typically for En8 material the surface hardness achieved by plasma nitriding is about 550 HV.

Sewing machines run at very high speeds, up to 10,000 cycles per minute and intermittently, so parts must be extremely resistant to wear. Conventionally, this has been achieved by immersing the steel components in a cyanide bath or an equivalent gaseous medium. However, the toxicity of these media can result in undesirable environment effects and the process time is relatively long.

Plasma nitro-carburizing, a high-tech surface technology, causes nitrogen and carbon molecules to diffuse into the surface of the steel, creating a hard layer composed largely of iron carbonitride. It offers a more environment-friendly approach. The process involves applying an electric charge to the objects to be hardened, while they are immersed in an atmosphere containing nitrogen, hydrogen and carbon. The gas atoms become excited and react with the surface of the charged engineering components.

Treatment is normally carried out by hanging parts from a stainless steel cathode in the centre of a cylindrical stainless steel chamber at around

570 volts and 550°C. Depending on requirements, each batch may be treated for up to 12 hours, resulting in a carbonitride diffusion layer up to half a millimeter deep. A hardness of up to 1000 Vickers units can be obtained on stainless steels, coupled with a reduced friction coefficient.

Also, with plasma nitriding and plasma nitrocarburizing the running-in period of the spinning ring is no longer required, leading to a significant increase of production yield. The extreme wear resistance and self-lubricating properties of the nitrided layer yield a higher life time of both the spinning ring and the corresponding cursor. Furthermore the quality of the yarn is improved and remains stable for a longer period. The treated tooling and forms do not snag or soil the delicate fibers. Hence, it can be concluded that plasma-aided diffusion processes like plasma nitriding and plasma nitro-carburizing play an important role in the textile industry by improving the quality of the final product, productivity and the life of the key components.

1.5.6 Plasma sterilization for cleaning the textile used in health industry

Plasma sterilization is fast evolving into a promising alternative to standard sterilizing techniques (Rossi et al., 2009; Moison et al., 2002). The process is usually at room temperature and hence poses no dangers associated with high temperatures (unlike autoclaves), doesn't involve any chemicals and hence is non-toxic (unlike EtOH), time of treatment is fast. It is versatile and can sterilize almost any material and any shape with nooks and crannies. Medical textiles are one of the most rapidly expanding sectors in the technical textile market. Medical textiles are all those textile materials used for medical and biological applications and are used primarily for first aid, clinical and hygienic purposes. Some examples of medical textiles are surgeons wear, wound dressings, bandages, artificial ligaments, sutures, artificial liver/ kidney/lungs, nappies, sanitary towels, sterilization pouches and protective nonwovens, vascular grafts/heart valves, artificial joints/bones, eye contact lenses and artificial cornea.

During the last decade, the bactericidal effects of plasmas have been extensively studied, using different gas mixtures and electrical sources as they offer a safe and low temperature of action, compared to present treatments such as autoclaving that do not allow sterilization of complex medical devices sensitive to temperature and humidity. Among low-temperature plasmas, different classifications can be established. Taking into account the working pressure, plasmas can be classified into low pressure and atmospheric pressure plasmas, whereas if the species involved in the treatment are taken into account, one can distinguish between plasma discharges and plasma post-discharges.

Plasma discharges consist of a state of mixed ions, radicals, electrons, excited molecules; UV and visible radiation that preserves electrical neutrality, while in post-discharges only the most stable species, such as neutral atoms, are present.

Plasma sterilizer as shown in Fig. 1.17 takes advantage of plasma and hydrogen peroxide vapor technology. As an sterilant, a hydrogen peroxide chemical is relatively non-toxic; however, hydrogen peroxide is a strong oxidant and these oxidizing properties allow it to destroy a wide range of pathogens, and it is used to sterilize heat or temperature sensitive articles such as rigid and flexible endoscopes. The biggest advantage of hydrogen peroxide as a sterilant is the short cycle time. Whereas the cycle time for ethylene oxide (EtO) may be 10–15 hours, the reaction of plasma and hydrogen peroxide vapor allows for a much shorter cycle time. Finally, it has been found that bacterial spores can also be significantly eroded by Ar/O_2 plasma discharge as demonstrated in Fig. 1.18. This finding clearly demonstrates, contrary to the statements made previously, that etching can significantly contribute to the sterilization process; etching can explain fast inactivation of bacterial spores in those plasma discharges where only limited intensity of UV radiation is produced.

Figure 1.17(a) Rf plasma-based plasma sterilizer works with H_2O_2 plasma (http://www.stormoff.ru/en/catalog_659.html)

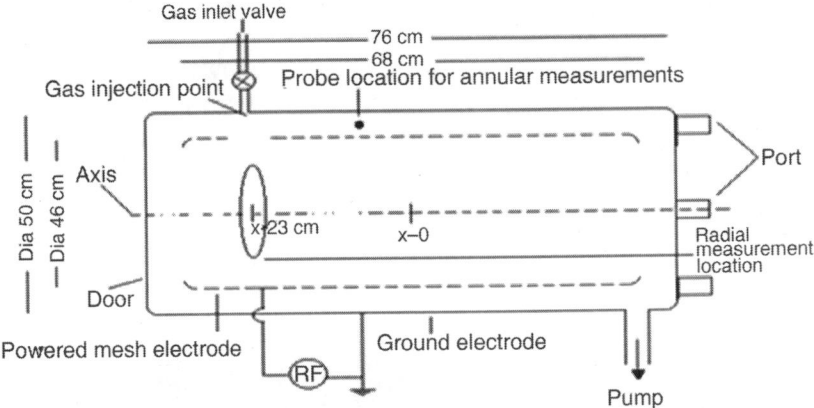

Figure 1.17(b) Rf plasma-based plasma sterilizer works with H_2O_2 plasma (http://www.stormoff.ru/en/catalog_659.html)

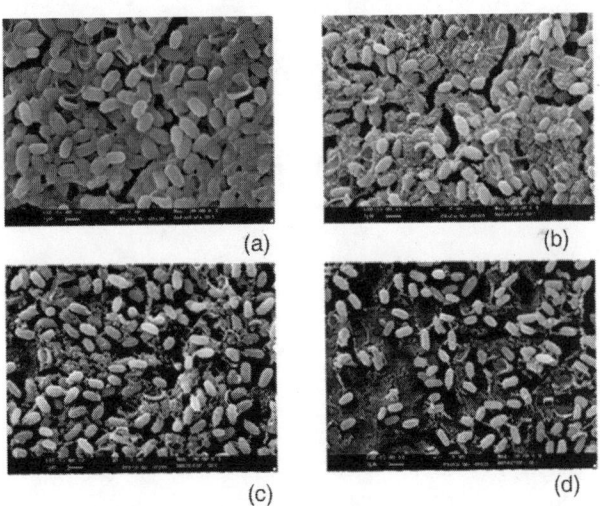

Figure 1.18 SEM images of (a) untreated G. stearothermophilus spores and spores treated for (b) 30 s (c) 60 s and (d) 120 s (Ar/O_2 20 : 1, 10 Pa, 200 W) (Rossi et al., 2009; Moisan et al., 2002)

1.5.7   Non-thermal atmospheric pressure plasma jet

Atmospheric Pressure Plasma Jets (APPJ), as shown in Fig. 1.19, offer a unique environment in plasma medicine, allowing treatment of soft materials, including biomaterials such as living tissues. The mechanism underlying this non-thermal discharge, however, remains unsettled that it has been often taken as resulting from dielectric barrier discharge or vaguely referred as streamer like. In this case the plasma jet of Ar (or compressed air) is formed between two metal electrodes at mid frequency range (50 to 150 kHz). The high electron temperature enhances the plasma chemistry processes while the plasma gas remains close to room temperature. APPJ are not confined by electrodes and can be adjusted from a few centimetres down to the sub-millimetre region. Advantages of APPJ are mentioned below:

(i)   Alters the surface properties without affecting their bulk property.

(ii) The temperature varies from 30°C to 70°C.

(iii) It is eco-friendly and no vacuum/water is needed for the process.

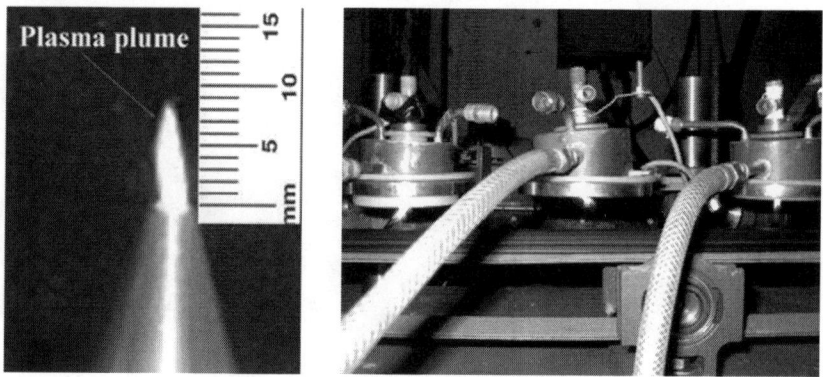

Figure 1.19 Atmospheric pressure non-thermal plasma jet: (a) Single Torch (b) Multiple Torches

1.5.8   Plasma power supplies

For industrial plasma applications, direct current (DC), alternating current (AC), radio frequency (RF), and microwave (MW) power supplies are used. Depending upon application and pressure conditions, appropriate power supply is selected. Different types of power supplies are used for plasma production at low pressure and at atmospheric pressure. For low pressure plasma DC, RF, microwave power supplies are used with voltage ratings used to be less than 1.0 kV and currents used to be in the range of a few amperes to tens of amperes. For applications such as plasma nitriding, plasma-assisted physical vapor deposition, plasma-enhanced chemical vapour deposition (PECVD), generally pulsed DC, radio frequency (RF) or microwave (MW) power supplies are used. The source frequency is selected based on substrate to be treated i.e. insulating, semiconductor or metallic; rate of etching or deposition which is governed by radical density and energy of charged particles that varies with source frequency. For non-thermal atmospheric pressure, plasma with high voltage, high frequency power supplies are used and current used to be in hundreds of milliampere. The voltage levels depend upon the inter electrode gap and type of electrode. In case of dielectric barrier discharge, if metal wire mesh electrodes are used, then breakdown can take place at lower voltages (Xinxin et al., 2006). Also Okazaki et al. (1993) have shown that for metal wire mesh, stable atmospheric pressure glow discharge is formed by using a 50 Hz frequency. In case of DBD, capacitance of the system is important for plasma formation and dissipated power. The plasma reactor capacitance depends upon reactor configuration, electrode size, gap between the electrodes, and the dielectric properties of the dielectric material. Amount of surface charge on the electrode is proportional to the capacitance of the reactor and discharge period. Hence for a given reactor dissipated power can be maximised at particular frequency (Kim and Song, 2006). In case of dielectric barrier discharge, the voltage level and frequencies are selected based upon the reactor type and required power levels.

1.6   Summary

The chapter describes the electric breakdown of gases, Paschen's law to produce plasma at a minimum voltage by adjusting pressure and gap between the electrodes and non-thermal and thermal plasma regimes in current–voltage characteristics. In textile industry, plasma technology seems to be clean and environment-friendly technique for surface modification of textile surface. Non-thermal atmospheric pressure plasma is becoming popular day by day and being widely used for surface treatment of textiles. The plasma system that works at atmospheric pressure can be effectively used for continuous treatment of textiles, which has been the basic need in textile industry because of huge production requirements. At atmospheric pressure, suitable electrode geometry and power supply have to be selected for discharge formation. Apart from textile surface modification, the machinery components used in textile manufacturing industries can also be modified by plasma treatment, i.e. nitriding and physical vapor deposition. Plasma nitriding increases surface hardness and wear resistance property of the steel component which enhances life of the component. Application of plasma techniques described here on actual textile fabric is demonstrated in subsequent chapters.

1.7   References

Basu A., Majumdar J.D., Alphonsa J., Mukherjee S., and Manna I. (2007). Plasma nitriding of a low alloy-high carbon steel, Institute of Metals, **60**(5), pp. 471–479.

Basu A., Duta, Majumdar J., Alphonsa J., Mukherjee S., and Manna I. (2008). Corrosion resistance improvement of high carbon low alloy steel by plasma nitriding, Materials Letters Issues, **62**(17–18), pp. 3117–3120.

Bogaerts A. E., Neyts R., Gijbels J., and Van der Mullen (2002). Gas discharge plasma and their applications, Spectrochimica Acta Part B 57, pp. 609–658.

Buyle G. (2009). Nanoscale finishing of textiles via plasma treatment, Materials Technology, **24**(1), pp. 1–50.

Chen F.F. (1983). Introduction to Plasma Physics and Controlled Fusion, second edition, volume1, Plenum Press, New York.

Chinta S.K., Landage S. M., Sathish Kumar M. (2012). Plasma Technology & Its Application In Textile Wet Processing, International Journal of Engineering Research & Technology, **1**(5), pp. 1–18.

Farkas S.Z., Filep E., Kolozsvary Z. (1980). Plasma nitriding improves service behaviour of textile machine components, Journal of Heat Treating, **1**(4), pp. 15–20.

Gorjanc Marija, Marija Gorenšek, Petar Jovančić, and Miran Mozetič (2013). "Multifunctional Textiles – Modification by Plasma, Dyeing and Nanoparticles" DOI: 10.5772/53376, ISBN 978-953-51-0892-4.

Gorjanc M., Bukosek V., Gorensek M., and Vesel A. (2010). The Influence of Water Vapor Plasma Treatment on Specific Properties of Bleached and Mercerized Cotton Fabric. Textile Research Journal, **80**(6), pp. 557–567.

Gries Th. (1993). PVD coatings for textile machine components, Surface and Coatings Technology, **62**(1), pp. 443–447.

Kunazawa, S. M., Kogoma, S., Okazaki, T. Morlwakl (1988). Stable Glow Plasma at Atmospheric Pressure, J. Phys. D: Appl. Phys., **21**, pp. 838–840.

Kim Gon-Ho, Song Sang-heon (2006). Optimum frequency of an atmospheric pressure dielectric barrier discharge for the photo-resistor ashing process, Journal of the Korean Physical Society, **49**(2), pp. 558–562.

Kogelschatz U. (2003). Plasma Chemistry and Plasma Processing, **23**(1).

Lieberman M and Allen J. Lichtenberg (1994). Principles of Plasma discharges and material Processing.

Lingling M., Xinmin H., Qufu W. (2011). Characteristics of Silver films deposited on the surface of PET fabric, Advanced Materials Research, **239–242**, pp. 2356–2360.

Moisan M., Jean B., Marie-Charlotte C., Jacques P., Nicolas P. and Bachir S. (2002). Plasma Sterilization: Methods and mechanisms, Pure Appl. Chem., **74**(3), pp. 349–358.

Manley T.C. (1943). The electric characteristic of the ozonator discharge. Trans. Electrochem. Soc. **84**, 83–96.

Okazaki S., Kogoma M., Vehera M., and Kumura Y. (1993). Appearance of stable glow discharge in air, argon, oxygen and nitrogen at atmospheric pressure using 50 Hz, J. Phys. D. Appl. Phys., **26**, pp. 889–892.

Raizer Y. (1991). Gas Discharge Physics, Springer-Verlag, Berlin.

Roth. J. (1995), Industrial Plasma Engineering, Vol. 1, IOP Publishing Ltd.

Rossi, F., Kylián O., Rauscher H., Hasiwa M., and Gilliland D. (2009). Low pressure plasma discharges for the sterilization and decontamination of surfaces, New J. Phys., **11**, 115017–1.

Shah J. N., and Shah S. R. (2013). Innovative Plasma Technology in Textile Processing: A Step towards Green Environment, Research Journal of Engineering Sciences, **2**(4), pp. 34–39.

Srikanth S., Saravanan P., Alphonsa J., and Ravi K. (2013), Surface Modification of Commercial Low-Carbon Steel using Glow Discharge Nitrogen Plasma and its Characterization, Journal of Materials Engineering and Performance, **22**(9), pp. 2610–2622.

Sterilization system STERRAD ASP (2014), http://www.stormoff.ru/en/catalog_659.html, May 6.

Vihodceva S., Kukle S., Blums J., and Zommere G. (2011). The Effect of the Amount of Deposited Copper on Textile Surface Light Reflection Intensity, Scientific Journal of Riga Technical University, 6, pp. 24–29.

Xinxin W., Luo H., Liang Z., Mao T. and Ma, R. (2006). Influence of wire mesh electrodes on dielectric barrier discharge, Plasma Sources Sci. Technology, pp. 15845–15848.

2

Atmospheric pressure plasma processing of textiles at FCIPT

S. K. Nema and P. B. Jhala

Abstract: The chapter briefly describes non-equilibrium (cold) plasmas and how they can be produced at atmospheric pressure. Four types of non-thermal plasmas which work at atmospheric pressure and commonly used for textile surface modification are discussed. These plasmas include corona discharge, atmospheric pressure plasma jet (APPJ), atmospheric pressure glow discharge (APGD) and dielectric barrier discharge (DBD). Cold plasma when interacts with polymer or textiles modifies their surface properties without altering the bulk properties of these materials. There are four chemical reactions that mainly take place when active species of non-equilibrium plasma interacts with the surface of textile surface: (i) plasma cleaning and surface activation, (ii) etching of surface, (iii) functionalization and grafting, (iv) deposition of coating or polymerization.

Further, the chapter covers different plasma and process parameters that govern gas phase and surface chemistry. It includes a few atmospheric pressure cold plasma technologies developed at FCIPT, Institute for Plasma Research, Gandhinagar, for various textile applications. In addition, we have covered important points which should be considered during scaling up of the plasma system. Finally, the future growth direction of FCIPT in textile applications is discussed.

Key words: Cold plasma, dielectric barrier discharge, atmospheric pressure glow discharge, textiles, surface modification, Angora wool, dry processing

2.1   Introduction

In 21st century, awareness among the people towards environment protection has significantly raised. The green process and environment-friendly product development has been the focus of market competitiveness. Currently, conventional processes which are in use to tailor the surface properties of textiles require large quantity of polluting chemicals, water and energy. Textile industry is looking for environment-friendly processes which can provide the desired surface properties and minimize the use of chemicals and water. In recent years, plasma surface modification technology is emerging very rapidly and has shown potential to replace some of the polluting processes of textile industry.

Plasma is the most active state of matter after solid, liquid and gas. Using appropriate device and power source, it is possible to generate cold or hot plasma. For textile surface modification, plasma at room temperature (cold plasma) is more appropriate. The cold plasma comprises of electrons, ions, radicals, metastables and UV radiation. Cold plasma is also known as non-equilibrium plasma, where electron temperature used to be much higher in comparison to ion or neutral temperature. Since electrons are very small in size, therefore, effective energy transfer does not take place. Consequently the plasma remains at room temperature.

In atmospheric pressure plasma, the electric field transmits energy to the gas electrons which are the most mobile-charged species. This electronic energy is then transmitted to the neutral species by collisions. These collisions can be divided in (i) elastic collisions, in this case there is no change in the internal energy of neutral species, however, their kinetic energy increases slightly; and (ii) inelastic collisions – when electronic energy is high enough, the collisions modify the electronic structure of the neutral species and result in the creation of excited species or ions. Most of the excited species have a very short lifetime and they come to ground state by emitting a photon; however, the excited 'metastable species' have a much long life. In helium plasma, large numbers of metastable species are formed. The metastable species interact with neutrals and results into penning ionization and dissociation. This plasma when interacts with polymer or textiles modifies their surface properties without altering the bulk properties of these materials. In the plasma, reactions occurs in gas phase as well as plasma species interact with the substrate surface. These reactions are given in Table 2.1 and plasma surface interactions are shown in Fig. 2.1.

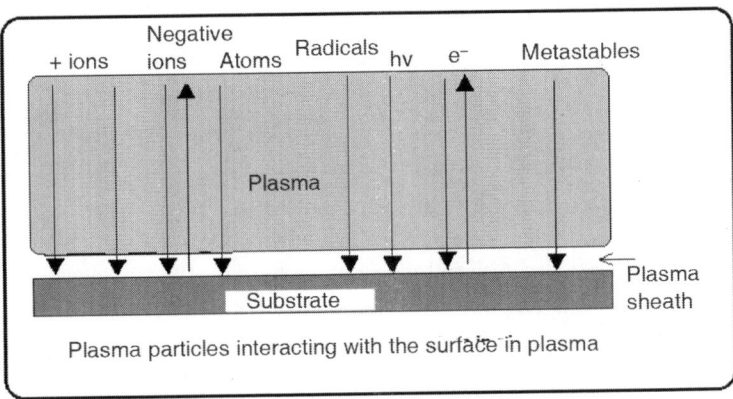

Figure 2.1 Interaction of plasma species with substrate surface

Table 2.1 Gas phase and surface reactions in plasma environment

Chemical reactions in plasma	Chemical equations
Electron impact reactions	
Ionization	$e + O_2 \rightarrow O_{2+} + 2e$
Excitation	$e + O_2 \rightarrow O_2^* + e$
Dissociation (radical formation)	$e + CH_4 \rightarrow .CH_3 + .H + e$
Dissociative ionisation	$e + CH_4 \rightarrow CH_{3+} + H + 2e$
Dissociative attachment	$e + SiCl_4 \rightarrow SiCl_3 + Cl^-$
Radiative recombination	$e + Ar^+ \rightarrow Ar + hv$
Three-body recombination	$e + H^+ + CH_4 \rightarrow CH_4 + H$
Reaction between atoms and chemical species	
Charge exchange	$Ar + O_2^+ \rightarrow Ar^+ + O_2$
Penning ionisation	$Ar^* + O_2 \rightarrow Ar + O_2^+ + e$
Penning dissociation	$Ar^* + CH_4 \rightarrow Ar + CH_3 + H$
Combination	$2 CH_3 \rightarrow C_2H_6$
Heterogeneous interactions of active species (with surfaces)	
Metastable deexcitation	$Ar^* + -CH_2-CH_2- \rightarrow Ar + -CH_2-CH_2-$ (polyethylene)
Adsorption	$CH_4 + (Solid) \rightarrow CH_4..(Solid)$
Sputtering	$Ar^+ + Metal/Polymer \rightarrow Ar + $ atoms of Metal/Polymer
Secondary electron emission	$Ar^+ + M \rightarrow M + e$
Radical formation at polymer surface (Hydrogen abstraction)	$-CH_2-CH_2- -CH_2-CH_2-$, CH_2 .CH $-.CH-CH_2-$
Functionalization	$.OH + .CH - CH_2 - \rightarrow HO-CH-CH_2 -$ polymer
Cross linking	Polymer chains Cross-linked chains

Surface activation of polymers and textiles takes place in plasma when it is exposed with air, oxygen, ammonia or nitrous oxide plasma. The primary reaction on plasma exposure is hydrogen abstraction. Hydrogen abstraction produces free radicals on polymeric chain. Different chemical species formed in the gas phase subsequently interact with its surface thus create functional groups on polymer's surface. Almost any fibre or polymeric surface may be modified to provide chemical functionality to specific adhesives or coatings, significantly enhancing the adhesion characteristics and permanency. For instance, polymers activated in such a manner show significantly high adhesive strength and permanency. For example, polyethylene, which consists of carbon and hydrogen, on exposure to plasma its surface is activated with the formation of radicals and functional groups such as hydroxyl, carbonyl, peroxyl, carboxylic, amino and amines based on the plasma-forming gas selected.

Cold plasma treatment of textiles is regarded as a dry, green process compared with traditional textile wet processing, and its potential for exploitation in the textile and clothing industries is currently the hot topic of study in a number of research and development centres worldwide. In the last decade, considerable efforts have been made by researchers and plasma technology suppliers across the world to develop both low-pressure and atmospheric pressure plasma systems and processes designed for industrial treatment of textiles and non-wovens to impart a broad range of functionalities. It has been observed that in textiles, biomedical components, polymers, wood, etc., finishing or surface modification is essential to obtain desired surface properties. This emerging technology has generated a considerable amount of interest for its use in the textile industry.

2.2   Status of atmospheric pressure plasma in the world

In the field of textiles, significant research work is going on since the early 1980s and many laboratories across the world are dealing with low-pressure plasma treatments of a variety of fibrous materials. Non-thermal plasmas are particularly suited to apply to textile processing because most textile materials are heat-sensitive polymers. It is a versatile technique, where a large variety of chemically active functional groups can be incorporated on the textile surface as well as it generates significant morphological changes. The possible actions of plasma treatment are improved wettability, adhesion of coatings, dyeability, printability, induced hydrophobic and/or oleophobic properties, changing physical and/or electrical properties, cleaning or disinfection of fibre surfaces, etc. The main advantage of plasma processing is that it is a dry treatment. Additionally, it is a very energy efficient and clean process. Therefore, the pre-treatment and finishing of textiles by non-thermal plasma technologies become very popular as a surface modification technique (Tomasino et al., 1995; Samanta et al., 2006; Hegemann, 2006; Radetic et al., 2007; Shishoo, 2007; Morent et al., 2008; Buyle, 2009; Kale et al., 2011). In general, the environmental benefits of plasma treatment can be summarized as reduced amount of chemicals needed in conventional processing, reduced BOD/COD of effluents, shortening of the wet processing time and significant amount of energy savings.

The recent trends of the textile industry are aimed to improve the environmental performance by reducing consumption of energy, water and chemicals used in processing as well as reducing environment impacts of end product and waste by using less toxic and biodegradable reactant (Dawson, 2012). Due to this, the textile industries have become more interested in plasma applications as novel finishing technology that can significantly reduce environmental impacts; however, industrial application of plasma pre-treatment to reduce environmental impact of conventional wet processing is still a challenge. The most plasma treatment of polymeric material have been conducted using low-pressure plasmas, which being batch process does not meet requirement of continuous processing of textiles. Also, requirement of creating and sustaining the vacuum/low pressure conditions leads to limitations on productivity and energy efficiency when applied at industrial scale processing (Samanta et al., 2006; Radetic et al., 2007; Morent et al., 2008; Kale et al., 2011). The development and use of atmospheric plasmas discharges has expanded enormously since the late 1990s, and thus new horizon opened for application of plasma technology for textile processing, renewing the interest of the scientific community (Kogelschatz et al., 1997; Napartovich, 2001; Fridman et al., 2005; Tendero et al., 2006). With atmospheric pressure plasmas, similar surface treatment can be achieved in a continuous process at reduced processing cost.

Depending upon type of gas chosen, there are two types of plasma treatment: surface activation and plasma polymerization (deposition), which are commonly used for changing the surface properties of textiles (Hegemann, 2006; Shishoo, 2007; Morent et al., 2008). A high functionality can be obtained by deposition of ultrathin coatings by plasma polymerization, however, the organic–inorganic precursor used has strong environmental impact. While treatment, using non-polymerizing gas, plasma primarily results in activation of polymer surface by the generation of functional groups.

The plasma treatment has shown very promising results and has improved various functional properties in plasma-treated textiles. Lately, many EU-financed projects within the 4th, 5th and 6th Framework Programme have had the objective of developing and demonstrating the feasibility of plasma-based industrial processes to meet the needs of the textile industry and offer tools for product development and innovation. As a part of EU Project Leapfrog CA, an extensive literature analysis and patent survey was carried out in 2005 in the area of plasmas and plasma-induced functionality of textiles. Hundreds of articles have been written on these subjects and very large numbers of patents have been granted in the field of plasma treatment of fibres, polymers, fabrics, non-wovens, coated fabrics, filter media, composites, etc., for enhancing their functions and performances. This survey has pointed out the potential use of plasma treatments of fibres, yarns and fabrics for various types of functionalization such as anti-felting/shrink-resistance of woollen fabrics, enhancement of hydrophilicity to improve wetting and dyeing, adhesive bonding, increasing hydrophobicity, anti-bacterial fabrics, etc.

This shows that plasma technology, when developed at a commercially viable level, has strong potential to offer new functionalities in textiles. In recent years, considerable efforts have been made by many plasma technology suppliers to develop both low-pressure and atmospheric pressure based plasma machinery and processes designed for industrial treatment of textiles and nonwovens to impart a broad range of functionalities. The standard and custom-designed plasma systems being offered are (i) low-pressure plasma systems for in-line and batch treatment: Europlasma (Belgium), P2i (UK), Mascioni (Italy); (ii) atmospheric-pressure plasma system for on-line continuous treatment: Dow Corning Plasma System (Ireland), Ahlbrandt (Germany), AcXys (France), APJeT (USA), Tri-Star (USA). For the first time, in an International Textile Machinery Exhibition (ITMA), 2007, at Munich, four manufacturers were showcasing atmospheric pressure and low-pressure plasma processing systems for textile applications namely Arioli (Italy), Grinp (Italy), HTP Unitex (Italy), Mageba (Germany) (Shishoo, 2007). Dow Corning has developed a technology to deposit highly functional thin coating using atmospheric pressure plasma of liquid precursors (Dow Corning, www. dowcorning.com/plasma).

In India too, the textile educational and research institutes have been carrying out research work in plasma textile applications at the laboratory scale contributing to the knowledge base in the field. The recent indigenous development of atmospheric pressure plasma processing prototype plant for Angora Wool (APPAW) by FCIPT-NID-CWDB has generated a lot of interest amongst the Indian textile technologists and researchers in the usage of plasma technology in textiles. IIT Delhi has developed atmospheric pressure uniform glow discharge in a laboratory reactor and studied the hydrophilic and hydrophobic coating deposited near atmospheric pressure glow discharge of different precursors on various textile fabrics such as cotton, silk, nylon and polyester. CSTRI also studied the effect of low pressure air plasma treatment on different types of silk such as Tasar, Muga, Eri. Wool Research Association (WRA) has developed itch-free woollens using plasma treatment either singularly or in combination with other processes. Also for applications in technical textiles, BTRA has imported a laboratory scale atmospheric pressure plasma machine from Italy. However, there is a growing need to look at various other potential plasma applications in textiles jointly with the textile educational and research institutes as well as industries in the country for the indigenous capacity building in technology development and machinery manufacturing.

2.3   Non-thermal plasma at atmospheric pressure

Atmospheric pressure non-thermal plasmas can be divided into the following four categories.

2.3.1   Corona discharge

Corona discharge is a process in which current flows from an electrode (kept at high potential) into a neutral fluid, usually air. It ionizes that fluid and creates a region of plasma around the electrode. The ions generated eventually pass charge to nearby areas of lower potential, or recombine to form neutral gas molecules. Corona discharge forms in Townsend discharge region where high electric field is produced through sharp points, edges or wires (Loeb, 1965). Coronas may be positive or negative. This is determined by the polarity of the voltage on the highly curved electrode. If the curved electrode is positive with respect to the flat electrode, it has a positive corona; if it is negative, it has a negative corona. Corona discharge is shown in Fig. 2.2 and its break down voltage and plasma density is mentioned in Table 2.2. Corona discharge has a number of commercial and industrial applications which include drag reduction over a flat surface, removal of unwanted electric charges from the surface of aircraft in flight, manufacture of ozone and sanitization of pool water, scrubbing particles from air in air-conditioning systems, removal of unwanted volatile organics, such as chemical pesticides, solvents, or chemical weapons agents, from the atmosphere.

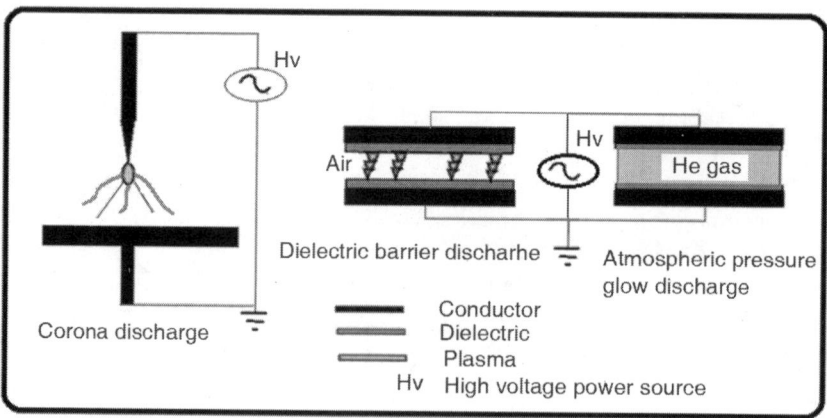

Figure 2.2 Atmospheric pressure discharges commonly used for textile treatment

Table 2.2 Breakdown voltage and plasma density value in different non-thermal plasma discharges (Schutze et al., 1998)

Source	Break down voltage (kV)	Plasma density (cm^{-3})
Low pressure discharge	0.2–0.8	10^8 to 10^{13}
Corona discharge	10–50	10^9 to 10^{13}
Dielectric barrier discharge	5–25	10^{12} to 10^{15}
Plasma jet	0.05–0.2	10^{11} to 10^{12}

2.3.2   Atmospheric pressure plasma jet (APPJ)

The atmospheric pressure plasma jet (APPJ) is a small (L <20 cm) RF plasma torch that generates non- thermal plasma jet and works at low power (Schutze et al., 1998; Tendero et al., 2006). The atmospheric pressure plasma jet is shown in Fig. 2.3. It consists of two concentric electrodes through which the working gas flows. By applying RF power to the inner electrode at a voltage between 100 and 150 V, the gas discharge is ignited. The ionized gas exits through a nozzle since the gas velocity used to be typically in the range of 5–12 m s^{-1}. The low-injected power enables the torch to produce a stable discharge and avoids the arc transition. In the jet, the gas temperature ranges from 25°C to 400°C, charged-particle densities are 10^{11} to 10^{12} cm^{-3}, and reactive species used to be present in high concentrations. Typical breakdown voltage and plasma density in plasma jet are shown in Table 2.2. Based on the impedance measurements and the emission spectra, it is estimated that the electron temperature inside the plasma jet averages between 1 and 2 eV (Schutze et al., 1998). Atmospheric pressure plasma jets are being used for large number of applications which include sterilization and blood clotting, cutting of unwanted tissues, surface modification of polymers, cleaning of metals (archaeological statues), deposition of SiO_2 and TiO_2 coating, production of fullerenes, etc.

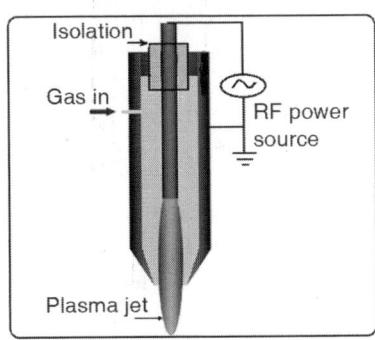

Figure 2.3 Plasma jet to produce atmospheric pressure discharge for textile treatment

2.3.3   Atmospheric pressure glow discharge (APGD)

Dielectric barrier discharges can be realized in two major operating modes, namely the filamentary mode and the homogeneous mode at atmospheric pressure, under certain conditions. Filamentary DBDs are characterized by transient microdischarges which are usually stochastically distributed in space and in time. These discharges result in inhomogeneous treatment and are less suitable for applications like surface treatments where uniformity is the main aspect. A homogeneous discharge is still advantageous than a filamentary discharge due to the uniformity rendered by the former. Unlike filamentary discharge, the discharge parameters of a uniform DBD can be controlled both spatially and temporally in number of ways. Besides the application prospects, studies are also focused on the phenomenal aspects of barrier discharges and the physics governing them. Unfortunately, the most common type of high-pressure DBD is non-uniform and filamentary. The existence of the uniform diffuse modes of DBD is usually strongly limited by the use of a specific gas mixture composition, dissipated power, operation frequency, etc. The transitions between discharge modes in the same experimental conditions have been profoundly investigated in nitrogen, rare gases and their mixtures with air and other gases.

The term atmospheric pressure glow discharge (APGD) was introduced in the 1990s by Kanazawa et al. (1988) who showed that for a proper selection of plasma forming gas (very often helium) and molecular admixtures, a stable diffuse discharge can be ignited in an AC-excited dielectric barrier set-up, free of streamers or filaments. Atmospheric pressure glow discharge has been depicted in Fig. 2.2. Further, Helium has a very low breakdown voltage and longer transition time T. Helium atoms also have some high-energy metastable states, as high as 20 eV. These states may help to extend the microdischarge points on an insulating plate, because the high energy He will easily ionise the mixed molecules. Generally, the ionisation potential is under 15 eV, and the dissociation energy of molecules is about 5 or 10 eV. The metastables of He having 20 eV energy exert large effects on the stabilisation of atmospheric pressure glow discharge and its applications. There are experimental evidences for the dissociation of oxygen molecules by He atomic plasma (Yokoyama et al., 1990).

The APGD is very useful for the homogeneous treatment of sensitive surfaces, since it can be operated at much lower voltages than the traditional DBD. Ayan et al. (2008, 2009) reported a uniform mode of dielectric barrier discharge in atmospheric air, where the discharge was generated by application of high voltage pulses with nanosecond duration. Below we describe new preliminary experimental results that allowed to formulate a hypothesis for the mechanism of generation of uniform DBD in atmospheric air by application of nanosecond pulses with fast rise time.

2.3.4   Dielectric barrier discharge (DBD)

Dielectric barrier discharge (DBD) is an electrical discharge between two electrodes separated by an insulating dielectric barrier. It is shown in Fig. 2.4. The process normally uses high voltage alternating current, ranging from lower RF to microwave frequencies. DBD is characterized by the presence of at least one insulation layer in the discharge gap between two metal electrodes. When a high AC voltage (typically tens of kilovolts in the frequency range of $50-10^4$ Hz) is applied to this configuration, the gas breakdown occurs. Various reactions that take place in gas phase and at surface are given in Table 2.1, and typical break down voltages and plasma densities in various non-thermal plasma discharges are shown in Table 2.2. In air the DBD normally operates in filamentary mode, i.e. it is constituted by large number of tiny filaments distributed uniformly over the entire dielectric layer. To ensure stable plasma operation, the gap which separates the electrodes is limited to a few millimeters wide. Plasma gas flows in the gap. The discharge is ignited by means of a sinusoidal or pulsed (Salge, 1996) power source. Depending on the working gas composition, the voltage and excitation frequency, the discharge can be either filamentary or glow (Massines et al., 1998; Yokoyama et al., 1990). A filamentary discharge is formed by micro-discharges or streamers that develop statistically on the dielectric layer surface.

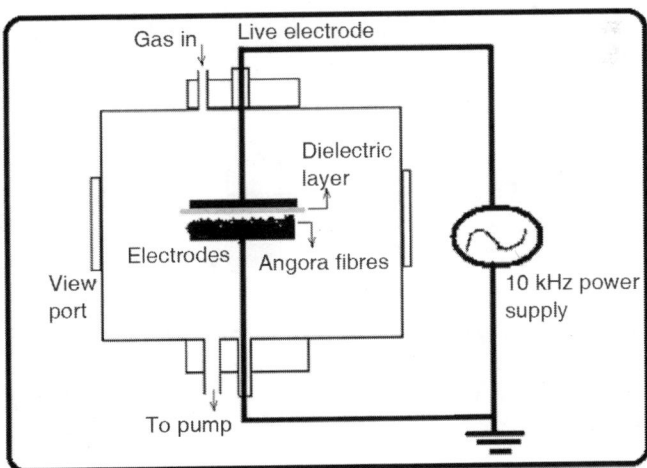

Figure 2.4 Block diagram for DBD System for Angora wool treatment

At atmospheric pressure, gases breakdown in a plane parallel gap with insulated electrodes that normally occur in a large number of individual tiny breakdown channels, referred to as micro-discharges (streamers). An individual filamentary discharge is initiated when a high voltage is applied between the electrodes such that the electric field in the open gap equals or exceeds the breakdown strength of the ambient gas. The term "microdischarge" refers to a bright, thin plasma filament that is observed in DBD gap space. Electrons emission from the surface of the dielectric is stimulated by UV photoemission accelerated in the electric field to energies that equal or exceed the ionization energy of the gas and create an avalanche in which the number of electrons doubles with each generation of ionizing collisions. The high mobility of the electrons compared to the ions allows the electrons to move across the gap in durations measured in nanoseconds. The electrons leave behind the slower ions and various excited and active species that may undergo further chemical reactions. When the electrons reach the opposite electrode, the electrons spread out over the insulating surface, counteracting the positive charge on the instantaneous anode. This factor, combined with the cloud of slower ions left behind, reduces the electric field in the vicinity of the filament and terminates any further ionization along the original track in time scales of tens of nanoseconds.

The use of helium as plasma gas seems to favor a glow discharge. High energetic He metastable species (Yokoyama et al., 1990) and Penning effect (Massines et al., 1998; Yokoyama 1990) provide uniform glow discharge. The dielectric layer plays the following important roles: (i) it limits the discharge current and avoids glow to arc transition that enables to work with a continuous or pulsed mode; and (ii) it distributes random streamers on the electrode surface and ensures a homogeneous treatment. The streamer creation is due to the electrons accumulation on the dielectric layer.

Dielectric barrier discharge process is becoming popular for the treatment of textiles at atmospheric pressure. The treatment can be used to modify the surface properties of the textile to improve wettability, absorption of dyes, and for sterilization. DBD plasma provides a dry treatment that doesn't generate waste water or require drying of the fabric after treatment. DBD can be characterized as non-thermal, weakly ionized plasma. The active plasma species, such as, UV photons, excited species and radicals react with the polymer surface and alter its surface energy. Surface modification of polymers occurs through surface oxidation and scission as well as by surface roughening.

2.4   Influence of process and plasma parameters and scaling up of system

The process parameters play a very important role in tailoring the surface properties of polymers and textile materials. These parameters include source frequency, power density, process pressure, plasma forming gas, residence time, etc. Similarly the plasma parameters which include electron temperature, plasma density, electron and ion energy distribution change the concentration of reactive species and therefore the overall treatment of the surface. It is shown in Fig. 2.5. Among various process parameters, role of excitation frequency is important since it influences the behaviour of the electrons and the ions, for example electron energy and ion energy distribution function, electron temperature and plasma density and free radicals produced in plasma. Reece Roth et al. (1995) reported that at one atmosphere plasma, applied frequency typically in tens of kHz range, traps the ions in between the plasma zone. The small change in frequency traps electron instead of ion and glow plasma change to filamentary plasma. The atmospheric plasma sources can be classified considering their following excitation mode.

- DC (direct current) and low frequency discharges
- RF discharges which are ignited by radio frequency waves
- Microwave discharges

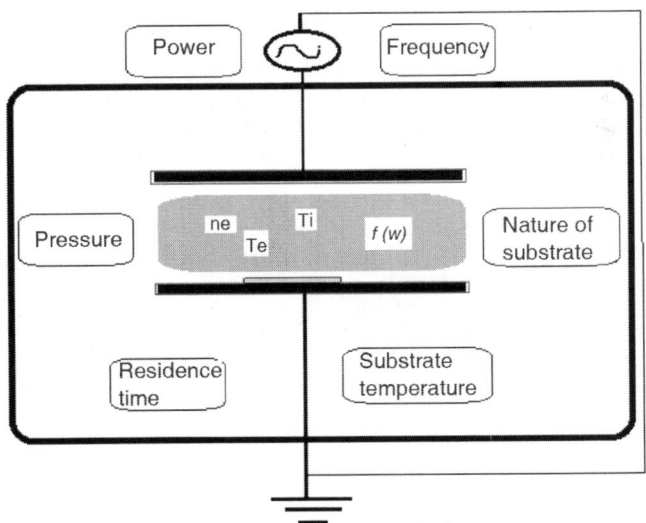

Figure 2.5 Process and plasma parameters that control the chemistry during the treatment

Power density has an influence on energy of ions and plasma density. At higher power density, surface etching becomes more dominant reaction. The process is at below atmospheric pressure or at atmospheric pressure decides the applied voltage for plasma breakdown. At low pressure (in the range of 10 torr to 10^{-2} torr), plasma can be produced by applying few hundred volts potential difference between the electrodes, whereas at atmospheric pressure few thousands of volts potential gradient between the electrodes is required for plasma breakdown. At higher pressure, ion–neutral collisions are dominant; and therefore, at same power coupling level, low pressure plasma provides better etching or surface activation. Plasma-forming gas is selected based on the type of surface treatment required. For surface activation and etching, gases such as Argon, Helium, Nitrogen, etc., are selected as a plasma-forming gas whereas oxygen, ammonia, air, CO_2, etc., are selected for functionalization. Gases or precursors such as CH_4, Hexamethyl disiloxane, $CF_2=CF_2$, etc., are introduced in plasma to deposit coatings. Residence time of polymers or textile surface in plasma is another important parameter for getting appropriate surface modification. Therefore, to expose the textile surface for the desired residence time, either plasma is produced over a large electrode area or processing speed of textile is adjusted as per the requirement. One can tailor the surface energy, number of functional groups per unit area, chemical composition of surface- and polymer-coating properties by varying process as well as plasma parameters.

Scaling up of plasma systems involves understanding of important plasma parameters, plasma chemistry, use of same power density, and surface to be modified should remain in plasma and should be exposed in plasma for the same residence time. Source frequency plays important role in changing plasma parameters such as electron energy distribution function, electron and ion temperature and densities. The dielectrics employed in these systems used to be polymers, glass, ceramics and quartz. Polymer dielectrics get punctured during long-operation cycles as well as also add impurities due to etching; therefore, ceramics and glass are more suitable dielectric materials. Atmospheric pressure uniform glow discharge is obtained using helium as a plasma-forming gas; however, it makes the process economically unviable because helium is costly gas. Therefore, the plasma reactor should be designed to recover helium or else one should use cheaper gases such as air, oxygen, etc. It is observed that filamentary discharge is formed when air and oxygen is used to produce dielectric barrier discharge. In order to make barrier discharge in air or oxygen nearly uniform, one should select appropriate power source, electrode design and gas flow mechanism. Plasma surface interaction with textiles generates many changes on its surface, few of them are discussed in the following sections.

2.5   Plasma textile interaction

When plasma is produced between the two electrodes and textile surface is placed inside the system for exposure, different interactions begin. In the gas phase, electrons, ions, excited neutral atoms and molecules interact and generate variety of active species. Different reactions that take place in plasma environment are given in Table 2.1. The active species present in plasma interact with textile surface and produce significant changes. It is shown in Fig. 2.1. The four chemical reactions that mainly take place when plasma active species interact with the surface of textile material are described below.

2.5.1   Plasma cleaning and surface activation

Ultra-violet light generated in the plasma is very effective in the breaking of most organic bonds of surface contaminants. This helps to break apart oils and grease. A second cleaning action is carried out by the energetic oxygen species created in the plasma. These species react with organic contaminants to form mainly water and carbon dioxide that are continuously removed (pumped away) from the chamber during processing. Plasma cleaning is a proven, effective, economical and environmentally safe method for critical surface preparation. Plasma cleaning with oxygen plasma eliminates natural and technical oils and grease at the nano-scale and reduces contamination.

UV radiation and active oxygen species from the plasma breaks the bond of silicones and oils present at the surface. Active oxygen species (radicals) from the plasma bind to active surface sites all over the material, creating a surface that is highly active to bonding agents. Plasmas treatment provides an effective surface activation prior to printing, lacquering or gluing. Similarly, glass and ceramics surface can be plasma activated. Surfaces should be used immediately after plasma treatment; plasma-treated surfaces may recover their untreated surface characteristics with prolonged exposure to air.

For example aging of polypropylene surface does not occur up to several weeks after the plasma treatment. Plasma activation is suitable for automotive components, medical plastics, aerospace components, rubber, etc.

2.5.2   Etching of surface

Generally, active plasma species interact with the surface exposed to the plasma. The plasma forming gas selected and power density decide which reaction would be more dominant, i.e. sputtering or chemical etching. When inert gas such as Ar or He is used as a plasma-forming gas, sputtering is more dominant reaction; whereas when oxygen, air or SF_6 is selected for plasma etching, chemical etching is more dominant than physical etching. This means, a chemical reaction takes place between the surface (the surface to be etched) atoms and reactive species from the plasma and form a product molecule, which is subsequently removed from the surface. The polymer or textile surface (mostly insulating) becomes negatively biased due to electron bombardment; therefore, there is always some sputtering takes place. The main steps in the chemical etching process are:

- formation of the reactive species in plasma
- arrival of the reactive species at the surface to be etched
- adsorption of the reactive species at the surface
- chemisorption of the reactive species at the surface, i.e. a chemical bond is formed
- formation of the product molecule
- desorption of the product molecule
- removal of the product molecule from the reactor.

Plasma etching is used to roughen a surface on the microscopic scale. After etching, the surface area is greatly increased, making the material easily wettable. Etching is used before printing, gluing, painting, adhesive bonding and dyeing and is particularly useful for processing of polyester, PTFE, large number of other polymers and textile materials.

2.5.3   Functionalization and grafting

The active species and UV radiation formed in plasma (as shown in Table 2.1) interact with polymer and textile surface. When the selected plasma-forming gas is either inert (e.g. Argon, He, etc.) or non-polymer forming (e.g. O_2, N_2, Air, etc.), then free radicals and functional groups are formed at the polymer or textile surface exposed to the plasma. The functional groups include – OH, –COOH, >C=O, –NH$_2$, etc. In grafting, an inert gas such as argon is employed as plasma-forming gas, the active plasma species interact with the surface and create many free radicals on the material surface subsequently, a monomer capable of reacting with the free radical is introduced into the chamber, which makes covalent bond with the free radicals and grafted. The grafting is possible at low pressure as well as at atmospheric pressure plasma. Typical monomers used for grafting are acrylic acid, allyl amine and allyl alcohol. Another important method to modify the fibre, to increase the uptake of dyes and finishes or to impart unique functionality, is performed through cold plasma (Sarmadi et al., 1995).

2.5.4   Deposition of coating or polymerization

When materials are subject to plasma treatments, the subsequent and significant reactions are based on free radical chemistry. The glow discharge plasma is efficient at creating a high density of free radicals by dissociating molecules through electron collision and photochemical processes. These gas-phase radicals have sufficient energy to disrupt chemical bonds in polymer surfaces on exposure, which results in formation of new chemical species.

Plasma polymerization refers to the formation of polymeric materials under the influence of plasma (partially ionized gas). In plasma polymerization process, thin polymer coating is deposited on a substrate surface. Therefore, this process alters the surface properties without affecting the bulk properties of the polymer substrate. To initiate plasma polymerization, monomer vapor/gas is introduced into the process chamber. Plasma is generated between the electrodes in which the excited electrons collide with the monomer molecules and ionize them. The monomer molecules break apart (fractionate) creating free electrons, ions, excited molecules and radicals. These radicals adsorb, condense and polymerize on the substrate surface. The electrons and ions cross link, or create a chemical bond with the substrate surface or with the already deposited molecules and create a denser film. The plasma polymerization process is done using low temperature or cold plasma. Comparison of conventional polymerization with plasma polymerization is shown in Fig. 2.6.

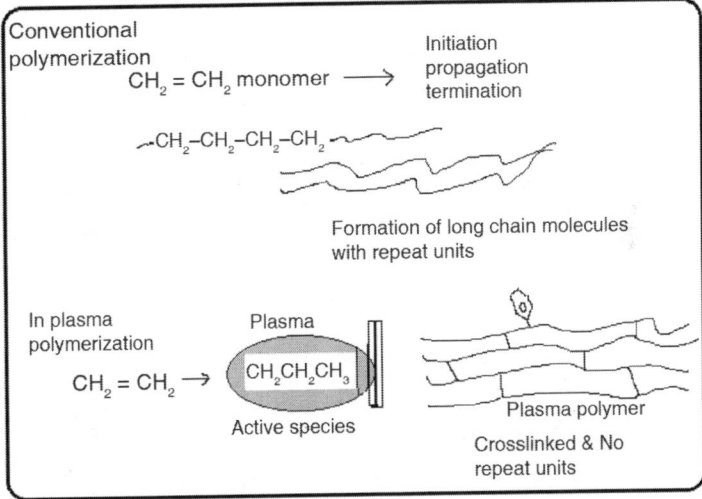

Figure 2.6 Comparison of conventional and plasma polymerisation

The type of discharge gas determines the stability of the glow discharge. Helium gives rise to a stable and homogenous glow discharge, whereas nitrogen, oxygen and argon require higher voltage for ionization and can cause the transition to a filamentary glow discharge. By changing electrode geometry, it is possible to obtain nearly homogenous, yet somewhat filamentary, glow discharge (Akishev et al., 2002). At atmospheric pressure, the reactive gases leading to the coating formation are diluted in a main gas, which is usually He or Ar in order to get a homogeneous discharge. The nature of the particular carrier gas largely influences the physics of the discharge and in particular the physics of the homogeneous DBD, i.e. atmospheric pressure glow discharge (APGD) which is generally obtained with noble gases.

2.6   Atmospheric pressure plasma technologies developed at FCIPT for textiles

Facilitation Centre for Industrial Plasma Technologies (FCIPT) is a division of Institute for Plasma Research, Gandhinagar, and is involved in the development of various plasma-based technologies for different applications. FCIPT has got expertise in the development of thermal and non-thermal plasma production devices and reactors. Considering the importance of green technologies in textiles, FCIPT has taken up projects to solve some of the problems that are existing in textile industries. Among various plasma technologies developed here, surface modification of Angora wool using atmospheric pressure plasma to improve cohesion among wool fibres is one of them.

Angora wool is extremely warm, soft and silky to touch. As it is slippery fibre to spin, its surface is modified by glow discharge plasma. Plasma treatment assists in increasing the friction and cohesion among the fibres and assists in spinning of 100% Angora yarn without any difficulties. Thereby it avoids problems like static charge, shedding of fibres and fibrosity. FCIPT, Institute for Plasma Research (IPR), carried out a pioneering research work by way of developing an Innovative Eco-friendly Atmospheric Plasma Processing System for Angora Wool (APPAW) under the sponsorship of Department of Science and Technology, Govt of India. It is the first proto-type system developed in the country, commissioned in Kullu. In this system, plasma is generated at atmospheric pressure using air as plasma-forming gas.

2.6.1   Plasma treatment of Angora wool to enhance cohesion among fibres

Angora fibre has chemical structure of protein. It consists of spindle-shaped cortical cells. It is one of the world's warmest, soft and light natural fibres. However, the smooth surface (low friction coefficient) and low-bonding force among the fibres lead to shedding as well as make the weaving of pure Angora wool difficult. In order to improve surface properties of Angora wool and to improve its processing, feasibility study for plasma surface modification of Angora fibres was taken up. The study show that the reactive species present in the discharge region interact with the surface of the materials and modify some physical and chemical characteristics such as roughness due to etching and functional groups formed due to oxygen species etc.

At FCIPT, dielectric barrier discharge was produced at atmospheric pressure for the treatment of Angora wool. Initially He, and later He and air mixture was used as a plasma-forming gas. The schematic of the experimental set up is shown in Fig. 2.7. The set up consists of 60 cm long rectangular chamber having width 30 cm. It has two rectangular parallel plate of stainless steel of dimension 16 × 7.5 cm, which works as electrodes. Both the electrodes were covered with the dielectric sheet of polypropylene of thickness 150 micron. The inter electrode gap was maintained at 6 mm in all the experiments. The applied voltage and current were measured with the help of high voltage probe (Tektronix P6015A, 1000X) and current transformer of model-IPC-CM-100-MG, respectively. Tektronis (TDS 224, 100 MHz) digital oscilloscope was used to record the voltage and current waveform. High voltage power supply of 10 kHz, 5 kV (rms) and 1 Amp rating was employed to generate the plasma. Initially, the Angora fibres were finely spreaded on the ground electrode and the chamber was evacuated to 1 m bar. Subsequently, the chamber was filled with He or He + air gas and the chamber pressure was raised up to 1 atmospheric pressure. The plasma was generated using high voltage power supply. In He, the discharge forms at lower breakdown voltage. It is due to the trapped electron and large number of He metastable atoms formed in the discharge. Indeed electrons are stored in the positive column of the discharge. Moreover, in the plasma the metastable atoms maintain a production of electron via penning ionization (Joko and Koga, 1990). He + air plasma provides active O radicals along with metastables for surface modification.

Figure 2.7 Plasma at 1 atmospheric pressure produced using He as a plasma-forming gas

We studied the chemical and morphological changes on the surface of Angora fibre when it is exposed to atmospheric pressure helium, and helium + air glow discharge (DBD). The morphological changes were determined by Scanning Electron Microscope (Fig. 2.8). The chemical changes were seen using X-ray photoelectron spectroscopy which is given in Table 2.3. We observed that He + air plasma treatment significantly modifies the chemical composition of the fibre surface. From Table 2.3, it is clear that the carbon concentration reduces whereas oxygen concentration increases at the surface after plasma treatment. The C_{1s} band of the untreated fibre can be fitted with two peaks. The hydrocarbon signal shows a peak at 285.0 eV. The second peak at 288.6 eV represents oxidized carbon on the fibre surface attributed to oxidized carbon species i.e. ketone [-(C=O)-] or acetal [-(O-C-O)-] carbon (Hesse et al., 1995; Molina et al. 2005). Increase in O content is also attributed to increased oxidation of S atoms present in the Angora fibre. From Table 2.2, it is clear that the S_{2p1} band of He + air plasma treated fibres can be fitted with two peaks, i.e. 164.0 eV for S present in macromolecule and 168.5 eV represent oxidized sulphur (-S-O bonding) (Chastain et al., 1995). The XPS results reveal the strong oxidation level at the surface of (He + air) plasma exposed fibre. In addition, coefficient of friction of untreated and plasma exposed Angora fibre was measured using fibre friction measuring device developed by National Institute of Design, Ahmedabad. The coefficient of friction increases after plasma treatment and it is shown in Table 2.4.

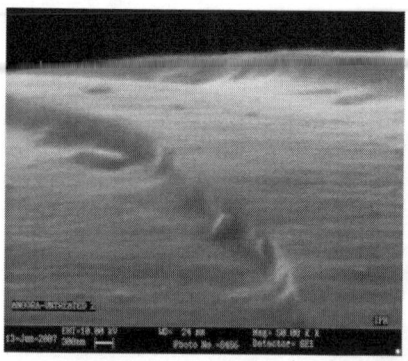
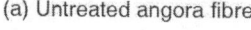
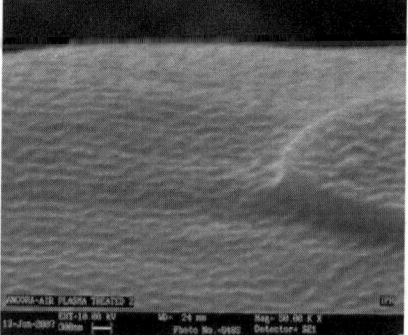

(a) Untreated angora fibre   (b) Plasma treated angora fibre

Figure 2.8 SEM image of (a) untreated and (b) plasma-treated Angora wool

Table 2.3 X-ray Photoelectron Spectroscopic Study of Angora Wool (Danish et al., 2007)

Band (Position eV)	Untreated (%)	(He + Air) plasma treated (%)
C_{1s} (285)	59.79	52.88
C_{1s} (288.6 ± 0.1)	22.50	25.54
O_{1s} (532.2 ± 0.1)	9.85	15.25
N_{1s} (400.1 ± 0.1)	5.61	4.40
S_{2p1} (164.0 ± 0.2)	2.25	1.04
S_{2p1} (168.5)	–	0.89

Table 2.4 Coefficient of Friction of untreated and plasma-treated Angora wool

Sample identification	Coefficient of Friction (μ)
Untreated Angora wool	0.08–0.1
He + Air plasma treated Angora wool	0.2–0.25
Air plasma treated Angora wool	0.2–0.3

It is evident from the XPS study that helium and air plasma treatment has increased oxygen concentration at the surface of Angora fibres, whereas the change in surface morphology is clearly seen in SEM micrograph of untreated and plasma-treated fibres. The increase in coefficient of friction value from 0.1 to more than 0.2 improves processing of wool and reduces shedding significantly.

2.6.2   Development of proto-type and commercial plasma system for Angora wool treatment

Based on the above findings, FCIPT along with Central Wool Development Board planned to develop proto-type plasma system for continuous treatment of Angora wool to improve its processing and establish the system in Kullu. There were a few constrains such as non-availability of costly gas He in Kullu, low inline voltage and fluctuations in power, treatment cost if He is used as plasma forming gas, etc. All these constrains were overcome one by one and finally the prototype system was successfully established in the Angora cottage industry at Kullu in 2009 for spinning of 100% Angora yarn and making newer products with the financial support from Department of Science and Technology, New Delhi, Govt. of India. It is illustrated in Fig. 2.9. Since then there has been demand from Angora farmers for establishing large size plants amongst Angora industry clusters of the country. In order to meet the growing demand of plasma-treated Angora fibre from the cottage industry in the country, the technology has been licensed to a leading textile machinery manufacturer, InspirOn Engineering Private Ltd., Ahmedabad, for commercialisation. The company has manufactured two industrial scale plants for plasma processing of Angora fibres and commissioned at two different locations first at Himalayan Institute for Environment, Ecology & Development (HIFEED), Ranichauri, Uttarakhand, and the second one at Kullu Handloom & H/C Weavers Cooperative Society, Kullu (Himachal Pradesh) in 2012 under sponsorship of DST, Govt. of India. DST has sponsored a project to install the third plasma system at Chungthang, Sikkim by May 2014.

Figure 2.9 Proto-type air plasma system for the treatment of Angora wool, commissioned at Kullu in 2009

This development has opened the door for using plasma technology in addressing the technological problems of the Angora cottage industry successfully. This will help in value addition and will provide access to better markets at home and abroad for Angora products. The techno-economic viability of the plasma treatment is excellent because the cost of treatment is a small fraction of the Angora wool cost. The industrial scale plasma system has following features:

- Air plasma production at atmospheric pressure on a large area (1.0 × 0.5 m, between 9 set of electrodes)
- 40" Angora Plasma Treatment system is developed with the linear speed of 4 m/min.
- The system is successfully synchronized with existing 40" Carding machine and the Sliver system.
- Study of the movement of the belt with respect to temperature, humidity and current are made.
- With the study mentioned above, the change in the belt material is made.

The carding machine had a speed of 5.7 m/min and to maintain desired residence time for plasma treatment, the speed of the machine was brought down to 4.5 m/min. The Carding Machine was synchronized with our Angora plasma treatment system. The mechanism built is such that Angora wool would easily transfer from carding machine to Angora plasma system. High voltage power supply output voltage was kept 0–10 kV rms at 50 Hz. Variable, and the maximum current that can be attained is 1000 mA (rms, load dependent). Suction pump was employed to remove small quantity of ozone that is formed during the treatment in process chamber. The plasma reactor is designed in such a way that the Angora web is exposed to the air plasma for appropriate time so that desired surface properties (coefficient of friction) could be obtained.

During design the electrical safety interlock were taken into account. The arrangement is made to control the movement of the belt mechanically. Also sensors were mounted to indicate the belt movement on both the sides (i.e. left or right). Applied voltage across the electrodes was measured using 1000X voltage probe and current in the circuit is measured using current transformer. Measured voltage and current waveform is shown in Fig. 2.10. The current waveform shows multiple number of peaks (current filaments) for a given voltage waveform.

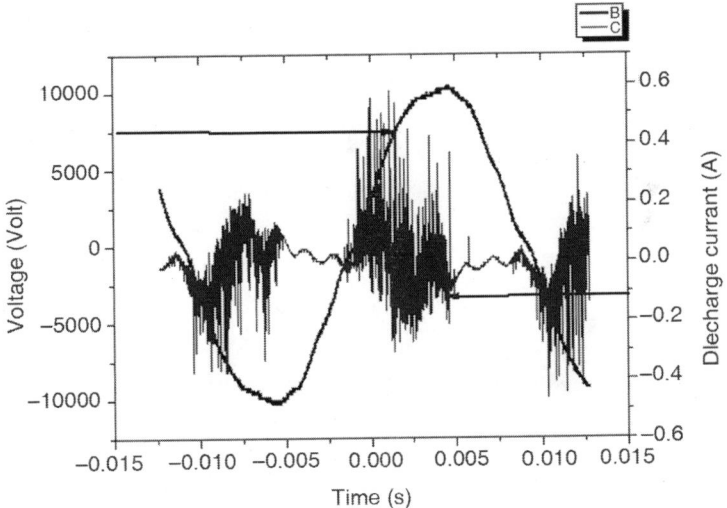

Figure 2.10 Voltage and current wave form of dielectric discharge generated using 50 Hz power source

We observed that the humidity and temperature affect the performance of the system, i.e. discharge current changes with the change in humidity and temperature. A systematic measurement of relative humidity was carried out and how the discharge current varies is determined. It is shown in Fig. 2.11. The change in discharge current with the variation in environment temperature is also recorded. The graph clearly reveals that discharge current increases with the increase in relative humidity in the environment. Hence discharge voltage has to be reduced at higher humidity in order to keep discharge current same for all treatments.

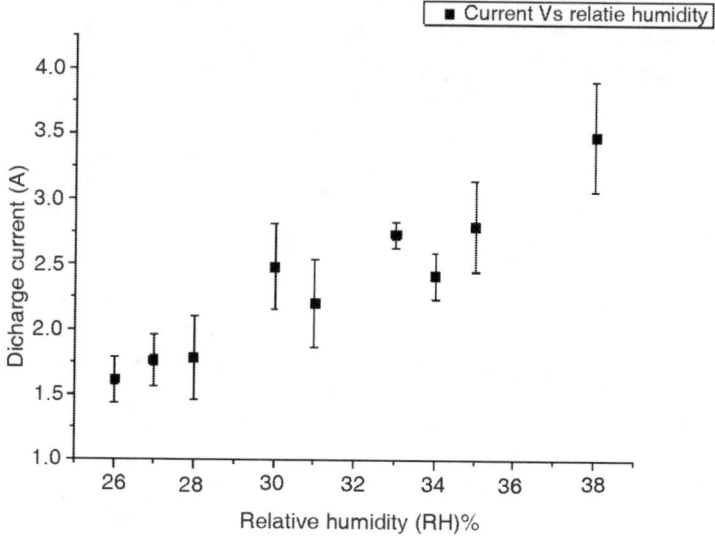

Figure 2.11 Graph showing discharge current dependence on relative humidity

Uniformity of plasma treatment was determined by placing small pieces of polyester at different locations on the electrode. The dielectric discharge was kept on for 1 minute. After plasma treatment, the surface of plasma-treated samples was tested by measuring water contact angle on all the samples using Goniometer. All the samples have shown contact angle in the range of 35–50°C which clearly depicts the uniformity of the plasma treatment.

The developed system is being used by HIFEED, Ranichauri, and Kullu Weavers Society at Kullu for the treatment of Angora wool. The plasma treatment has eliminated the use of other wool for blending with Angora for easy processing; hence, 100% Angora wool products are being made.

2.6.3   Plasma surface modification of polyester fibre and fabric

An experimental plasma reactor and power source was designed and fabricated at FCIPT which is illustrated in Fig. 2.12, from the financial support of DST. The plasma system was used to modify the surface of polyester fabric and fibres to enhance dye uptake properties. Dielectric barrier discharge was created in NH_3, air and O_2 gases. Polyester surface was exposed to the plasma of NH_3/O_2 for different time duration and at different power density, and dye uptake properties of polyester fibre and fabric studied. Scanning electron micrograph of untreated and plasma-treated polyester fabric is shown in Fig. 2.13. Plasma treatment though not affects the bulk properties of polymers; however alteration in surface crystallinity induced by plasma can alter the surface-related properties. Since dyeing of polyester is more related to surface than the bulk properties, it is possible that alteration in surface crystallinity induced by plasma may influence the dyeing properties of polyester fabric. The formation of functional groups and amorphization of polyester surface has been observed by its exposure to low pressure plasmas. ATR-FTIR spectroscopy study has revealed that surface amorphization takes place due to plasma exposure (Dave et al. 2013).

Figure 2.12 Laboratory scale plasma reactor for processing of cotton/polyester fabric

(a) Untreated polyester fabric   (b) Plasma treated polyester fabric

Figure 2.13 SEM image of (a) untreated (100KX) and (b) plasma-treated polyester fabric (100KX)

Dye uptake properties of plasma-treated polyester fabric are studied with six commercially available natural dyes. These natural dyes are (i) eco-alizarin (EA), which is root extract of Indian madder plant (*Rubia cordifolia*), contains various anthraquinones chiefly alizarin and purpurin as coloring constituents; (ii) eco-smoke grey (SG), which is powdered root barks of alkanets, chemically belongs to Naphthoquinones classes of compounds; (iii) eco-hill brown (HB) is mixture of root extracts of alkanet, Indian madder and pomegranate (*Punica granatum*) rind; (iv) dye eco-granet brown (GB) is powered wood extracts of Black Catechu (Acacia catechu) which is traditionally used as tanning agent for leather and dyeing the cotton and wool. The chief constituents of the plant are catechin and catechutannic acid. (v) Eco-turkey red (TR) is anthraquinone-based dye from root of Indian madder plant (*Rubia cordifolia*), and (vi) dye eco-orange (EO) is mixture of roots of Indian Madder plant with Pomegranate rind.

Initially, dyeing of plasma-treated polyester fabric is carried out at temperature of 80°C with eco-alizarin and eco-hill brown. Further, in order to decrease dyeing temperature polyester with natural dyes and achieve good dye uptake, various dyeing techniques including use of different solvents, mordents are studied. The solvent-assisted dyeing using little amount of solvents such as alcohol, acetone along with water to prepare dye bath has been found as a best technique to increase dye uptake for plasma-treated polyester. With solvent-assisted dyeing techniques, dyeing of plasma-treated polyester with above-mentioned natural dyes is achieved using cold extracts of natural dyes at room temperature. Study has demonstrated dyeing time as low as 1 hour gives good dye uptake; however, dyeing time of 12 to 16 hours is required for optimum results. Significant improvement in dye uptake has taken place, which is shown in Table 2.5 (Dave et al., 2013).

Table 2.5 Improvement (%) in color depth for low pressure plasma treated dyed samples using alcohol-assisted natural dyeing

No.	Name of dye	%Improvement (alcohol-assisted dyeing)	
		15 min Low pre. Air plasma	15 min Low pre. NH3 plasma
1	Hill brown	272	186
2	Eco-orange	367	137
3	Smoke grey	244	64
4	Turkey red	197	60
5	Eco-alizarin	152	144
6	Garnet brown	94	11

For colour measurement, spectral reflectance factors (R) of the dyed samples are measured using a Datacolor 3F-500 reflectance spectrophotometer (Datacolor International Ltd, UK) interfaced to a computer. Four different areas of each sample are measured and the average colour value is automatically calculated and saved by the computer. K/S values (representing absorption) are calculated from reflectance factors R at the desired wavelength by the software using the following Kubelka-Munk equation. The percentage increase in colour depth (% I) is obtained from the difference between the (K/S) value measured for plasma treated sample and untreated sample dyed using same dyeing procedure according to Eqs. 2.1 and 2.2.

$$\left(\frac{K}{S}\right) = \frac{(1-R)^2}{2R} \qquad \text{...(2.1)}$$

$$\%I = \frac{\left(\dfrac{K}{S}\right)treated - \left(\dfrac{K}{S}\right)untreated}{\left(\dfrac{K}{S}\right)untreated} \times 100 \qquad \text{...(2.2)}$$

Increase in K/S value with different dyes as results of low pressure plasma treatment of polyester fabric is shown in Fig. 2.14. The study reveals that DBD plasma treatment has changed polyester surface and increased dye uptake properties with natural dyes.

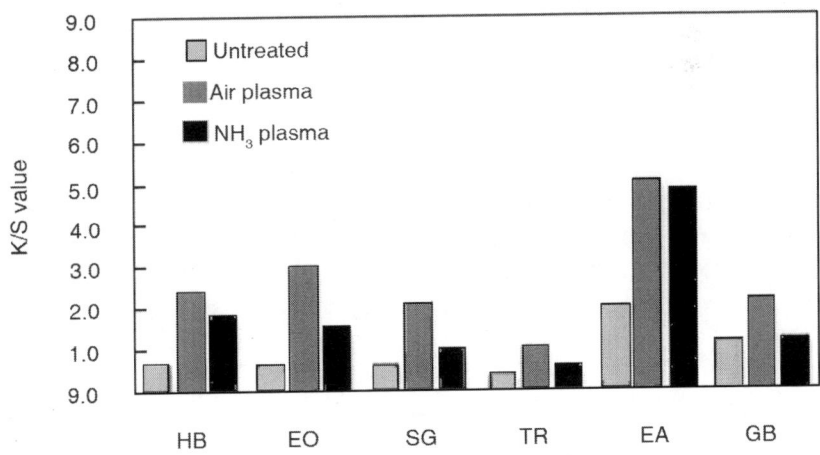

Figure 2.14 Increase in K/S value as results of low pressure plasma treatment of polyester fabric

2.6.4   Fading of Jeans by pencil plasma torch

Denim is the most preferred clothing of today's youth. Various items of denim like trousers, shirts, skirts, jackets, belts and caps, etc., are available in the market. To give distressed denim look, many types of washing techniques are used like stone washing, cellulose treatment, acid washing, chlorination, ozone bleaching, etc. But all these conventional processes have several disadvantages which are mentioned below.

- Stones could cause wear and tear of the fabric.
- Problem of environmental disposition of waste of the grit produced by the stones.
- High labor costs are to be incurred as the pumice stones and its dust particles produced are to be physically removed from the pockets of the garments and machines by the labors.
- Denim is required to be washed several times in order to completely get rid of the stones.
- The process of stonewashing also harms the big expensive laundry machines.

To minimize such drawbacks, stonewashing of denim is carried out with the aid of enzymes. The method of giving the denim a stonewash look by use of enzyme like cellulase is known as Enzymatic Stonewashing. Here cellulases are used to provide that distressed worn out look to the denim fabric. However there has been a problem of back staining like blue threads becoming bluer and white threads becoming blue in case of enzymatic stonewash. Another method, which is being conventionally employed, is bleaching. Here a strong oxidative bleaching agent such as Sodium Hypochlorite or Potassium Permanganate is used during washing. This process is difficult to control when desired level of bleaching is reached. Moreover these harsh chemicals can cause damage to cellulose fibre and also to health. This also results in corrosion of the textile machineries. Another route for denim fading is laser technology, which is water-free and computer-controlled process for denim fading, and is used for creating patterns on denim. However, this is relatively very costly and used limitedly. It has been observed that the conventional processes which are being used to create fading effect on denim, take lot of time, use large quantity of water, chemicals and pollute the environment.

At Facilitation Centre for Industrial Plasma Technology (FCIPT), a process has been developed to modify the surface of denim fabric to create the worn out effect using atmospheric pressure plasma torch. In this work we have used a novel process to create worn out effect on denim using non-thermal AC pencil plasma torch that operates at atmospheric pressure. The atmospheric pressure plasma modifies the surface of denim to create permanent effect.

The active radical of plasma react with the dye molecules and cause fading. The plasma process causes de-colorization of denim fabric in short time span (<60 seconds). The process can be controlled to create desired fading and pattern. It is continuous, online treatment at atmospheric pressure. The process is environment friendly. The pencil torch works at high voltage and low current. The schematic diagram of the torch is shown in Fig. 2.15 and working torch is shown in Fig. 2.16. AC plasma torch consists of a tungsten rod that works as powered electrode and cylindrically planner copper cup as grounded electrode. This plasma torch operates at 50–100 kHz, and average power output of plasma torch is in the range of 0.1 to 1.0 kW. The advantages of this process are mentioned below:

- The desired colour fading is achieved in a short time span and in a controlled manner.
- The effect is permanent and is retained even after several washes.
- The strength of cellulose fibre is not affected.
- It is a dry and environment-friendly process, so a lot of water can be saved.
- Labour cost can be reduced greatly.
- Online treatment of denim can be done in short time span in a controlled manner.

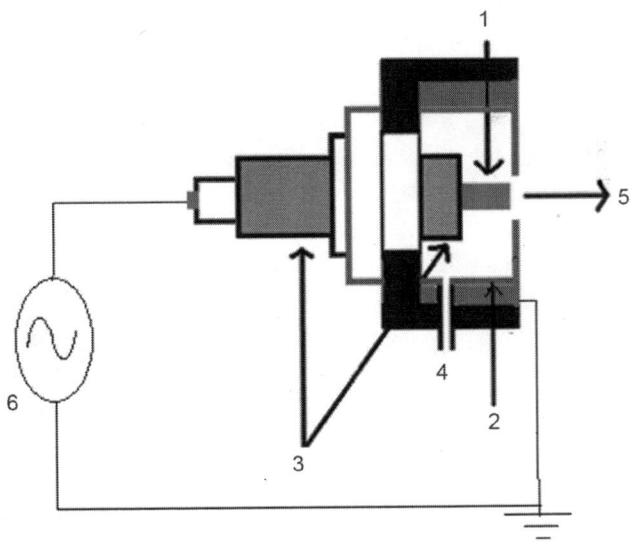

Figure 2.15 Schematic diagram of Pencil Torch [(1) Live Electrode, (2) Ground Electrode, (3) Insulator, (4) Gas Inlet, (5) Gas Outlet, (6) Power Supply]

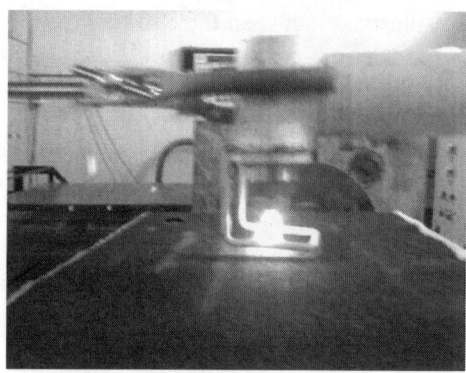

Figure 2.16 Pencil plasma torch in operation for fading of Jeans

2.6.5   Adhesion improvement by plasma treatment

Technical textiles are one of the fastest growing sectors in global textile industry. In India too, it is an emerging area with great potential for growth. Coating and lamination are twin technologies that can completely transform all appearance, handle, properties and performance of technical textiles. Consequently, their application has grown rapidly worldwide. Textile companies are continuously looking for innovative coating and laminating technologies that are economical, flexible and versatile to make novel products with high-added values.

In order to improve adhesion between textile substrate and deposited coating or laminate layer, plasma treatment is an environment-friendly and economical alternative. Plasma accomplishes this through an improved wetting or surface activation of textiles and enhancing substrate coating or lamination affinity. In the present project, a prototype plasma reactor has been developed for fabric treatment and commissioned at MANTRA, Surat. The salient features of the plasma reactor are as follows:

- Dielectric barrier discharge based plasma reactor uses air as a plasma-forming gas and works at atmospheric pressure.
- The power source used for plasma formation operates at low frequency (50 Hz).
- Continuous plasma treatment of 0.5 m wide fabric with speed of 0.5 to 4 m/min is given.
- Plasma power density and exposure time can be varied to obtain better adhesion strength.

The surface of the substrate plays an important role in getting higher adhesion strength between the substrate and the material applied onto it, i.e. adhesive or coating or laminated layer. The adhesion strength depends upon many factors, such as wetting of the substrate, its morphology, polarity, formation of chemical bonds between the substrate and coating. Conventional methods used to achieve good adhesion, include treatments with solvent-based and aqueous solutions, flame treatment and corona treatment. However, these methods have several disadvantages, such as material consumption and environmental contamination, reproducibility problems in the case of flame or corona treatment.

Plasma exposure changes surface characteristics which include surface cleanliness; surface roughness and formation of various polar groups. These properties increase the wettability. The surface can be provided with chemical functional groups resulting into a chemical interaction between the substrate and the coating or the adhesive. In textile industries particularly for technical textiles, fabric is coated or laminated for the applications like interior decoration, transport, sports and leisure wear, safety and protection, etc. Fabrics such as polyester, nylon, etc., are coated with PU or PVC material. In this process appropriate adhesion between substrate, i.e. fabric and the coating material, is important for achieving the right quality in coated or laminated fabric. Therefore, the fabric surface is activated by a dry plasma process before coating, which enhances the adhesion properties. FCIPT developed a prototype plasma reactor for fabric treatment, which is commissioned at MANTRA, Surat. It is shown in Fig. 2.17. Plasma process has been optimised for the surface modification of polyester/nylon fabric to improve its adhesion with PU/PVC coating or lamination. SEM micrograph of untreated and plasma-treated polyester fibres is shown in Fig. 2.18. Subsequently, adhesion strength is determined for untreated and plasma-treated fabrics which were laminated with PU/PVC.

Figure 2.17 Plasma treatment system integrated with coating/lamination unit at MANTRA, Surat

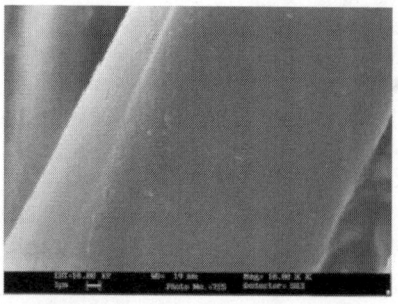

(a) Untreated polyester fibre

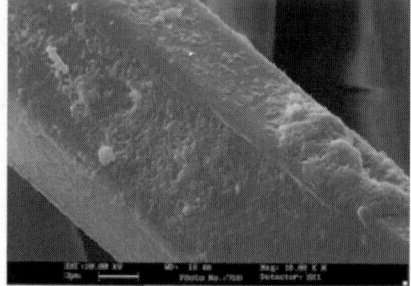

(b) Plasma treated polyester fibre

Figure 2.18 SEM micrograph of (a) untreated and (b) plasma-treated polyester fibres for coating/lamination

At MANTRA, the polyester fabric was treated in plasma for 1 minute and at different voltages. The treated polyester fabric was coated with PVC. The adhesion strength between fabric and coating was measured for both plasma-treated and -untreated fabric by tensile strength tester (Lloyd's Instrument). Testing for strength of adhesion was carried out as per ASTM D 751:98 method. The results clearly indicate that there is 30–40% improvement in adhesion strength, which has taken place after the plasma treatment. Results are shown in Fig. 2.19.

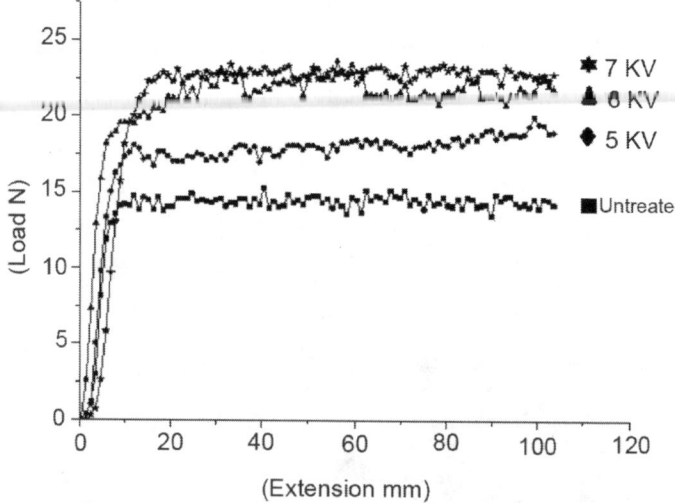

Figure 2.19 Adhesion strength at different applied voltage to form plasma at MANTRA, Surat (ASTM D 751:98 method)

2.7   Future directions

FCIPT, Institute for Plasma Research, has demonstrated its strength by developing a few industrial scale systems which are running smoothly in the field. While designing and developing any new technology, we concentrate on easy availability of consumables, economic viability, environment friendliness of the process. Encouraged with the response from textile industries, FCIPT would concentrate in following research areas in the future:

- Shrink proofing of wool by plasma route would be an area which will have demand in India in near future; therefore, FCIPT would keep its focus in this direction. Plasma exposure of wool followed by enzyme treatment to eliminate pricking seems another important research area which will be in demand in the country. FCIPT will collaborate with some of the leading organization involved in enzyme treatment of wool.
- Design and development of plasma-treatment equipments with higher coupling of power to generate intense atmospheric pressure plasma using high frequency power sources. This in turn would reduce treatment time and would significantly improve textile processing which used to be the requirements in textile industry.
- Surface activation of different polymeric materials, powders for improving adhesion and for composite applications. High frequency intense plasma systems would be designed for meeting textile industries requirements.
- Reactive dyeing would be another area where plasma functionalization can play very important role. FCIPT would be targeting its research in this direction in near future.
- Deposition of uniform coatings at atmospheric pressure without mixing helium or argon gases along with precursors. This would include superhydrophobic and hydrophilic coatings, bactericidal coatings, nano-silver and nano- titania coatings on the surface of textiles to provide unique functionality.
- Development of high speed pencil torches for designing tattoos on denim and fading of denim would be another area of research here.

2.8   Summary

The atmospheric pressure plasma and corona technologies can be effectively used for in-line continuous treatment of textiles. Most of the plasma processes used to be dry and consume gases and electricity therefore do not pollute the environment. Dielectric barrier discharge is suitable for continuous (in-line) treatment of two-dimensional textile materials of thickness typically in the range of 0.01 mm to 1 mm because of lower inter electrode gap. Three-dimensional objects can be treated using APP jets. It is possible to use multiple plasma jets and one can expose 3D objects uniformly to plasma plume using robotics. Plasma treatment activates textile surface, changes surface morphology and produces functional groups which in turn improve adhesion. Plasma treatment improves hydrophilic and hydrophobic properties, dye uptake properties and assist in grafting of molecules to achieve desired functionality. FCIPT has successfully demonstrated the improvement in cohesion properties of Angora fibres by treating the wool in air plasma. The process requires air and electricity and is eco-friendly and cost effective. The process economics are discussed in Chapter 12. Similarly, air plasma treatment of polyester surface has increased the adhesion strength more than 30%. The plasma-based technologies are becoming popular because one can vary plasma environment to obtain desired surface properties, as well as they fall in the category of green technologies.

2.9   References

Akishev Y., Grushin M., Napartovich A. and Trushkin, N. (2002). Plasmas and Polymers, **22**, p. 261.

Ayan H., Fridman G., and Gutsol A. F. (2008). IEEE Trans. Plasma Sci. **36**, p. 504.

Ayan H., Staack D., and Fridman G. (2009). J. Phys. D: Appl. Phys. **42**, p. 202.

Babayan S. E., Jeong J. Y., Tu V., Park J. J., Selwyn G. S., and Hicks R. F. (1998). "Deposition of silicon dioxide films with an atmospheric pressure plasma jet," Plasma Source Sci. Technol., **7**, pp. 286–288,

Biederman H. (2004). Plasma Polymer Films, Imperial College Press.

Buyle G. (2009). Nanoscale finishing of textiles via plasma treatment, Materials Technology 24, pp. 46–51.

Chastain J. et al. (1995). Handbook of X-Ray Photoelectron Spectroscopy, Physical Electronics Inc., Minnesotta, pp. 40–61.

Danish N., Garg M.K., Rane R.S., Jhala P. B., and Nema S. K. (2007). Surface modification of Angora rabbit fibres using dielectric barrier discharge; Applied Surface Science **253**, p. 6915.

Dawson T. (2012). Progress towards a greener textile industry, Coloration Technology **128**, pp. 1–8.

Dave H., Ledwani L., Chandwani N., Kikani P., Desai B., and Nema S. K. (2013). Surface Modification of Polyester Fabric by Non-thermal Plasma Treatment and Its Effect on Coloration Using Natural Dye, *Journal of Polymer materials: An International Journal*, **30**, p. 291.

Dogai R., and Kushner M. J. (2003). J. Phys. D: Appl. Phys., **36**, p. 666.

Dow Corning "Plasma Solutions" Application Note (2014). (http://www.dowcorning.com/content/publishedlit/01-3137-01.pdf ; accessed May 2, 2014

Fridman A., Chirokov A., and Gutsol A. (2005). Non-thermal atmospheric pressure discharges. Journal of Physics D: Applied Physics **38**, R1.

Hegemann D. (2006). Plasma polymerization and its applications in textiles. Indian Journal of Fibre & Textile Research **31**, pp. 99 –155.

Hesse A., Thomas, H., and Hocker, H. (1995). Textile Res. J. **65**, pp. 355–361.

Jeong J. Y., Babayan S. E., Tu V. J., Park J., Hicks R. F., and Selwyn G. S. (1998). "Etching materials with an atmospheric-pressure plasma jet," *Plasma Source Sci. Technol.*, **7**(3), p. 282.

Joko K. and Koga J. (1990). Proc. 9th Int. Wool Text. Res. Conf., pp. 19–26.

Kale K.H. and Desai A.N. (2011). Atmospheric pressure plasma treatment of textiles using non-polymerising gases. Indian Journal of Fibre & Textile Research **36**, pp. 289–299.

Kanazawa S., Kogoma M., Moriwaki T., and Okazaki S. (1988). J. Phys. D **21**, p. 838.

Kogelschatz U., Eliasson B., and Egli W. (1997). Dielectric-Barrier Discharges: Principle and Applications. J. Phys. IV France 07, C4-47-66.

Loeb L. (1965). "Electrical Coronas Their Basic Physical Mechanisms. University of California Press.

Ma G., Liu B., Li C., Huang D., and Sheng J. (2012). Appl. Surf. Sci. **258**, pp. 2424.

Massines F. and Gouda G. (1998). A comparison of polypropylene-surface treatment by filamentary, homogeneous and glow discharges in helium at atmospheric pressure, J. Phys. D: Appl. Phys. **31**, pp. 3411–3420.

Molina et al. (2005). Appl. Surf. Sci. **252**, 1417–1419.

Morent R., Geyter N. D., Verschuren J., Clerck K.D., Kiekens P., and Leys C. (2008). Non-thermal plasma treatment of textiles. Surface and Coatings Technology **202**, pp. 3427–3449.

Napartovich A.P. (2001). Overview of Atmospheric Pressure Discharges Producing Nonthermal Plasma. Plasmas and Polymers **6**, pp. 1–14.

Radetic M., Jovancic P., Puac N., and Petrovic, Z.L. (2007). Environmental impact of plasma application to textiles. Journal of Physics: Conference Series 71, 012017.

Rane R., Vaid A., Patil C., Mukherjee S. and Nema S.K. (2013). Development of large scale

atmospheric pressure plasma system for Angora wool treatment, Technical Report IPR/ TR- 258/2013.

Reece Roth J. (1995). Industrial Plasma Engineering, Institute of Physics Publishing, **1**.

Reece Roth J. (2001). Industrial Plasma Engineering, Institute of Physics Publishing, **2**.

Salge J. (1996). Plasma-assisted deposition at atmospheric pressure, Surf. Coat. Technol. **80**, pp. 1–7.

Samanta A.K. and Agarwal P. (2009). Application of natural dyes on textiles. Indian Journal of Fibre & Textile Research **34**, pp. 384–399.

Samanta K., Jassal M., and Agrawal A.K. (2006). Atmospheric pressure glow discharge plasma and its applications in textile. Indian Journal of Fibre & Textile Research 31, pp. 83–98.

Sarmadi, A., Ying, T., and Denes, F. (1995). European Polymer Journal, **31**(9), p. 847.

Schutze A., Jeong J.Y., Babayan S.E., Park J., Selwyn G.S., and Hicks R.F. (1998). The atmospheric pressure plasma jet : a review and comparison to other plasma sources, IEEE Trans Plasma Sci. **26**(6), pp. 1685–1693.

Shishoo R. (2007), Plasma Technologies for Textiles, CRC Press & Woodhead Publishing Ltd.

Tendero C., Tixier C., Tristant P., Desmaison J., Leprince P. (2006). Atmospheric pressure plasmas : A review, Spectrochimica Acta Part B, **61**, pp. 2–30.

Tomasino C., Cuomo J.J., Smith C.B., and Oehrlein G. (1995). Plasma Treatments of Textiles. Journal of Industrial Textiles **25**, pp. 115–127.

Yashuda H. (1985), Plasma Polymerization, Academic Press, Orlando.

Yokoyama T., Kogoma M., Moriwaki T, and Okazaki S. (1990). The mechanism of the stabilization of glow plasma at atmospheric pressure, J. Phys. D: Appl. Phys. **23**, pp. 1125–1128.

3

Atmospheric pressure plasma polymerization of HMDSO on different textile fibres and their studies

Shital S. Palaskar and A. N. Desai

Abstract: Plasma polymerization refers to formation of polymeric material under the influence of plasma (partially ionized gas). Plasma polymerization today is gaining recognition as an important process for the formation of entirely new kind of material, which has unique properties such as excellent adhesion to substrate and strong resistance to most of chemicals. In this chapter, the studies on effect of plasma polymerization of hexamethyldisiloxane on natural and synthetic fibres are discussed.

Key words: Plasma polymerization, HMDSO, cotton, polyester, polyester/cotton bended fabric, nylon

3.1   Introduction

Plasma technology, being a dry, clean, and environment-friendly process, has gained considerable attention for surface modification of textile materials to impart various functional properties. Plasma surface modification allows control of fiber surface chemistry and morphology without affecting bulk properties (Radetic, 2009). Plasma can alter textile properties in various ways (e.g., etching, cleaning, activation, grafting and polymerization). Historically, polymers formed under plasma conditions were recognized as an insoluble deposit that provided nothing but difficulty in cleaning the apparatus. This undesirable deposit, however, had very important characteristic that are sought after in the modern technology of coating. Plasma polymerization is an attractive and innovative alternative to conventional techniques for deposition of thin polymer films with specific functional properties. Plasma polymerization offers several advantages including polymerization in a clean environment, formation of pinhole-free films, and the ability to tailor films with specific thickness and chemical functionality (Ameen, 1994).

The term 'plasma polymerization' encompasses different approaches to impart surface-specific functional properties to the textiles. One of the approaches is plasma-assisted grafting, which comprises surface activation of the textile with plasma followed by grafting with a monomer (Zhang, 1997). When a textile substrate is bombarded with highly energetic plasma species, formation of free radicals on the surface takes place. Free radicals formed during plasma activation play an important role in the grafting process (Wang, 2007). Other approach refers to direct polymerization of the monomer on the textile substrate, where monomer is fed directly into the plasma zone. In order to increase the homogeneity and stability of the plasma, carrier gases like helium or argon are often used as energy carriers along with the monomer (Hegemann, 2005).

Plasma-induced polymers differ significantly from those produced by conventional synthesis techniques in their chemical structure and degree of cross-linking (Zhang, 2004). Unlike conventional polymerization, plasma polymerization process tends to form irregular three-dimensional structures, where chemical structure of the plasma polymer may be quite different from conventional polymer which has been derived from the same monomer. Moreover, conventional polymerization needs a double bond or a functional group to polymerize. On the other hand, plasma can induce polymerization by breaking single bonds (Yang, 2009). Plasma polymerization is a very complex and consequently non-specific phenomenon where different processes like activation, formation of free radicals, recombination of radicals, etc., simultaneously take place. However, it is possible to control the properties of plasma polymer films with use of well-defined reactor geometry enabling homogeneous plasma conditions (Hegemann, 2003).

The chemical nature of monomers used as precursors, predominantly affects the ultimate properties of the film/coating formed by the plasma polymerization. Techniques like plasma polymerization can induce different chemical functionalities to the textiles depending on the nature of the monomers used. Many researchers (Cireli, 2007, Ameen, 1994, Oktem, 2000) have studied plasma-assisted grafting using monomers like acrylic acid, acryl amide and similar chemicals on different textile substrates to impart functional properties.

The extent and type of plasma interaction with substrate and its effect on chemical, morphological, physical, and performance properties of polymers are principally governed by monomer chemical composition.

Important plasma parameters include discharge power, treatment time, and gas flow (Kale, 2011). Plasma polymerization of organosilicon monomers has been extensively explored for various applications such as coatings for scratch resistance, chemical barrier, optical barrier, and corrosion resistance (Theirich, 1996; Korzec, 1995; Saloum, 2008; Vautrin, 2000; Lamendola, 2010). Hexamethyldisiloxane (HMDSO) and tetraethylorthosilicate (TEOS) are commonly-used organosilicon precursors. Plasma polymerization of HMDSO has been used for imparting water repellency to various textile substrates such as polyester, wool, cotton and polypropylene. In this chapter we discuss about the plasma-enhanced chemical vapour deposition (PECVD) of Hexamethyldisiloxane (HMDSO) for imparting water repellency to the different textile materials.

3.2   Experimental

For the experimental studies, we have used HMDSO as a monomer precursor for plasma polymerization. The experiments were carried out using atmospheric pressure plasma reactor, PLATEX-600 (make GRINP S.R.L., Italy) on commercially available cotton, polyester, polyester/cotton (P/C) blended and nylon ready-for-dyeing (RFD) fabrics.

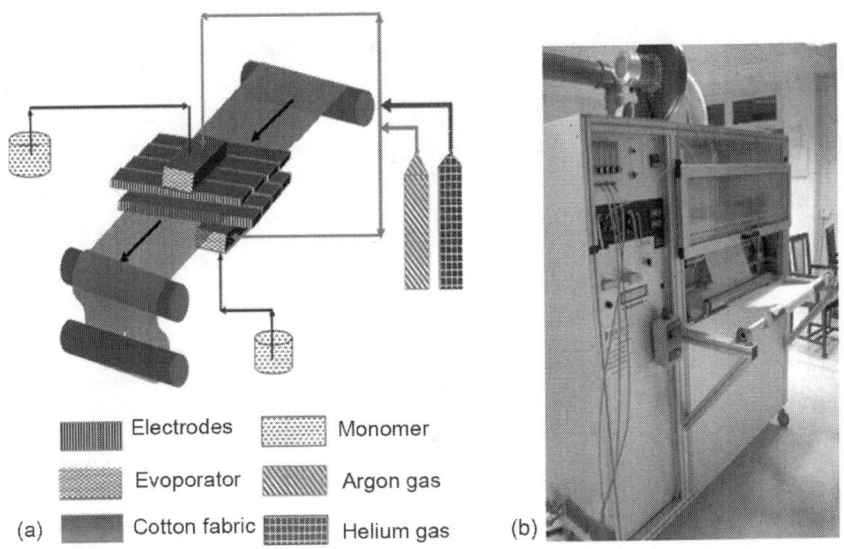

Figure 3.1 Atmospheric pressure plasma treatment unit PLATEX 600 at BTRA
(Make: GRINP S.R.L., Italy)

The schematic diagram of the plasma reactor at BTRA is shown in Fig. 3.1. The system operates in continuous mode where online treatment of fabric is possible. The length of plasma zone (total length of electrodes in the direction of fabric movement) is 12 cm and width is 55 cm. The minimal possible gap of 0.5 mm was kept in-between the top and bottom electrodes since previous research (Kale, 2010) showed better efficiency of plasma treatment at narrower inter-electrode gap. Carrier gases like helium and argon are often added to the polymerizing gas plasma mixture to increase homogeneity and stability of plasma. Helium has certain properties like high-energy metastable state, thermal conductivity, and chemical inertness which reduce instabilities in the plasma. Though argon is widely used for deposition of plasma polymers, in this study, we used helium, argon and mixture of both for generating uniform and homogeneous non-thermal plasma. In our study, liquid monomer (HMDSO) at a fixed flow rate of 0.65 ml/min was fed continuously to each evaporator. At the evaporator, monomer (HMDSO) was vaporized to a gaseous form at a temperature of about 150°C and then the monomer in vapour phase was mixed with carrier gas. This mixture of inert gas and vaporized monomer was then fed to the electrode system where electrical power was applied to the electrodes to generate plasma.

Various experiments were carried out on different fibres to capture the data on various plasma process parameters. The discharge power density for plasma generation was kept in the range of 3.0 W/cm^2 to 7.6 W/cm^2 in different experiments. In each experimental series, the plasma exposure duration was varied from 30s to 320s. The plasma is generated from the mixtures of HMDSO (0.65 ml/min), with helium, argon and He + Ar mixture.

Experiments were carried out at different power as mentioned below and their corresponding plasma density (power per unit area) is also given here.

Power	2000 W (2 kW)	3000 W (3 kW)	4000 W (4 kW)	5000 W (5 kW)
Power density	3.0 W/cm^2	4.5 W/cm^2	6.0 W/cm^2	7.6 W/cm^2

3.3   Characterization

The contact angle (CA) measurements were carried out using sessile drop method on "Easy Drop" standard drop shape analysis system of KRUSS GmbH, Germany, equipped with high speed camera IEEE1394b interface. CA was measured by sessile drop fitting method (Young Laplace fitting). In addition, water repellency of treated samples under different discharge conditions was evaluated as per American Association of Textile Chemist and Colorist (AATCC) test method 22-2005.

The surface morphological changes in samples after plasma treatment were investigated by scanning electron microscopy (SEM). The surface chemistry of fabrics was studied by attenuated total reflectance-Fourier transform infrared (ATR-FTIR) spectroscopy.

The IS: 11056-2006 method was employed to determine the air permeability of the plasma-treated fabric. Experiments were conducted at an air pressure of 100 Pa through the fabric area of 5 cm^2 on a SDL ATLAS air permeability tester. The water vapour transmission (WVT) of the plasma-treated fabric was determined by ASTM-E 96 (2005) using water as a test liquid. The WVT through the test specimen of surface area of 18 cm² for 2 hours was determined. The values of water vapour transmission were calculated as follows:

$$WVT \ (mg\,/\,cm^2\,/\,h) = \frac{Weight\ change\ (mg)}{Time\ (h) \times test\ area\ (cm^2)}$$

Thermal characterization – The inorganic residue of the untreated and HMDSO plasma-treated sample was calculated as per IS 199:1989 method. An oven-dried sample was weighed and transferred to crucible and then subjected to complete ashing at about 750°C in the muffle furnace. Mass of the residue was determined and reported as % residual ash content as below.

$$\%\ Ash\ content = \frac{Weight\ of\ residue\ (g)}{Oven\ dry\ weight\ of\ the\ speciment\ (g)} \times 100$$

Moreover, tensile properties were evaluated by tensile strength measurement as per ASTM D 5035. The durability of the treated samples was studied by subjecting the samples to repeated no. of washing cycles as per AATCC test method 61(2A)-2007.

3.4 Results and discussion

Results obtained are mentioned in different sub-sections.

3.4.1 Contact angle

Contact angle measurement is one of the established methods to determine the hydrophobicity or hydrophilicity of the textiles and polymer films. Since the untreated P/C and nylon sample gradually absorbed the water droplet placed on it, their CA value is considered as zero. The influence of treatment time and discharge power on the surface wetting properties of all fabric samples are evaluated by CA measurements. The CA values of different fabric samples treated at 3.0 W/cm² discharge power density were plotted against treatment time.

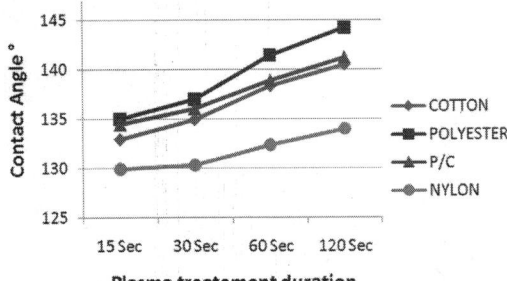

Figure 3.2 The effect of treatment time on contact angle values of plasma-treated samples at discharge power 2000W

Figure 3.2 shows the influence of plasma treatment time on the CA of different fibres samples. It can be seen from the figure that increase in plasma exposure time has resulted in higher CA values. The CAs of cotton samples treated for 15 s, 30 s, 60 s and 120 s are 132.9°, 134.9°, 138.3°, and 140.5°, respectively. Similar trend is exhibited by the polyester, P/C and nylon samples for all treatment duration. Moreover, the highest CA values were obtained for the polyester samples which can be attributed to inherent hydrophobic nature of the polyester samples which further becomes super-hydrophobic due to deposition of HMDSO plasma polymer (HMDSO pp).

During plasma processes, the monomer molecules undergo selective chemical bond breaking and recombination processes to form macromolecules (Denes, 1999). The improved water repellency in HMDSO plasma-treated samples can be attributed to the introduction of silicon atoms on the fiber surface (Hong, 2008). In addition, there is also an increase in the CA values with the increase in discharge power as can be seen from Fig. 3.4. For example, at a fixed exposure time of 60 s, CA values of polyester 141.4°, 145.7°, 149.5° and 150.1° are obtained for samples treated with discharge power densities of 3.0 W/cm^2, 4.5 W/cm^2, 6.0 W/cm^2, and 7.6 W/cm^2, respectively. Figure 3.4 illustrates the enhanced hydrophobic character of the plasma-treated fabric when exposure time and discharge power densities are increased. Although there is an increasing trend in the CA with increase in discharge power density, yet the changes in CA values of 6.0 W/cm^2 and 7.6 W/cm^2 plasma-treated samples are very small. Therefore, it appears that the saturation of hydrophobic properties after discharge power density of 6.0 W/cm2 take place.

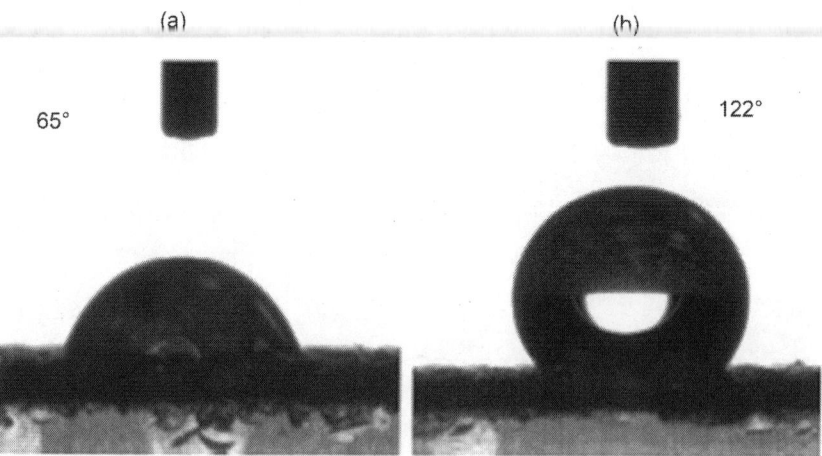

Figure 3.3 Typical photographs of contact angles on (a) untreated nylon and (b) HMDSO plasma-treated nylon

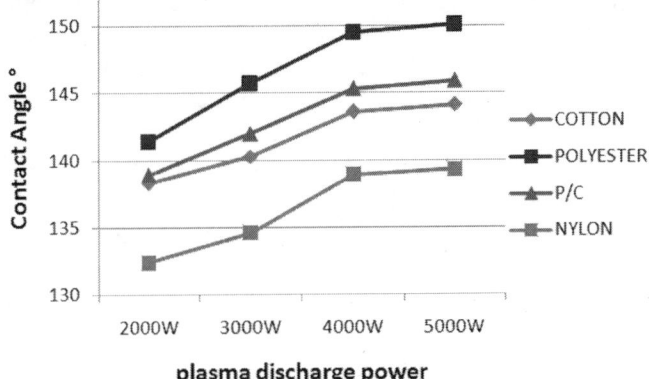

Figure 3.4 The effect of discharge power on contact angle values of plasma treated samples at treatment time 60 s

3.4.2   Water repellency by AATCC22 spray test

For many applications, CA measurement does not fulfill the requirement of water repellency. Since this study deals with imparting super hydrophobic properties to the textiles by the novel approach of plasma polymerization, more severe tests like AATCC22 spray test is used to determine the water repellent properties. In the spray test, water is sprayed on the taut textile fabric surface in controlled conditions. The results of AATCC22 spray test for samples treated under different discharge conditions are depicted in the histogram (Fig. 3.5).

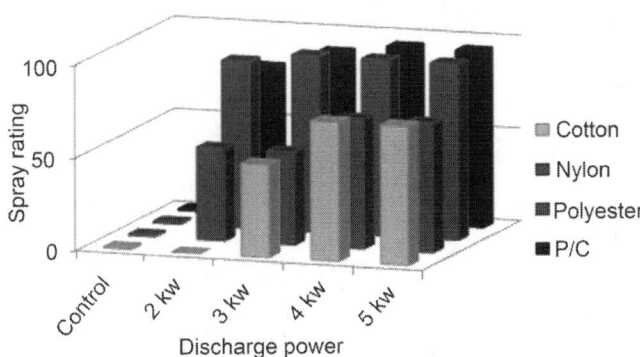

Figure 3.5 AATCC22 spray rating of HMDSO plasma-treated fabrics

The untreated fabric samples were assigned with "0" spray rating due to complete wetting of both upper and lower surfaces. Irrespective of treatment time used, there is a noticeable increase in the spray rating with increase in plasma discharge power. For example, polyester samples treated with plasma exposure time of 60 s for power densities of 3.0 W/cm^2, 4.5 W/cm^2, 6.0 W/cm^2, and 7.6 W/cm^2 have exhibited spray rating values of 80, 90, 95 and 95, respectively.

These observations of improved resistance to wetting with water are in accordance with the results reported by (Young-Y, 2008) on polyester fabric. In addition, at a fixed discharge power, there is an increase in spray rating with higher exposure time. For example, spray ratings of P/C samples treated in 4.5 W/cm^2 discharge power density for 15 s, 30 s, 60 s, and 120 s are 50, 70, 90, and 95, respectively. It can be noted from Fig. 3.5 that the effect of water repellency is getting levelled after discharge power density of 6.0 W/cm^2 which are in agreement with results of CA measurements. The chemical changes on the surface of HMDSO plasma-treated fabric samples at different discharge powers are elaborately discussed in FTIR spectroscopic analysis section. It is interesting to note that treatment with discharge having power density of 6.0 W/cm^2 for 60 s has shown spray rating of 95, which indicates very good water repellent properties. The treatment time of 60 s appears to be appropriate for the purpose of commercial scale textile processing. The enhanced hydrophobicity of plasma-treated samples treated with higher discharge power for longer durations may be attributed to higher amount of deposition of plasma polymer. Increase in deposition thickness of plasma polymer with increase in treatment time is reported by Morent (2009).

3.4.3   Effect of different gases on the water repellency

The cotton sample was treated with HMDSO for 2 min with the discharge power density 4.5 W/cm^2 using different gases to investigate the effect of gas chemistry on hydrophobicity. Table 3.1 shows the spray rating of cotton samples treated with different gases.

Table 3.1 Effect of different gases on hydrophobicity

Type of plasma gas	Spray rating
Helium	90
Argon	80
Helium + Argon	70-80
Helium + Oxygen	70
Argon + Oxygen	50

It can be seen from Table 3.1 that helium is the most effective gas for generation of homogenise plasma, as stated earlier that helium has certain properties like high-energy metastable state, thermal conductivity, and chemical inertness which reduce instabilities in the plasma. Argon is having more atomic weight (39.9) as compared to helium; therefore, argon is more aggressive and intensive etching may be responsible for the less deposition due to which spray rating is reduced.

In addition, it can be seen from the Table that introduction of oxygen reduce the hydrophobicity of the treatment, which may be attributed to formation of more hydrophilic group on the surface. When oxygen is introduced in the HMDSO polymer, Si-CH$_3$ groups which are responsible for hydrophobic nature of polymer convert into the Si-O like structure which is hydrophilic in nature. Therefore, water repellency may be reduced by incorporation of oxygen gas.

As mentioned, AATCC spray test is a commonly employed technique to determine the efficiency of the water repellent finish applied to textile fabric. The photographs taken while carrying out the spray test on untreated and plasma-treated cotton samples are shown in Fig. 3.6. The difference in the wetting behaviour of cotton fabric after HMDSO plasma polymerization can be easily noticed in Fig. 3.6.

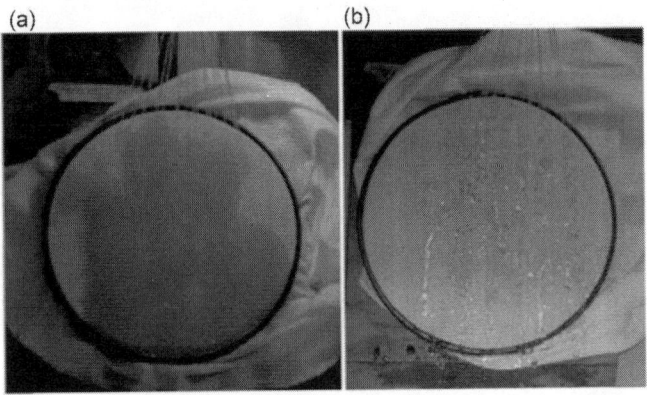

Figure 3.6 Typical image captured during AATCC spray test on (a) untreated cotton, and (b) HMDSO plasma-treated cotton sample (Kale, 2010)

3.4.4   Surface morphology by scanning electron microscopy

SEM was used to study the surface morphological changes in the treated samples. SEM of polyester/cotton-blended fabric treated with different discharge conditions is depicted in Fig. 3.7. The SEM of untreated sample showed clean and smooth surface morphology (Fig. 3.7a). The SEM photographs of samples treated in plasma having 6.0 W/cm^2 power density for 15 s, 30 s, 60 s, and 120 s are shown in Fig. 3.7(b–e), respectively.

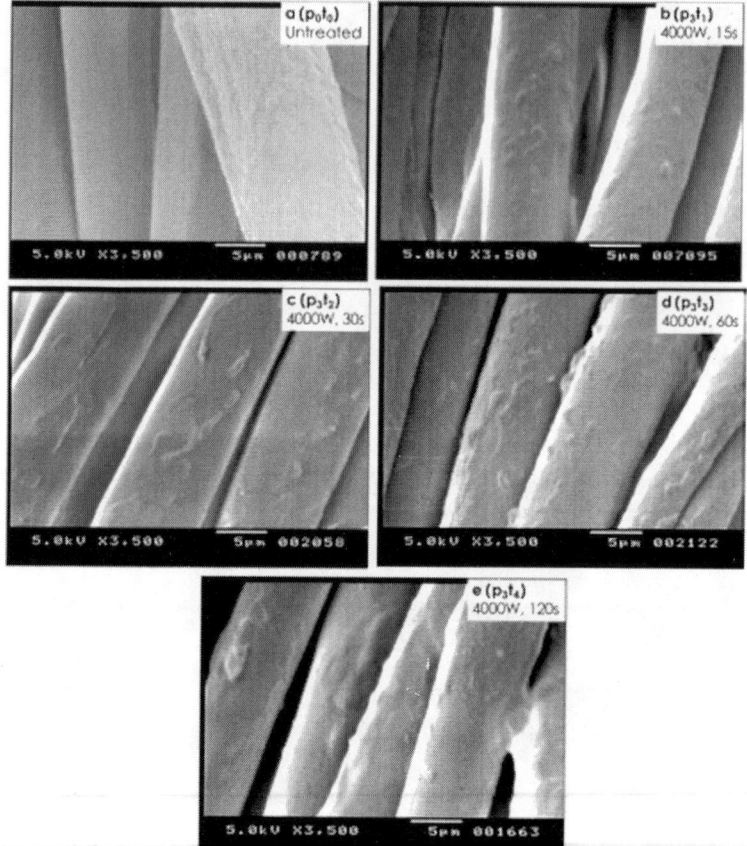

Figure 3.7 SEM photographs of (a) untreated sample, and samples treated with 4000 W for (b) 15 s, (c) 30 s, (d) 60 s, and (e) 120 s (Palaskar, 2011)

All samples treated with HMDSO exhibit deposition of plasma polymer on their surfaces. In addition, it may be pointed out that there is an increase in the deposition of plasma-induced polymer with respect to treatment time. The sample treated for 120 seconds showed higher deposition of plasma polymer on its surface. A similar kind of enhanced deposition with increase in plasma exposure time was observed in other samples treated in discharge with power densities of 3.0, 4.5 and 7.6 W/cm^2.

Figure 3.8 illustrates the SEM photographs of plasma-treated cotton samples treated for different discharge power. Increase in deposition of HMDSO plasma polymer on its surface can be observed with increase in discharge power.

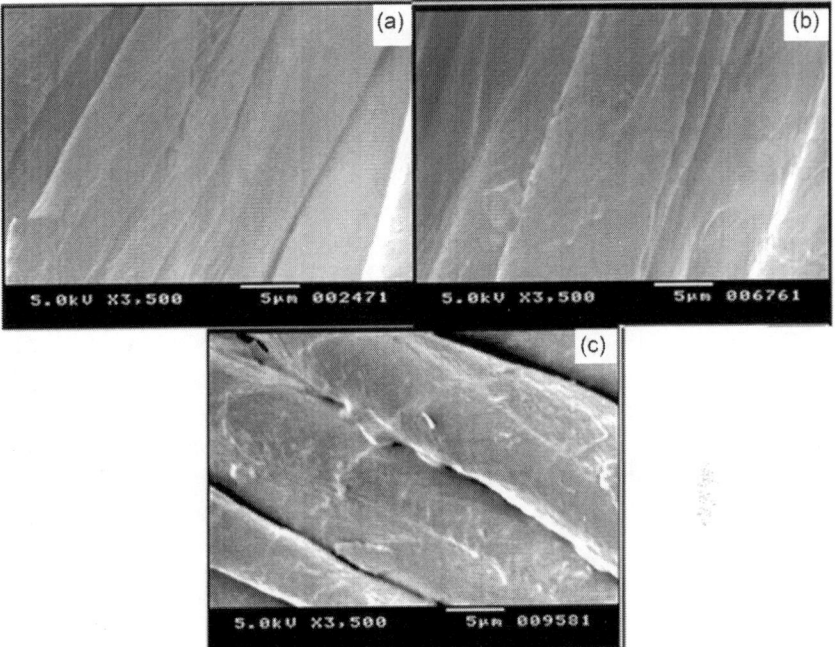

Figure 3.8 SEM images of 100% cotton samples (a–untreated, b–2000W and c–4000W HMDSO plasma treated) (Kale, 2010)

3.4.5   FTIR spectroscopic analysis

The FTIR spectra of untreated cotton and HMDSO plasma-treated cotton samples are shown in Fig. 3.9 and Table 3.2. The ATR-FTIR spectra of untreated cotton exhibits characteristic peaks at 1030 cm^{-1}, 1055 cm^{-1}, 1107 cm^{-1} and 1316 cm^{-1}, which can be assigned to a C–O stretch, asymmetric in-plane ring stretch, asymmetric bridge C–O–C and CH wagging vibrations, respectively (Chung, 2004). It can be seen from Fig. 3.9 that the FTIR spectra of all HMDSO plasma polymerized cotton samples exhibit characteristic absorbance peaks at 1256 cm^{-1}, 840 cm^{-1} and 795 cm^{-1}. The absorbance peak at the frequency band 840 cm^{-1} can be assigned to the Si-C rocking vibrations in the Si-CH$_3$ groups. The peak near the band frequency of 795 cm^{-1} can also be assigned to Si-CH$_3$ rocking vibrations. Another characteristic peak at wave number 1256 cm^{-1} can be attributed to the characteristic sharp band of the CH$_3$ group attached to Si due to the symmetric bending in Si–CH$_3$.

Table 3.2 Characteristic absorption peaks in the IR spectra of HMDSO plasma treated

S. No.	Band	Assignment
1	1233 cm^{-1}	Asymmetric stretching of aromatic ester
2	1256 cm^{-1}	Symmetric CH$_3$ deformation
3	721 cm^{-1}	C=O out-of-plane bending and ring CH out-of-plane bending
4	839 cm^{-1}	Si–C rocking vibrations in the Si-CH$_3$ groups
5	793 cm^{-1}	Si–C rocking vibrations in the Si-CH$_3$ groups

The presence of these peaks indicates that the final polymer formed by HMDSO plasma polymerization has retained the Si-CH$_3$ groups in its structure. The peak at 2958 cm-1 is also observed in plasma-treated samples, which can be assigned to CH$_3$ asymmetric stretching. The occurrence of this peak is prominent especially in the case of the treatment times of 320 and 240 s. Characteristic peaks related to silicon containing groups (1256 cm^{-1}, 840 cm^{-1} and 795 cm^{-1}) are completely absent in the FTIR spectra of untreated cotton.

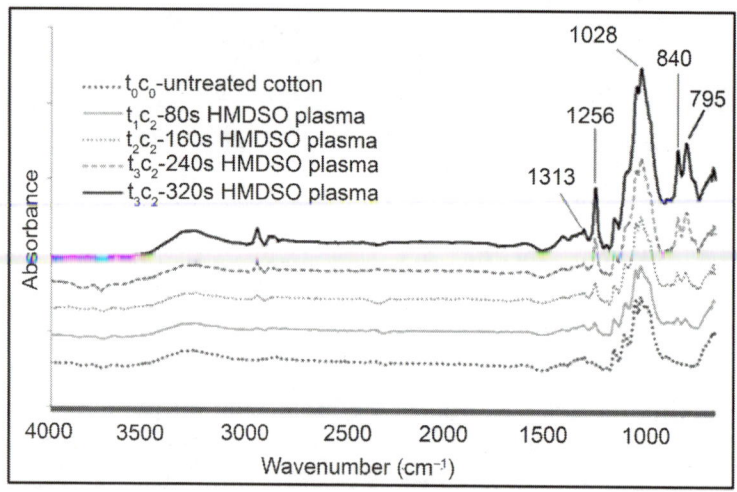

Figure 3.9 FTIR spectra of HMDSO plasma-treated cotton at various exposure times (Kale, 2010)

Figure 3.9 shows that the absorbance intensities at 1256 cm^{-1}, 840 cm^{-1} and 795 cm^{-1} increase significantly with an increase in the plasma exposure time from 80 s to 320 s. Higher absorbance intensity at longer plasma exposure time indicates a higher number of CH$_3$ groups bonded to Si. The presence of CH$_3$ groups in the structure of the plasma polymer confers water repellent properties to the cotton substrate. Spray test results have revealed the water repellent properties of HMDSO plasma polymer deposited on the HMDSO plasma-treated fabric.

Figure 3.10 shows the FTIR spectra of polyester/cotton blended fabric. The fabric contains polyester as a major component with a blend percentage of 67 (%). Therefore, the FTIR spectrum of untreated P/C blended fabric (Fig. 3.10a), predominantly exhibits peaks at 1709 cm^{-1}, 1233 cm^{-1} and 721 cm^{-1} pertaining to polyester component of the blend. The peak at 1709 cm^{-1} can be assigned to stretching vibration of C=O group in ester and the peak at wave number 1233 cm^{-1} can be assigned to asymmetric stretching of aromatic ester (Gupta, 2010). The peak at 721 cm^{-1} can be attributed to aromatic C-H out of plane vibrations.

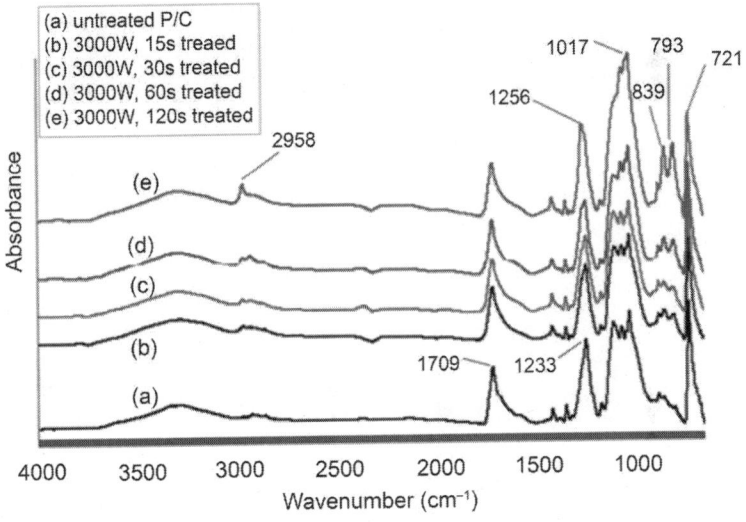

Figure 3.10 ATR-FTIR spectra of untreated and plasma-treated P/C samples with 3000 W for 15, 30, 60, and 120 s (Palaskar, 2011)

The ATR spectra of all the HMDSO plasma-treated samples show distinct peaks at the wave number 839 cm^{-1} and 793 cm^{-1}. The occurrence of these peaks can be attributed to Si-C rocking vibrations in the Si-CH$_3$ groups (Socrates, 2001).

3.4.6   Quantitative analysis of the FTIR spectra

Quantitative FTIR spectroscopic analysis is done by way of analyzing the ratio of absorbance intensity at specific characteristic bands and to discuss the effect of treatment time and discharge power on chemical nature of plasma polymer. The intensity of peak at 721 cm^{-1} was found to be least affected by the different plasma treatment conditions. Therefore, it was considered as a reference peak for absorbance ratio calculations. The ratio of absorbance intensity was calculated as follows:

$$\text{Absorbance ratio} = \frac{\text{Absorbance int ensity of characteristic peak}}{\text{Absorbance int ensity of reference peak}}$$

3.4.7   Effect of treatment time on surface chemistry

Here absorbance ratio refers to the ratio of corrected heights of the peaks. The effect of treatment time on the absorbance ratios of characteristic peaks viz. A_{2958}/A_{721}, A_{1017}/A_{721}, A_{839}/A_{721}, A_{793}/A_{721} are shown in Fig. 3.11. It is evident from Fig. 3.11 that with increase in duration of plasma treatment, there is a gradual increase in absorbance ratios of all the characteristic peaks.

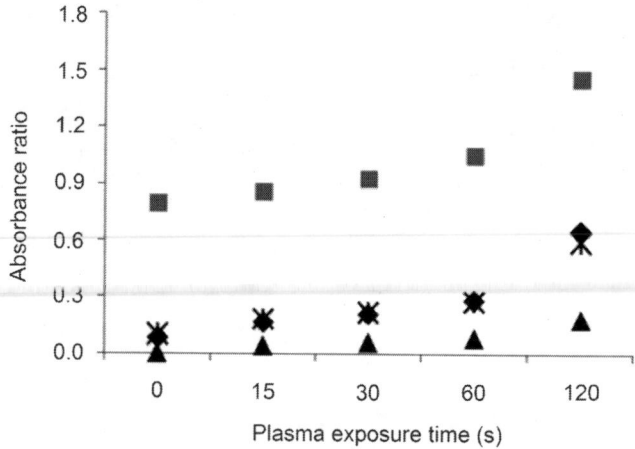

▲ A2958/A721 ■ A1017/A721 ✘ A839/A721 ◆ A793/A721

Figure 3.11 Absorbance ratios of characteristic peaks at 2958 cm^{-1}, 839 cm^{-1}, 793 cm^{-1}, 1017 cm^{-1} (discharge power 3000 W, time of treatment 15 s, 30 s, 60 s, and 120 s).

The ratios A_{839}/A_{721} obtained for plasma treatment time of 0 s (untreated), 15 s, 30 s, 60 s, and 120 s are, 0.10, 0.18 0.22, 0.27, and 0.59, respectively. Similarly, absorbance ratios A_{793}/A_{721} and A_{2958}/A_{721} showed considerable increase with respect to treatment time. The peak at 1017 cm^{-1} also show increase in absorbance ratio at longer treatment times. For example, absorbance ratio A_{1017}/A_{721} for untreated sample is 0.79, which then gradually increased almost by two times to the value of 1.457 for plasma exposure time of 120 s. It is pertinent to note that Morent (2009) has reported strong absorption band at 1020 cm^{-1} which can be assigned to Si-O- Si stretching vibrations.

In this study, the monomer HMDSO contains Si-O-Si in its backbone. Therefore, the increase in absorbance intensity of the peak at 1017 cm^{-1} may be assigned to Si-O-Si groups in the plasma polymer deposited on the surface of the polyester/cotton blended fabrics. The FT-IR spectra of samples treated with discharge of power density 7.6 W/cm^2 for 15 s (p4t1), 30 s (p4t2), 60 s (p4t3), and 120 s (p4t4) are shown in Fig. 3.12. Similar to the treated samples in plasma having power density 4.5 W/cm^2, the ATR spectra of samples treated with 7.6 W/cm^2 plasma exhibit characteristic peaks at 2958 cm^{-1}, 1017 cm^{-1}, 839 cm^{-1}, and 793 cm^{-1}. Moreover, the absorbance ratios of these characteristic peaks namely A_{2958}/A_{721}, A_{1017}/A_{721}, A_{839}/A_{721}, and A_{793}/A_{721} showed similar trend at longer treatment times (Fig. 3.13). However, it is interesting to note that the ratio A_{1709}/A_{721} remains practically unaltered for all the treatment times. This behaviour of the peak at 1709 cm^{-1} signifies the fact that HMDSO plasma treatment does not alter composition of carbonyl group in the fabric. From quantitative analysis of FTIR spectra, it can be inferred that the increase in absorbance intensity or absorbance ratio is most probably due to higher deposition of HMDSO plasma polymer at longer treatment times.

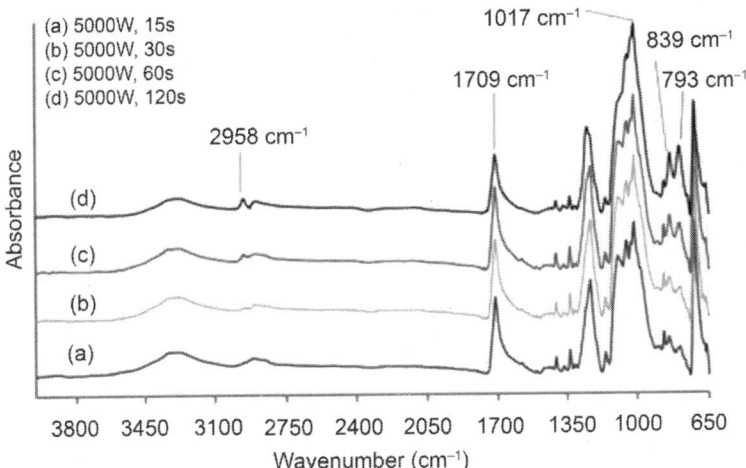

Figure 3.12 ATR–FTIR spectra of plasma-treated P/C samples with 5000 W for 15, 30, 60, and 120 s.

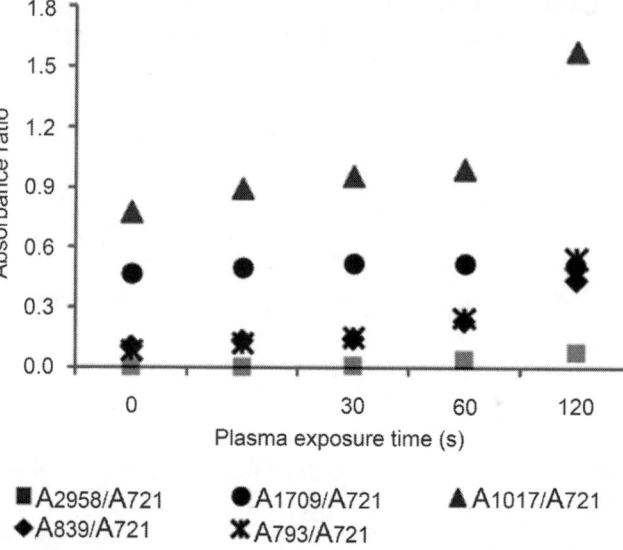

Figure 3.13 Absorbance ratios of characteristic peaks at 2958 cm⁻¹, 839 cm⁻¹, 793 cm⁻¹, 1017 cm⁻¹ (Discharge power 5000 W, time of treatment 15 s, 30 s, 60 s, and 120 s).

The mechanism of HMDSO plasma polymerization was reported by (Jiao, 2005). HMDSO molecules are initially dissociated via electron impact to $(CH_3)_3$-Si-O-Si-$(CH_3)_2{}^+$ ions and CH_3 radicals. In subsequent steps, $(CH_3)_3$-Si-O-Si-$(CH_3)_2{}^+$ ions further polymerize with HMDSO source gas molecules to form large activated ion complexes. Worbel (1980) proposed that after ionization of the HMDSO molecule, Scheme 3.1 reaction takes place.

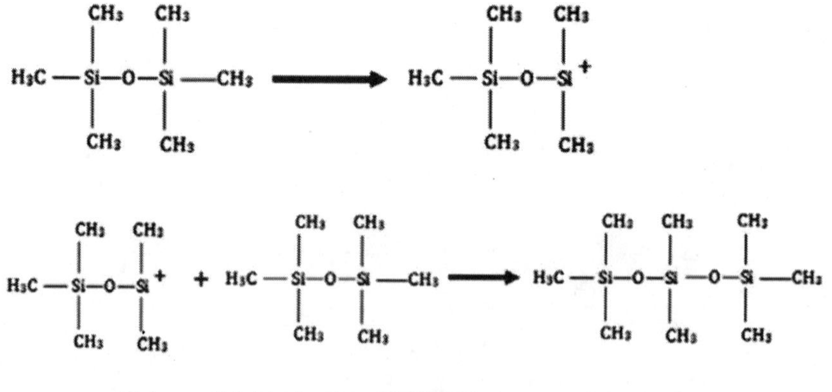

Scheme 3.1 Mechanism of HMDSO plasma polymerization

Fanelli et al. (2011) have also studied the dissociation of HMDSO, which forms different liner and cyclic compounds due to fragmentation of the monomer and condensation of the same unit (Fanelli, 2011). They have identified various species including pentamethyldisiloxane, tetramethyldisiloxane, ethoxtrimethylsilane and heptamethyltrisiloxane. The presence of species containing the dimethylsiloxane (-Me_2SiO-) repeat unit is indicative of the oligomerization processes. Therefore, it may be postulated that chain propagation takes places on the Si-O backbone, resulting in deposition of polymer by the plasma.

3.4.8   Effect of discharge power on surface chemistry

The change in absorbance ratio of characteristic peaks with respect to power is shown in Fig. 3.14. The ratios are obtained for samples treated for 120 s using plasmas having power densities 3.0 W/cm^2, 4.5 W/cm^2, 6.0 W/cm^2 and 7.6 W/cm^2. It could be seen from Fig. 3.14 that initially with increase in discharge power density from 3.0 W/cm^2 to 6.0 W/cm^2, there is a gradual increase in the absorbance ratio of all the characteristic peaks at 2958 cm^{-1}, 1256 cm^{-1}, 1017 cm^{-1}, 839 cm^{-1} and 793 cm^{-1}. Increase in deposition rate of plasma polymer at higher discharge powers is reported by Fang (2011). They have attributed it to more fragmentation of a monomer providing better chances of polymerization. However, beyond 6.0 W/cm^2, further increase in discharge power density (i.e., to 7.6 W/cm^2) showed decrease in the absorbance ratios A_{2958}/A_{721}, A_{1256}/A_{721}, A_{839}/A_{721}, and A_{793}/A_{721}.

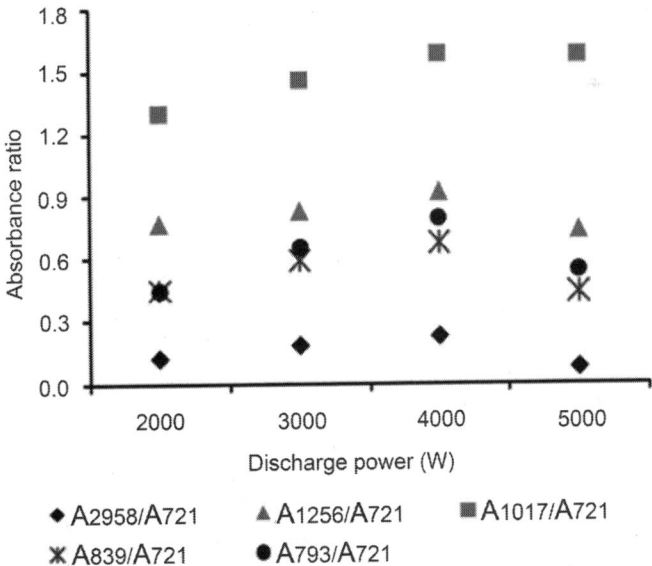

Figure 3.14 Effect of discharge power on the absorbance ratios of characteristic peaks at 2958 cm^{-1}, 839 cm^{-1}, 793 cm^{-1}, 1017 cm^{-1}

It may be pointed out that the peaks at 839 cm^{-1} and 793 cm^{-1} represent organic character of a plasma polymer film due to CH$_3$ groups. Decrease in absorbance ratios of these peaks may be due to reduction in organic character of plasma polymer at higher discharge power density of 7.6 W/cm^2. Whereas, it is interesting to note that unlike other characteristic peaks, the absorbance ratio of peak at 1017 cm^{-1} shows continuous increase even at plasma of 7.6 W/cm^2 power density. The peak at 1017 cm^{-1} may be assigned to inorganic nature of plasma polymer due to Si-O-Si groups. Therefore, further analysis was carried out to get an idea about the organic/inorganic nature of plasma polymer deposited on the surface of P/C blended fabric.

3.4.9   Organic–inorganic nature of plasma polymer

Data on absorbance ratio signify the importance of different process parameters on the deposition rate and chemical nature of plasma-induced polymer. The Si-O group represents the inorganic nature of the films, while the Si-(CH$_3$) represents the organic nature of plasma polymer. The ratios organic and inorganic IR bands reflect the fundamental differences in the chemical structure of HMDSO plasma polymer (Zuri, 1996). To evaluate the effect of discharge power on the chemical nature of plasma polymer, samples treated for 120 s with different discharge powers are taken into consideration. The peak at 1017 cm^{-1} for all the samples was normalized to the absorbance value of 1.5. The normalized FTIR spectra of samples treated for 120 s with different discharge powers is shown in Fig. 3.15. The ratio of organic to inorganic part of the plasma polymer is calculated as follows:

$$R = \frac{\text{Sum of corrected area under peaks (839 cm}^{-1} + 793 \text{ cm}^{-1})}{\text{Corrected area under peak 1017 cm}^{-1}}$$

Increase in the ratio R indicates increase in organic nature, where as decrease in ratio R shows more inorganic nature of the plasma polymer deposited on the fabric surface.

The effect of discharge power and treatment time on the ratio R is shown in Fig. 3.16. It can be noticed that initially there is increase in the R value with increase in discharge power density from 3.0 W/cm^2 to 6.0 W/cm^2. Schmachtenberg et al. (2006) have reported increase in the organic character of plasma polymer at higher discharge powers. However, further increase in discharge power led to drop in the ratio R, indicating thereby the loss in the organic character of the plasma polymer. It may be inferred from the above results that up to 6.0 W/cm^2, plasma polymer show a tendency to retain its original CH$_3$ groups in the structure. On the other hand, further increase in power has caused decline in organic structure due to very high fragmentation of a monomer.

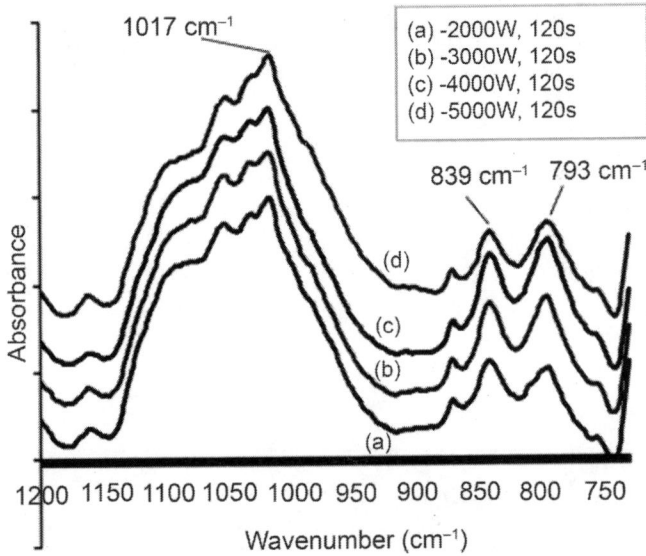

Figure 3.15 Normalized FTIR spectra of samples treated for 120 s with discharge power of 2000 W (p1t4), 3000 W (p2t4), 4000 W (p3t4), and 5000 W (p4t4)

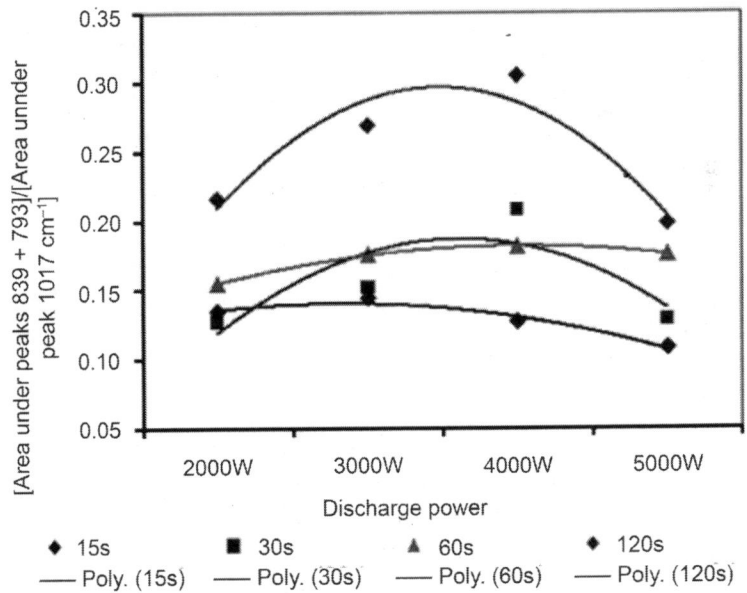

Figure 3.16 Effect of discharge power and treatment time on the organic–inorganic nature of the HMDSO plasma polymer

3.4.10   Air permeability

The values of air permeability of untreated and HMDSO plasma-treated cotton fabric are given in Table 3.3. It is seen from the Table that there is a gradual decrease in the air permeability with an increase in the plasma treatment time. Statistical significance of the decrease in air permeability was checked with the help of a t-test with a significance level of 0.05. It is clear from the t-test that the change in the air permeability, after the plasma treatment, is statistically significant when compared to the untreated cotton. This can be attributed to the deposition of plasma polymer on the surface of cotton fabric, as seen in SEM images.

Table 3.3 Air permeability and water vapour transmission of HMDSO plasma-treated cotton

Plasma treatment tine	Air permeability (LPM)	Variance	t_{table}	$t_{calculated}$	WVT (mg/ cm$_2$/h)
Control cotton	15.5	0.23	1.725		3.00
Plasma treatment 80s	13.26	0.15		11.59	3.00
Plasma treatment 160s	13.1	0.81		10.51	2.96
Plasma treatment 240s	12.86	0.76		11.85	2.78
Plasma treatment 320s	21.1	0.51		17.6	2.73

3.4.11   Water vapour transmission (WVT)

Water vapour transmission of a fabric is an important factor which decides the comfort properties of the product. WVT of untreated cotton as well as HMDSO plasma polymerized cotton is given in Table 3.3. The WVT obtained for untreated cotton sample is 3 mg/cm^2/h. For a sample (HMDSO, 80 s), the WVT remains unchanged, showing no effect on the comfort level of the cotton sample. Overall values of WVT show that comfort properties of the HMDSO plasma-polymerized cotton samples are not considerably affected.

3.4.12   Thermal analysis

The thermo gravimetric analysis was carried out to calculate the % residue in both untreated and plasma-treated samples. The results of residual ash content of the samples are given in Table 3.4. The % residual ash content for all HMDSO plasma-treated samples is higher than that of untreated sample. A higher amount of non-volatile residue after HMDSO plasma treatment was also observed by Young-Yeon (2008). Furthermore, an increase in the plasma exposure time has yielded a gradual increase in the residual ash content.

Table 3.4 Residual ash content of the HMDSO plasma-treated samples

Sample	Ash content (%)			
	Cotton	**Polyester**	**P/C**	**Nylon**
Control	0.21	0.41	0.22	0.18
Plasma treatment 80s	0.29	0.6	0.39	0.25
Plasma treatment 120s	0.41	0.7	0.46	0.27
Plasma treatment 240s	0.53	0.93	0.55	0.34
Plasma treatment 320s	0.68	1.26	0.66	0.37

3.4.13   Fabric strength and elongation

The effect of APPECVD process on the tensile properties of the samples was evaluated in the warp direction. Effect of discharge power on fabric tensile strength is shown in Fig. 3.17a. The tensile strength of untreated cotton is 7.36 N/mm. Figure 3.17 shows that the tensile properties of cotton remain unaffected due to treatment with discharge conditions. Tensile strength of cotton fabric treated even at higher power density (7.6 W/cm^2) does not show statistical significance when compared to untreated one. The elongation-at-break (Fig. 3.17b) of the untreated cotton sample is 9.9%. The elongation-at-break for samples treated in plasmas of power densities 3.0 W/cm^2, 4.5 W/cm^2, 6.0 W/cm^2 and 7.6 W/cm^2 are 9.25, 9.89, 10.03 and 10.43%, respectively. Figure 3.17b shows that variation in the discharge powers has no significant effect on the elongation-at-break of cotton fabric. Discharge power can be calculated by multiplying power density with electrode area which is 660 cm^2.

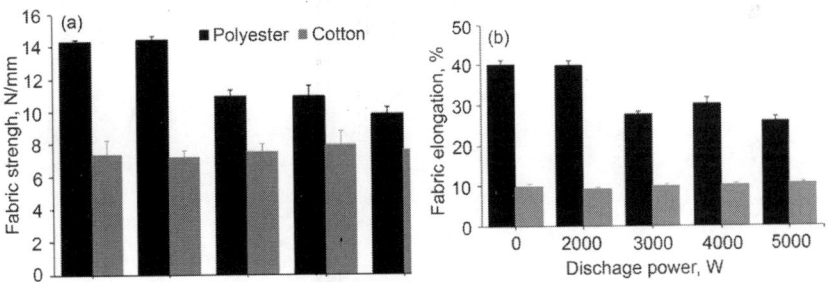

Figure 3.17 Effect of plasma discharge power on (a) tensile strength and (b) elongation-at-break of polyester and cotton fabrics (Kale, 2012)

The effect of discharge power on the tensile properties of polyester fabric is also depicted in Fig. 3.17a. The initial strength of untreated polyester fabric is 14.29 N/mm. For lower discharge power density of 3.0W/cm^2, tensile strength of polyester fabric is 14.43 N/mm, showing hardly any effect on tensile strength. However, further increase in the discharge power densities to 4.5 W/cm², 6.0 W/cm² and 7.6 W/cm² shows tensile strength values of 10.93, 10.93 and 9.85 N/mm, respectively. It indicates that higher discharge powers considerably deteriorate the tensile properties of the polyester sample during plasma polymerization. The percentage elongation-at-break of the plasma-treated polyester samples (Fig. 3.17b) shows significant reduction due to the plasma treatment.

The deterioration in tensile properties of polyester fabric at higher discharge powers may be attributed to formation of weak spots on fibre surface due to excessive etching process. Though etching could not be observed in SEM photographs of plasma-treated polyester, it might have been camouflaged due to deposition of plasma polymer on the surface. Etching process may contribute to the loss in tensile properties of plasma-treated samples; at the same time, the deposition process may not contribute to increase in tensile properties. In addition, polyester being a thermoplastic fibre, possibility of thermal degradation at very high power needs to be separately investigated.

3.4.14   Effect of washing cycles on the durability of the treatment

Durability to washing is a very vital factor in many textile applications. Finished fabrics should withhold their hydrophobicity when they are subjected to certain number of washing cycles. Therefore, to study the washing fastness in nylon HMDSO-pp, a sample (HMDSO plasma, treatment time: 60 s) was subjected to repeated number of washing cycles. Hydrophobic performance of the samples was assessed by the measurement of contact angle of the samples after every five washing cycles. Figure 3.18 illustrates the effect of number of washing cycles on the contact angle values.

Prior to washing, the nylon fabric showed CA =133.8°, 128.4° and 122.8° for test liquids water (H_2O), ethylene glycol (20%) ($C_2H_6O_2$) and formamide (CH_3NO), respectively. Gradual decrease in contact angles could be noticed with an increase in the number of washing cycles. After 25 washing cycles, contact angles were 120.4°, 117.8° and 112.5°, respectively. Figure 3.18 clearly shows a gradual decline in the hydrophobic character of the nylon fabrics with an increase in the number of washing cycles. However, it is interesting to note that the values (measured with H_2O, $C_2H_6O_2$ and CH_3NO) even after 25 washing cycles are greater than those of the untreated fabric. Therefore, it may be inferred that HMDSO-pp exhibits good durability to washing on nylon samples.

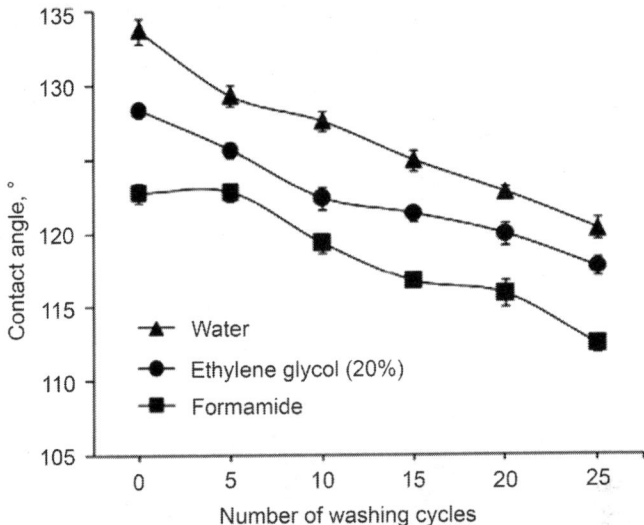

Figure 3.18 Contact angle of nylon fabric after repeated number of washing cycles. (Monomer: HMDSO, power = 4000W, treatment time = 1 min) (Kale, 2012)

Table 3.5 shows the spray rating of cotton, polyester, P/C and nylon fabrics treated with HMDSO plasma for 2 min at 4.5 W/cm^2 power density, after repeated numbers of washing up to 20 washing cycles. It can be seen from the Table that PECVD of HMDSO is more durable to synthetic material than natural fabrics.

Table 3.5 Effect of washing on different fabrics (Spray rating)

Sample	Spray rating			
	Cotton	Polyester	P/C	Nylon
Without wash	90	100	90–100	90–100
After 5 wash	50	80–90	50	80–90
After 10 wash	0	80–90	0	80–90
After 15 wash	0	80	0	80
After 20 wash	0	70–80	0	70–80

3.5   Summary

Plasma polymerization of hexamethyl disiloxane at different power density has shown change in organic and inorganic nature of deposited polymer at atmospheric pressure. There is enhancement in hydrophobic character of the plasma-treated fabric when exposure time and discharge power densities are increased. Power density and deposition time plays important role in changing various physical and mechanical properties of plasma polymer coated fabric. Water vapour transmission study showed that comfort properties of the HMDSO plasma polymerized cotton samples are not considerably affected. It has been observed that plasma polymer coating of HMDSO is more durable to synthetic material than natural fabrics.

3.6   References

Ameen A.P., Short R.D., and Ward R.J. (1994). The formation of high surface concentrations of hydroxyl groups in the plasma polymerization of allyl alcohol Polymer, **35–20**, pp. 4382–4391.

Chung C., Lee M., and Choe E.K. (2004). Characterization of cotton fabric scouring by FTIR ATR spectroscopy. Carbohydr Polym, **58**, pp. 417–420.

Cireli A., Kutlu B., and Mutlu M. (2007). Surface modification of polyester and polyamide fabrics by low frequency plasma polymerization of acrylic acid. J Appl Polym Sci, **104**, pp. 2318–2322.

Denes A.R., Tshabalala M.A., Rowell R., Denes F., Young R.A. (1999). Hexamethyldisiloxane plasma coating of wool surface for creating water repellent characteristic, Holzforschung, **53**(3), pp. 318–326.

Fanelli F. (2011). www.ispc-conference.org/ispcproc/papers/594.pdf

Fang J., Chen H., and Yu X. (2001). Studies on plasma polymerization of hexamethyldisiloxane in the presence of different carrier gases. J Appl Polym Sci, **80**, p. 1434.

Gupta B., Srivastava A., Grover N., and Saxena S. (2010). Modification of Polyester Monofilament by Plasma-Induced Graft Polymerization of Acrylic Acid, Indian J Fibre Text Res, **35**, pp. 9–19.

Hegemann D., Brunner H., and Oehr C. (2003). Evaluation of deposition conditions to design plasma coatings like SiOx and a-C:H on polymers. Surface Coat Technol, **174**, p. 253.

Hegemann D. and Hossain M.M. (2005). Influence of non-polymerizable gases added during plasma polymerization. Plasma Processes Polym, **2**, pp. 554–562.

Ji Y.Y., Hong Y.C., Lee S.H., Kim S.D., and Kim S.S. (2008). Formation of super-hydrophobic and water-repellency surface with hexamethyldisiloxane (HMDSO) coating on polyethyleneteraphtalate fiber by atmospheric pressure plasma polymerization, Surf Coat Technol, **202**, pp. 5663–5667.

Jiao C.Q., DeJoseph C.A., Garscadden A. (2005). Ion chemistries in hexamethyldisiloxane, Journal of vacuum Science and Technology, **A23**(5), pp. 1295–1304.

Kale K.H. and Palaskar S.S. (2010). Resume of Papers, 51st Joint Technological Conference of ATIRA, BTRA, NITRA & SITRA (NITRA, Ghaziabad), **33**.

Kale K.H. and Palaskar S.S. (2010). Atmospheric pressure plasma polymerization of hexamethyldisiloxane for imparting water repellency to cotton fabric. Textile Research Journal, **81**(6), pp. 608–620.

Kale K.H., Palaskar S.S., Hauser P. J., and El-Shafei A. (2011). Atmospheric pressure glow discharge of helium- oxygen plasma treatment on polyester/cotton blended fabric, Ind. J of fibre and textile research, **36**, pp. 137–144.

Kale K.H., Palaskar S.S., and Kasliwal P.M. (2012). A novel approach for functionalization of polyester and cotton textiles with continuous online deposition of plasma polymers, Indian Journal of Fibre & Textile Research, **37**, pp. 238–244.

Kale K. H. and Palaskar S. S. (2012). Structural studies of plasma polymers obtained in pulsed dielectric barrier discharge of TEOS and HMDSO on nylon 66 fabrics. The Journal of The Textile Institute, **103**(10), pp. 1088–1098.

Korzec D., Theirich D., Werner F., Traub K., and Engemann J. (1995). Remote and direct microwave plasma deposition of HMDSO films: comparative study. Surface and Coating Technology, **74–75**, pp. 67–74.

Lamendola R. and R. d'Agostino (2010). Process control of organosilicon plasmas for barrier film preparations. Pure & Applied Chemistry, **70**(6), pp. 1203–1208.

Morent R., Geyter N.D., Jacobs T., Vlierberghe S.V., Dubruel P., Leys C., Schacht E. (2009). Plasma-Polymerization of HMDSO Using an Atmospheric Pressure Dielectric Barrier Discharge, Plasma Process Polym, **6**, pp. 537–542.

Oktem T., Seventekin N., Ayhan H., and Piskin E. (2000). Modification of Polyester and Polyamide Fabrics by Different in Situ Plasma Polymerization Methods, Turk J Chem, **24**, pp. 275–285.

Palaskar S.S., Kale K.H., Nadiger G.S., and Desai A.N. (2011). Dielectric Barrier Discharge Plasma Induced Surface Modification of Polyester/Cotton Blended Fabrics to Impart Water Repellency using HMDSO, Journal of Applied Polymer Science, **22**(2), pp. 1092–1100.

Radetic M., Puač N., Jovančić P., Šaponjić Z., and Petrović Z. L. (2009). Plasma Induced Decolorization of Indigo Dyed Denim Fabrics Related to Mechanical Properties and Fiber Surface Morphology. Textile Research Journal, **79**(6), pp. 558–565.

Saloum S., Naddaf M., and Alkhaled B. (2008). Properties of thin films deposited from HMDSO/O2 induced remote plasma: effect of oxygen fraction. Vacuum, **82**, pp. 742–747.

Schmachtenberg E., Costa F.R., and Gobel S. (2006). Microwave assisted HMDSO/oxygen plasma coated polyethylene terephthalate films: Effects of process parameters and uniaxial strain on gas barrier properties, surface morphology, and chemical composition, J. Appl Polym Sci, 99, p. 1485.

Socrates G. (2001). Infrared and Raman Characteristic Group Frequencies-Tables and Charts, John Wiley and Sons Ltd: West Sussex, England.

Theirich D., Ningel K.P., and Engemann J. (1996). A novel remote technique for high rate plasma polymerization with radio frequency plasmas, Surface and Coating Technology, **86–87**, pp. 628–633.

Vautrin-Ul C., Boisse-Laporte C., Benissad N., Chausse A., Leprince P., and Messina R. (2000). Plasma-polymerized coatings using HMDSO precursor for iron protection Progress in Organic Coatings, **38**, pp. 9–15.

Wang C. and Chen J.R. (2007). Studies on surface graft polymerization of acrylic acid onto PTFE film by remote argon plasma initiation. Appl Surf Sci, **253**, pp. 4599–4606.

Worbel A.M. (1980). Polymerization of Organosilicones in Microwave Discharges. Journal of macromolecular science and chemistry, **A14**(3), pp. 321–337.

Worbel A.M. and Werthremer M.R. (1990). Plasma deposition treatment and etching of polymers, edited by R. D'Agostiono, Academic press, San Diego, Calif., USA, p. 163.

Yang S.H., Liu C.H., Su C.H., and Chen H. (2009). Atmospheric-pressure plasma deposition of SiOx films for super-hydrophobic application. Thin Solid Films, **517**, pp. 52–84.

Young Y.J., Hong Y.C., Lee S.H., Kim S.D., and Kim S.S. (2008). Formation of super-hydrophobic and water repellency surface with hexamethyldisiloxane (HMDSO) coating on polyethyleneteraphthalate fibre by atmospheric pressure plasma polymerization. Surf Coat Technol, **202**, pp. 5663–5667.

Zhang J. (1197). The surface characterization of mulberry silk grafted with acrylamide by plasma copolymerization. J Appl Polym Sci, **64**, pp. 1713–1717.

Zhang C., Wyatt J., and Weinkauf D.H. (2004). Carbon dioxide sorption in conventional and plasma polymerized methyl methacrylate. Thin Films Polymer, **45**, p. 7665.

Zuri L., Silverstein M., and Narkis M. (1996). Organic–inorganic character of plasma-polymerized hexamethyldisiloxane. J Appl Polym Sci, **62**, p. 2147.

4

In-situ plasma reactions for hydrophobic functionalization of textiles

Ashwini K. Agrawal, Manjeet Jassal, Prashanta K. Panda and Kartick K. Samanta

Abstract: Recent developments on hydrophobic functionalization of hydrophilic textile using atmospheric plasma have been discussed. Effective surface functionalization can make the textile stain repellent/self-cleaning without affecting its bulk properties. Atmospheric pressure plasma has been used for surface modification of textiles by in-situ reaction of precursors. Different types of gaseous and liquid precursors i.e. tetrafluoroethane, dodecyl acrylate, acrylonitrile, 1,3-butadiene and hexamethyl disiloxane along with helium, have been used for in-situ reaction on cellulosic textile substrates using glow plasma at atmospheric pressure. Attempts were made to understand the mechanism of functionalization with different precursor molecules. The surface was characterized using X-ray photoelectron spectroscopy, FT-IR spectrometer and measuring contact angles. The precursors or its fragments were found to react with the substrate under suitable conditions and resulted in durable hydrophobic functionality. The influence of different plasma process parameters on the fragmentation of precursors and their reaction with substrate has also been discussed. Using the above-mentioned fluoro and non-fluoro compounds, hydrophobic functionalization of textile substrates has successfully been established.

Key words: Non-thermal glow plasma, atmospheric pressure, cellulosic textiles, hydrophobic functionalization, HMDSO, acrilonitrile, 1,3-butadiene

4.1   Introduction

Plasmas play an essential role in many industrial applications, such as thermal spraying (coating), etching in microelectronics, metal cutting, welding, vehicle exhaust cleanup, fluorescent/luminescent lamp, etc. Recently, plasma techniques are utilized in lighting and large-screen televisions. The application of plasma technique has made roads into the plastic, polymer and textile field as well.

In light of environmental regulations and concerns, the textile industry has become more interested in application of plasma as a novel finishing technology that significantly reduces toxic-chemical pollution. Plasma technology has generated enormous interest as a solution for environmental problems in textiles, and there has been significant research interest in this area over the past decade. The recent research on application of plasma treatment in these areas covers superfine surface cleaning, surface activation and surface modification (such as desizing, wettability enhancement, water/soil repellency, printability, dyeability, shrink resistance, adhesion enhancement and sterilization, etc.) of textiles, filters, films and webs.

Hydrophobic finish on the hydrophilic textile makes it stain repellent and self-cleaned without affecting its bulk properties. Both low pressure plasma and atmospheric pressure plasma can be used for this application. Each one has its own advantages for a specific application. The atmospheric pressure plasma has been mostly used for cleaning/etching and modification of polymeric substrates by depositing polymeric layers. However, the generation of glow plasma and carrying out reactions of precursors on the surface of fabric/polymer to provide a durable chemical finish to the fabric are tricky.

In this chapter, the recent developments on hydrophobic functionalization using plasma treatment method have been discussed.

4.2   Surface properties and wettability

Surface roughness and its chemical composition are two key parameters for surface wettability. An adhesive force between surface and a liquid droplet depends on these two parameters. Surface roughness may be micro or nano in size. When the surface roughness and chemical composition together result in low surface energy, the adhesion force between solid surface and water droplet decreases significantly (Jafari et al., 2013). Hydrophobic functionalization on the surface of substrate can be achieved by tailoring these two parameters using plasma technology. Researchers have followed various ways to tailor the surface of the substrate using plasma technique, such as deposition, ex-situ grafting and in-situ reaction.

4.2.1   Hydrophobization of textile by plasma enhanced chemical deposition

In this technique, polymerization or fragmentation and recombination of precursor takes place in gas phase. As a result a polymeric/chemical layer deposits on the surface of the substrate. The deposited layer physically adheres to the surface of substrate. Physical adherence ultimately gives less wash durable effect on the substrate. Figure 4.1 shows the schematic of the process. Some of the recent studies based on this process are given in Table 4.1.

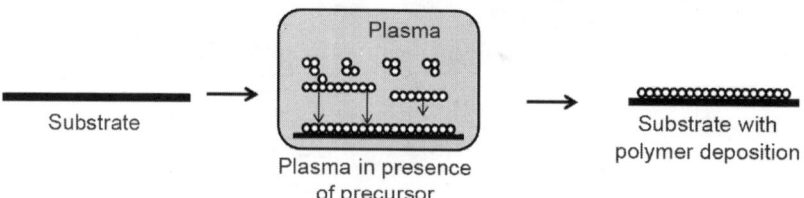

Figure 4.1 Schematic of the plasma-enhanced deposition process

Table 4.1 Recent studies on plasma-enhanced deposition.

S. No.	Precursor used	Result obtained	Reference
1	Silicone-modified surfactants	Excellent hydrophobic effect on nylon	Lin et al., 2006
2	Diamond-like carbon coating	Hydrophobicity on cotton	Caschera et al., 2013
3	Fe particle deposition	Water repellency on polyimide film	Bansode et al. 2012
4	Fluoropolymer aerosol dispersion treatment	Water repellency effect on PET fabric	Leroux et al., 2008
5	HMDSO polymer layer	Makes hydrophobic coating on polyester	Ji et al., 2008
6	PECVD of Heptadecafluoro-1,1,2,2-tetrahydrodecyl-1-trimethoxysilane	Ultra water repellent effect on PET surface	Teshima et al., 2005
7	Fluorocarbon nanoparticulate hydrophobic film deposition	Hydrophobic effect on cotton	Zhang et al., 2003
8	CF_4 and CHF_3	Teflon-like hydrophobic layer formation on carbon cloth	Lee et al., 2009
9	PECVD of HMDSO and TEOS	Water repellency on nylon 66	Kale and Palaskar, 2012a
10	PECVD of HMDSO and TEOS	Hydrophobicity on polyester	Kale and Palaskar, 2012b
11	C_2F_6 fluorocarbon coating	Water repellency effect on polyester	Farsari et al., 2011
12	1,1,2,2-tetrahydroperfluorodecyl acrylate and 1,1,2,2-tetrahydroperfluorododecyl acrylate	Makes polyester cotton blend fabric hydrphobic	Davis et al., 2011
13	HMDSO	Water repellent layer on polyester cotton blend fabric	Palaskar et al., 2011

4.2.2   Hydrophobization of textile by ex-situ plasma grafting

In this process, first a substrate is treated with plasma and then exposed to a reactive monomer for a long time to achieve grafting of the monomer on it.

Plasma treatment of the substrate activates its surface by generating radicals and other highly reactive sites. When this activated surface comes in contact with monomer solution, grafting of monomer takes place at the activated sites. Contrary to chemical deposition, this method results in a durable functionalization. In this method, the hydrophobic precursors with reactive vinly groups are used. The grafting of subtrates in a monomer bath leads to high values of add on. However, this changes the feel and texture of the fabric significantly, which is undesirable, and at the same time is not cost effective. Further this process also involves wet treatment steps, which makes the process environmentally unfriendly. A schematic of the method is shown in Fig. 4.2, and the recent literatures related to this method are summarized in Table 4.2.

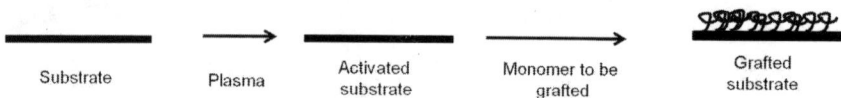

Substrate   Plasma   Activated substrate   Monomer to be grafted   Grafted substrate

Figure 4.2 Schematic of plasma-grafting process

Table 4.2 Recent studies on plasma grafting.

S. No.	Precursor used	Result obtained	Reference
1	1,1,2,2-tetrahydroperfluorodecyl acrylate (AC8)	Hydrophobicity on PAN	Hochart et al., 2003
2	Oleic acid	Hydrophobic effect on 100% cotton	Cabrales and Abidi, 2012
3	HMDSO	Water repellency on cotton	Nattinen et al., 2011
4	Dichloredimethylsilane	Hydrophobicity on 100% polyester	Jahgirdar and Tiwari, 2007
5	Vinyl laurate	Durable hydrophobic finish on cotton	Abidi and Hequet, 2005

4.2.3   Hydrophobization of textile by in-situ plasma treatment

This method does not require preactivation of the surface and any precursor (without reactive /functional groups) can also be used. Both precursor and substrate are activated at the same time inside the plasma chamber. The fragmentation and recombination of the precursor occur in gas phase and the fragmented radicals or the recombined radicals react with the created radical sites of the substrate. This method helps to establish a chemical reaction between the fragments of precursor and molecules of the substrate. This leads to a durable functionalization of the surface of the substrate. Figure 4.3 shows the schematic of the process and Table 4.3 summarizes some of the recent studies reported on in-situ plasma reaction. Some physical deposition may occur which can be avoided under the appropriate conditions.

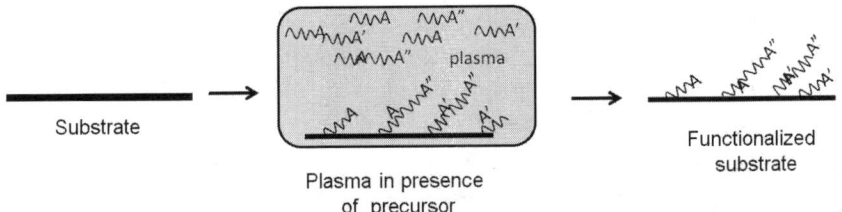

Plasma in presence
of precursor

Figure 4.3 Schematic of in-situ plasma reaction process

Table 4.3 Recent studies on in-situ plasma reaction.

S. No.	Precursor used	Result obtained	Reference
1	Oxygen	Hydrophobicity on cotton rating equal to fluorocarbon finish	Tsoi et al., 2011
2	Per fluoropropene	Hydrophobicity on cotton-polyester textile	Vohrer et al., 1998
3	Fluorocarbon precursors	Durable water repellency on nylon. Durability in case of saturated monomer was better than unsaturated monomer	Iriyama et al., 1990
4	SF_6	Hydrophobicity on silk	Chaivan et al., 2005
5	1,1,2,2 heptadecaflurodecyl acrylate	Hydrophobicity on PLA knitted fabrid	Khoddami et al., 2010

4.3   Challenges in hydrophobic functionalization of textiles using atmospheric pressure glow plasma

Despite the widespread and rapidly growing use of plasma modification, the application of plasma in textile manufacturing is still an area of research. From the above-mentioned literature, it can be seen that plasma reactions for polymerizing volatile, organic molecules to produce thin, pinhole free coatings are mostly reported for the vacuum systems. However, vacuum plasma technology has several limitations, which include the requirement for vacuum, batch treatment process and low deposition/reaction rates. As textile manufacturing deals with processing of large volume of materials at significantly high throughput in a continuous manner, the modifications carried out in homogeneous glow atmospheric pressure plasma is relatively more promising technology.

Plasma reactions at low pressure have distinct advantages. It allows better control of the degree of ionization, i.e. fragmentation of the starting materials. Also the generated radicals get sufficiently longer time to react with the radical of the textile. On the other hand, generation of plasma at atmospheric pressure, especially in presence of a precursor molecule, is challenging. Again the generation of glow-type cold plasma and its stabilization in the presence of precursor is a matter of concern at atmospheric pressure. Most of the studies do not distinguish between the treatments carried out in filamentary or glow plasma. It is necessary to investigate plasma reactions in glow regime as it can help in better control of radicalization and reaction of the precursor with textile substrates. Also, it results in uniform treatment of the textile without damaging the textile and the di-electric plates during the reaction. Formation of active species such as radicals and preserving them in active form at the atmospheric pressure are difficult because radicals are readily killed by recombination or in the presence of oxygen. Therefore, it is highly challenging to carry out plasma reactions inside atmospheric pressure glow plasma. A few studies that have reported the incorporation of various functionalities in textile substrates using atmospheric pressure plasma have not shown the stability of the imparted finishes to soap washing.

Further, majority of the research on plasma polymerization, whether at atmospheric or low pressure, has been carried out using gaseous precursors. It is challenging to carry out reactions in plasma using liquid precursors. Most of the studies using liquid monomer have been carried out by using the approach of post polymerization (i.e. grafting) after the radicals have been generated on the substrate by plasma treatment. It is important to understand the types of reactions a vapourized monomer would undergo inside an atmospheric pressure plasma reactor.

For the atmospheric pressure plasma technology to become economically viable, it is necessary to investigate the nature of generated radicals and the possible reactions that may take place in the plasma zone. Also, the effect of various plasma parameters on the reactions is indispensable. A thorough understanding of these would help in obtaining durable functional textile without adversely affecting the desirable textile properties. As mentioned earlier, plasma reaction is a complex process involving both constructive as well as destructive types of reactions resulting in various kinds intermediate and end products. Hence, the mechanism of plasma reaction of a particular precursor with a specific textile is not very clearly understood till date. In the literature attempts have been made to explain the plasma mechanism based on the analysis of the plasma modified textile. However, it is also necessary to understand the species generated inside the plasma zone before the mechanism can be elucidated. This is because many of the generated species in the plasma zone may not react with substrates and if they do, the attached big molecules cannot be directly detected by XPS or SIMS.

This chapter deals with understanding plasma reactions inside atmospheric pressure plasma for hydrophobic modification of textile surfaces.

4.4   Effect of different parameters on plasma reactions

Different parameters such as frequency, power density, electrode gap, carrier gas to precursor flow ratio and treatment time are key parameters to control the process and generate stable glow plasma.

(a) *Frequency:* Roth et al. (2005) reported in his study that applied frequency should be in a range, which would trap the ions in between the plasma zone. The small change in frequency traps electron instead of ion and glow plasma change to filamentary plasma so frequency should be controlled to trap the ions only.

(b) *Energy and power density:* The plasma process needs a minimum amount of energy for the ionization of gas and formation of glow plasma. The energy can be expressed in terms of power density (W/cc) and treatment time (s). The change in power density affects the glow nature of the plasma. Applied voltage (V) and gap between the electrodes for a given area jointly decide the power density of a plasma zone. Increase in the applied voltage increases the power density, whereas increasing gap between the electrodes, which changes the volume of the plasma, decreases the power density. These two factors should be precisely controlled to maintain the stability of glow plasma.

(c) *Role of carrier gas and precursors:* Electronegative gases such as helium easily ionize and form stable glow plasma; however, the introduction of reactive precursor in gaseous form or in vapor form inside a plasma zone turns it into filamentary one. Precursor molecules require a certain amount of time at a given voltage (i.e. energy) for their effective ionization. Ratio of precursor to carrier gases is, therefore, an important parameter in stabilizing a glow regime and carrying out desirable reactions. The generation of stable glow plasma in presence of precursors is a difficult task. Also the formation of glow plasma using electronegative gas other than helium is also a challenge.

(d) *Temperature of electrodes:* Plasma generates heat, and during introduction of precursor, the temperature of plasma tends to increase, as an attempt is made to effectively ionize the large molecules. Cooling of electrode is generally required to enhance the stability of glow. Increase in temperature affects the glow condition of the plasma so the cooling of electrode is generally required to enhance the stability of glow.

4.5   In-situ hydrophobic functionalization of textiles using gaseous precursor

In this section, hydrophobization of cellulosic textile using non-polymerizable fluorocarbon gaseous precursor (tetrafluoroethane) and non-fluorocarbon gaseous precursor (1,3-butadine) has been discussed.

4.5.1   Plasma treatment of cellulosic textile using tetrafluoroethane (TFE)

In-situ plasma reaction of TFE with cellulosic substrate was carried out in an indigenously developed atmospheric pressure cold plasma reactor. Figure 4.4 shows the schematic diagram of the set-up (Samanta, 2010).

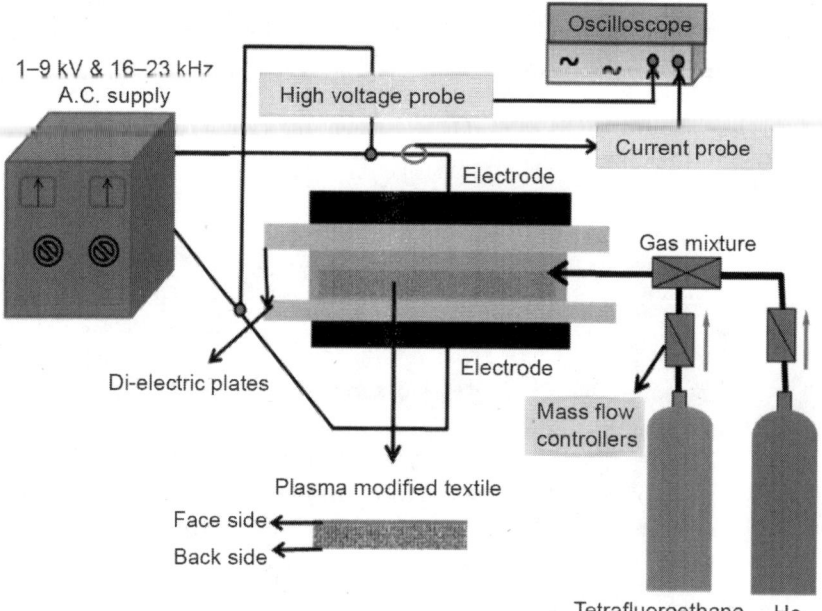

Figure 4.4 Schematic diagram of plasma reactor set-up

The plasma reaction was found to be highly sensitive to atmospheric contaminants. Therefore, the reaction was carried out in an enclosed chamber with a provision to distribute plasma gases uniformly. Before the reaction, the reactor was flushed with helium (He) gas for 10 min to remove contaminants such as air and oxygen trapped inside the reactor and the fabric substrate. Thereafter, the textile sample was treated with He plasma for 2 min followed by plasma reaction of He/TFE plasma 1–8 min.

Glow plasma is usually characterized by a smooth profile of I-V waveform. A number of experiments were conducted to understand the effect of various parameters on plasma formation and to generate the glow plasma in He/TFE gases. It was observed that suitable ratio of helium and precursor, discharge voltage and low discharge gap were important to generate stable glow plasma in presence of precursor. The I-V profile and image of He/TFE plasma are shown in Fig. 4.5(a) and (b).

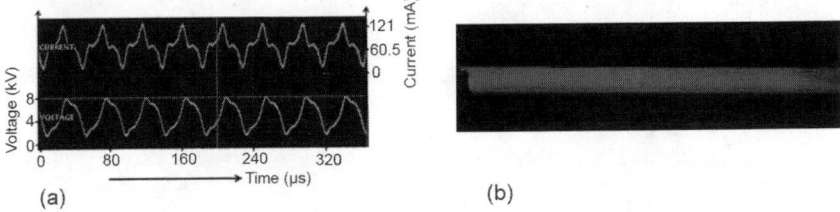

(a)   (b)

Figure 4.5 (a) I-V waveform of He/TFE plasma (b) He/TFE glow plasma

(a) *Effect of parameters*
This can be seen from Fig. 4.6 that fragmentation of TFE increases with increasing plasma discharge voltage. The frequency also seemed to have significant influence on the fragmentation of TFE. The increase in intensity of F line with increasing discharge voltage was more pronounced at higher frequency compared to the lower frequency.

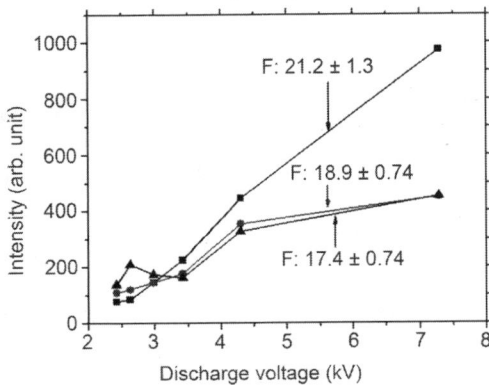

Figure 4.6 Effect of discharge voltage on the intensity of F 518 nm line at different frequencies

Figure 4.7 shows the effect of total gas flow rate and the ratio of He to TFE (gas ratio) on OES spectrum of He/TFE plasma. As the total gas flow rate was increased from 310 ml/min to 620 and then to 930 ml/min at a gas ratio of 30, the intensity of peaks in OES spectrum reduced while several of them completely disappeared. When the gas ratio was increased from 30 to 60 at the lowest flow rate of 305–310 ml/min, fragmentation of TFE increased significantly.

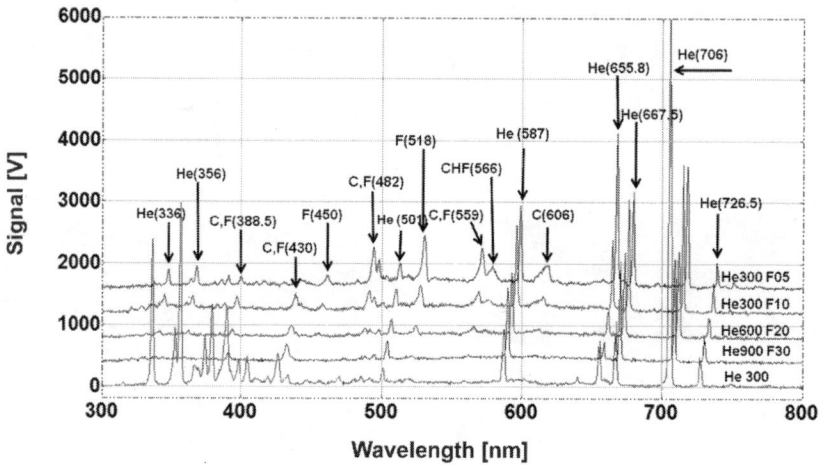

Figure 4.7 Effect of total gas flow rate and He to TFE (carrier to precursor) gas ratio on ionization of He/TFE on OES spectra at a discharge voltage of 7,28 kV and frequency of 18.9 ± 0.74 kHz

The presence of more number of emission lines and with higher intensity was possibly the result of higher degree of fragmentation of precursor gases to atomic species. It was assumed that when concentration of atomic species increase in the plasma zone, the amount of larger fragments may tend to decrease.

The effect of treatment time of TFE plasma on the cellulosic textile was evaluated by determining the water drop absorption test, contact angle and rolling angle. With increase in treatment time, the water contact angle on the treated fabric was found to increase significantly. The 8 min. plasma-treated sample showed excellent hydrophobicity with a water contact angle of 153°, rolling angle of 5° and absorbency time of more than 1 h. Similarly, the durability of the treatment to washing improved gradually and the best results of hydrophobic functionality and its durability were achieved at plasma treatment time of 8 min. Images of water drop on untreated and plasma-treated sample are shown in Fig. 4.8.

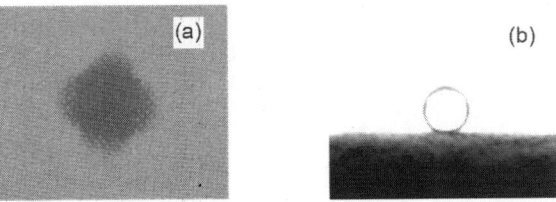

Figure 4.8 Shape of water droplet on (a) untreated fabric (b) TFE plasma-treated fabric (8 min plasma treated as prepared)

The GC-MS in combination with OES results indicated that when TFE was introduced in plasma zone, it first broke down into various fragments such as CF_3, CFH_2, CF_2, CFH, CH_2, etc., which further combined to form two types of species. The first type, with molecular masses such as 73, 93, 184 and 209 a.m.u., had long hydrocarbon backbone with fluorocarbon moieties attached to one of the ends. These molecules had high carbon to fluorine ratio. The likely structures of these species are shown in Fig. 4.9. The other type was principally represented by molecular species of 151 a.m.u., which was formed by a combination of TFE with smaller fluorocarbon moieties such as CF_2. These molecules were rich in fluorine. Clearly both kinds of molecules are likely to impart high level of hydrophobicity to the fabrics.

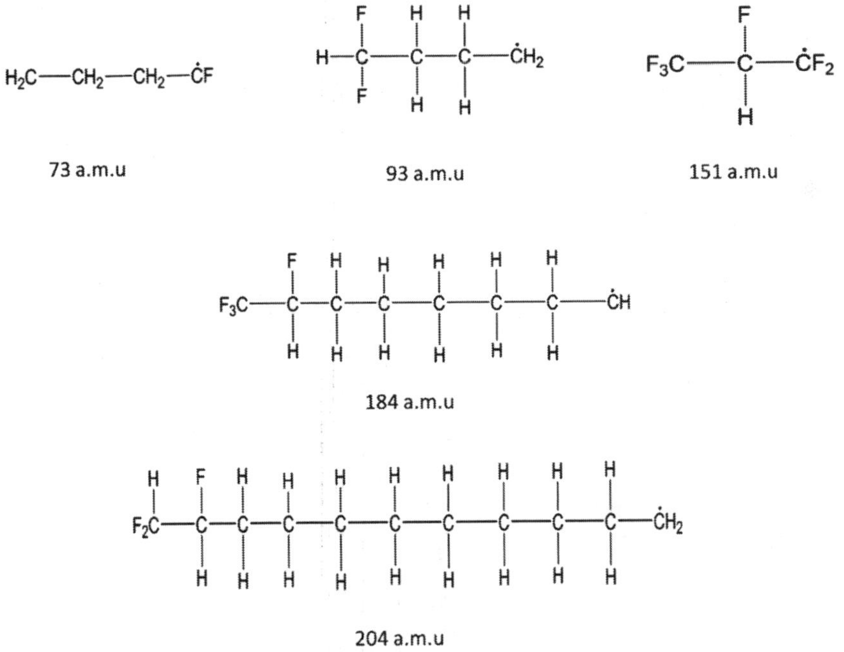

$$H_2C\!-\!CH_2\!-\!CH_2\!-\!\dot{C}F$$

73 a.m.u

$$H\!-\!\underset{F}{\overset{F}{C}}\!-\!\underset{H}{\overset{H}{C}}\!-\!\underset{H}{\overset{H}{C}}\!-\!\dot{C}H_2$$

93 a.m.u

$$F_3C\!-\!\underset{H}{\overset{F}{C}}\!-\!\dot{C}F_2$$

151 a.m.u

$$F_3C\!-\!\underset{H}{\overset{F}{C}}\!-\!\underset{H}{\overset{H}{C}}\!-\!\underset{H}{\overset{H}{C}}\!-\!\underset{H}{\overset{H}{C}}\!-\!\underset{H}{\overset{H}{C}}\!-\!\underset{H}{\overset{H}{C}}\!-\!\dot{C}H$$

184 a.m.u

$$F_2C\!-\!\underset{H}{\overset{H}{C}}\!-\!\underset{H}{\overset{F}{C}}\!-\!\underset{H}{\overset{H}{C}}\!-\!\underset{H}{\overset{H}{C}}\!-\!\underset{H}{\overset{H}{C}}\!-\!\underset{H}{\overset{H}{C}}\!-\!\underset{H}{\overset{H}{C}}\!-\!\underset{H}{\overset{H}{C}}\!-\!\dot{C}H_2$$

204 a.m.u

Figure 4.9 Possible structures of the molecules formed in He/TFE plasma

The relative fraction of the molecules in GC suggested that very large molecules such as 209 and 184 were being formed in very small quantity, while species of a.m.u. 73, 93 and 151 were formed in moderate to high quantity.

(b) Surface analysis

X-ray photoelectron spectroscopy (XPS) survey spectra of the untreated and 8 min plasma-treated soap-washed samples is shown in Fig. 4.10. The untreated sample showed only C (1s) and O (1s) peaks in the spectrum. On the other hand, in the He/TFE plasma-treated samples (both as-prepared and soap-washed) there was a strong appearance of F (1s) peak at 688.2 eV binding energy in addition to C (1s) and O (1s) peaks as shown in the same figure. Presence of F (1s) peak in plasma-treated samples indicated that the fragments of TFE produced in the plasma zone could react to the cellulosic substrate. It was found that in the 1 min plasma-treated and soap-washed sample, the fluorine atomic percentage was as high as 29% and in 8 min plasma-treated sample, it was 39.60%. There was no significant change in F percentage when the sample was subjected to soap washing. This signifies that F containing molecules were chemically attached to the cellulosic substrate. It could also be seen that with the increase in plasma treatment time, both C and O percentages decreased. Degree of surface fluorination (F/C ratio) i.e. attachment of F atom to the surface of the cellulosic textile was zero in the untreated sample and was 0.57 in the 1 min plasma-treated sample and the ratio increased to 0.96 when the plasma treatment time was increased from 1 min to 8 min.

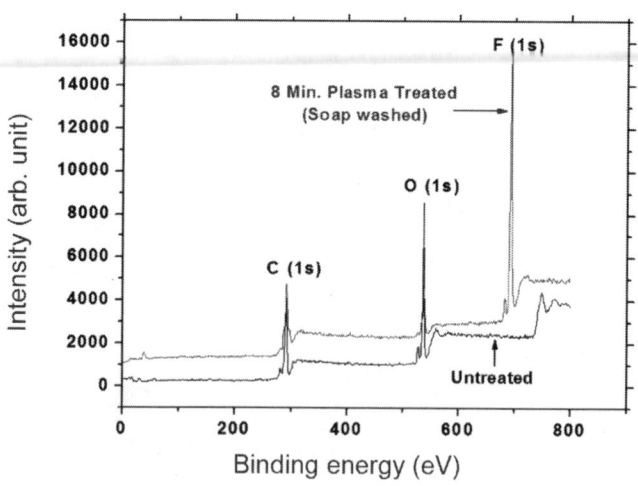

Figure 4.10 XPS survey spectra of the untreated and 8 min. plasma-treated soap-washed cellulosic samples

High resolution C (1s) spectra of the untreated and plasma-treated samples were used for deconvolution to find the type of chemical bonds present in different samples. In an untreated cellulosic sample, only four types of chemical bonds were necessary for deconvolution as shown in Fig. 4.11. These were $-C-C-$, $-CH$, $-COH$, $-C-O-C-$ and $-O-C-O$. Surface charging in the different cellulosic samples was corrected by considering the C(1s) peak to be located at 285 eV. This leads to consistent value for all other peaks, such as $-C-C-/-CH$ at 284.6 eV, $-COH$ at 286 eV, $-C-O-C-$ at 287 eV and $-O-C-O-$ at 288 eV. On the other hand, in all the TFE plasma-modified samples, three additional peaks corresponding to $-CF$ at 289.54 eV, $-CF_2$ at 291.5 eV and $-CF_3$ at 293 eV were needed to be included to complete the deconvolution. In deconvoluted spectra of all the samples, an additional peak corresponding to an unknown bond at lower binding energy level of around at 282.3 eV was found to be necessary to completely fit the deconvoluted spectra in that region. The peak was weak for the untreated sample and increased in intensity with the plasma treatment. This peak was tentatively assigned to the formation of graphite-like structure with double bonds or amorphous carbon. These species may get formed during the plasma treatment because of the interaction of high energy ions, electrons with the cellulosic polymer and/or due to the interaction of higher energy X-ray beam with cellulosic polymer during the XPS analysis.

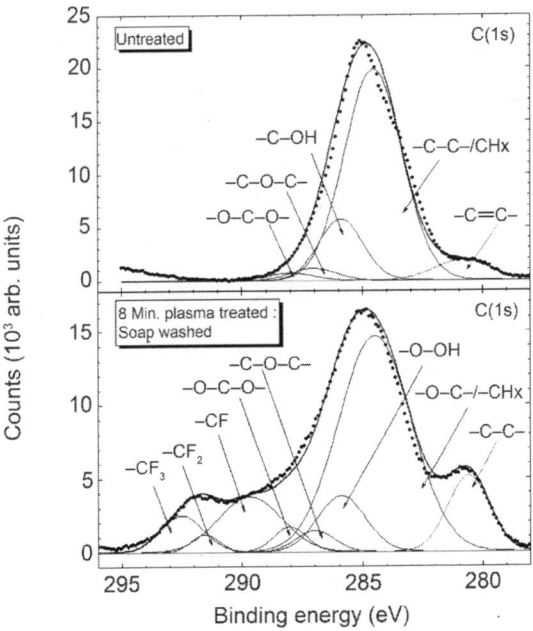

Figure 4.11 Deconvoluted of C (1s) spectra of the different cellulosic samples

In all the TFE plasma-treated samples, high percentage of fluorine compounds were present having $-CF$, $-CF_2$ and $-CF_3$ bonds. These are responsible for the development of hydrophobicity in the otherwise hydrophilic cellulosic substrates. The $-CF$, $-CF_2$ and $-CF_3$ bonds were present in both the as-prepared and soap-washed samples indicating the formation of covalent bonds between the fragments of TFE and cellulosic polymer. It was found that $-CF$ and $-CF_3$ bonds were the main contributing species towards the developed hydrophobicity. The presence of $-CF_2$ bond was very small. This was explained on the basis of the structure of TFE molecule which has one carbon atom with three fluorine atoms and another with one. It appeared that in plasma zone, the TFE molecule might be getting fragmented at the C$-$C bond giving out two species as $-CF_3$ and $-CH_2F$ and getting attached to cellulosic substrate.

With increasing plasma treatment time, the attachment of $-CF$ and $-CF_3$ molecules to the cellulosic polymer increased. This also correlated well with the improvement in hydrophobicity in terms of water absorbency time and contact angle. Upon soap washing, there was a slight decrease in the concentration of $-CF_3$ bond by 4.5%, which could be due to the removal of a small amount of condensate products deposited on the surface of the textile substrate.

There was a decrease of $-C-C-/-CH$ and $-C-OH$ bonds by 17.8%, and 4.8% when the sample was subjected to 1 min TFE plasma treatment; and a decrease of 18% and 6.7% when the sample was subjected to 8 min plasma treatment. Interestingly, there was no significant change of $-C-O-C-$ and $-O-C-O-$ bonds percentage with the plasma treatment. This indicates that the attachment of $-CF$ and $-CF_3$ bonds has happened primarily at the expense of $-C-C-/-CH$ and $-C-OH$ bonds of cellulose.

Secondary ion mass spectrometric analysis of the surface was carried out. Figure 4.12 shows the negative mass spectra of the untreated and 8 min plasma-treated samples. Negative mass spectrum of the untreated sample showed the presence of major species at mass values of 13 a.m.u. for CH^-, at 16 a.m.u. for O^-, and at 17 a.m.u. for OH^-. In addition to these, there were also peaks for C^- at 12 a.m.u., CH_2^- at 14 a.m.u., $C-C^-$ at 24 a.m.u. and $CH-C^-$ at 25 a.m.u. These smaller fragments C^-, CH^-, CH_2^-, O^-, HO^-, $C-C^-$ and $C-CH^-$ were broken fragments of the groups present in cellulose such as $-CH_2OH$, $-CHOH$, $-C-C$, etc., during SIMS analysis.

The spectrum of 8 min TFE plasma-treated and soap-washed sample showed a strong peak for F^- at 19 a.m.u. This confirmed the presence of fluorocarbon compounds on the surface of the washed samples. It appears that the attached $-CF$ and $-CF_3$ molecules were broken down into F^- and C^- atoms during the SIMS analysis. In addition to that peaks for $C-C^-$ molecule at 24 a.m.u. and CH−C⁻ at 25 a.m.u. (present in the untreated samples) were completely absent. This indicated that the prolonged plasma treatment (8 min) resulted in enhanced attachment of fluoro compounds. The relative lowering of C⁻ peaks was similar to the reduction of −C−C⁻ bonds observed in the XPS analysis.

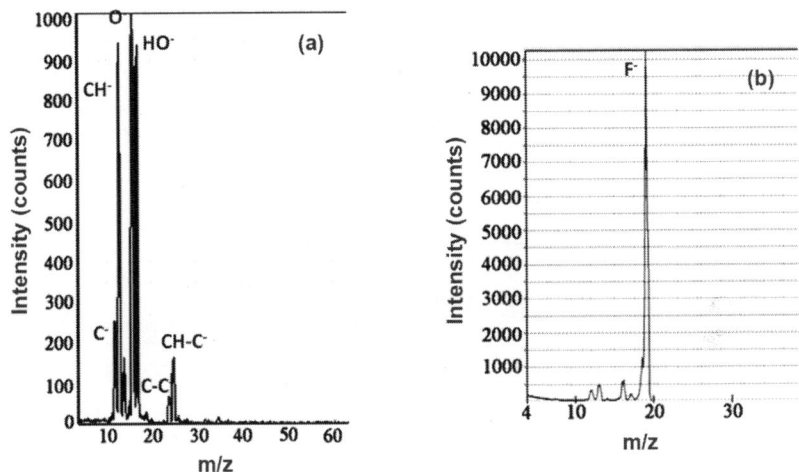

Figure 4.12 Negative mass spectra of the cellulosic samples: (a) untreated and (b) 8 min plasma treated soap washed

In order to understand the mechanism of action of He/TFE plasma on cellulosic textile, the plasma exhaust gases were analysed using GC-MS. The species from the plasma exhaust gases, which could be captured in the two organic solvents, isopropyl alcohol and acetone, were analysed in GC-MS. The mass spectra showed various masses at 73, 93, 151, 184 and 209 (Fig. 4.9). However, very small fragments of the TFE, which were detected in OES, could not be trapped in the solvents possibly because of their high volatile nature.

Based on the GC-MS analysis, the possible structures of the molecule that were formed in He/TFE plasma are mainly of two types: one with long hydrocarbon chains with one of the end with fluorocarbon moiety and the other highly fluorinated compounds formed by condensation of fluorine containing fragments. Both XPS and SIMS analysis showed that there was a significant decrease of the −C−OH, −C−C−/ −C−H bonds. Therefore, during the plasma reaction there is possibility of breaking of −C−OH, −CH₂−OH and −C−C−/ CH bonds of cellulose and at the same time attachment of the fluorocarbon moieties. However, if the moieties with long hydrocarbon chains (based on GC-MS fragments) were to attach to cellulose, then −C−H bond percentage would have remained high and fluorine percentage would have been much lower. Therefore, it appears that only highly fluorinated compounds have attached mainly by replacing the −C−OH bonds of cellulose.

XPS also showed the addition of CF_3 and CF as the two important bonds within the carbon peak. CF_2 has only a minor contribution. These bonds would appear/form/ if the precursor molecule would preferentially split into two parts (i.e. CF_3 and CFH_2) with the breakage of −C−C⁻ bond and react immediately with the cellulose. In addition, species with structure similar to that of 151 a.m.u. may also react with the cellulosic substrate. Figure 4.13 shows the possible sites at which the bonds in cellulose are likely break to react with various fragments of tetrafluoroethane such as •CF_3, •CFH_2 and •CF_3−CHF−CF_2−.

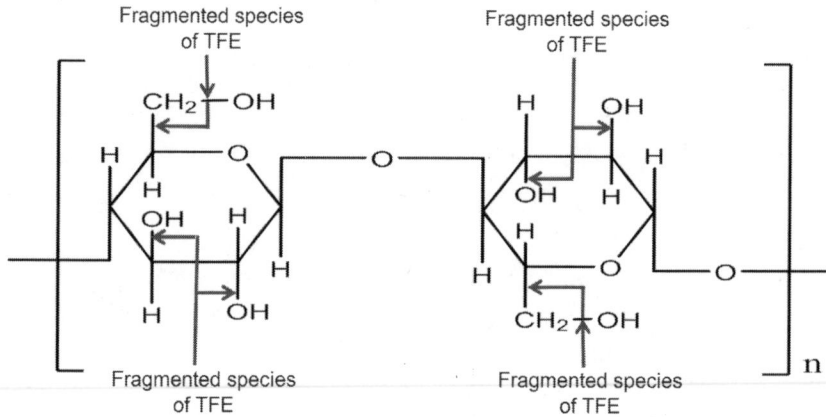

Figure 4.13 Possible sites where fragments of tetrafluoroethane may attach to cellulosic textile

4.5.2   Using 1,3-butadiene (BD) gas as a precursor (Samanta et al., 2012)

(a) Plasma generation and analysis
Reactor used for the experiments was same as discussed in previous section. A number of experiments were carried out to fix the parameters for the glow plasma generation, after that the ratio of carrier to reactive gas flow rate was kept at 5 and a discharge gap of 2.1 mm, stable helium/butadiene (He/BD) plasma could be generated over a specific range of voltage and frequency. The mixture of BD and He formed uniform glow plasma with a milkish white colour, which was noticeably different from the bluish purple colour of the He plasma. The smooth profile of I-V waveform, shown in Fig. 4.14, i.e. the absence of any short or long spikes, was a confirmation of the formation of glow plasma.

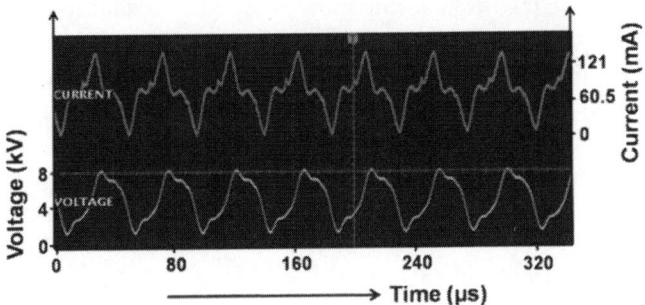

Figure 4.14 I-V characteristic of He/BD plasma

Figure 4.15 shows the emission spectra of He/BD plasma at two different carrier (He) to precursor (BD) gas ratios (total gas flow rate 510–540 ml/min) along with that of pure He. Pure He gas upon ionization exhibited eight major atomic lines at wavelengths of 388.6, 491.5, 501.2, 587, 655.8, 667.5, 706 and 726.5 nm, with a maximum emission at 706 nm. At He to BD gas flow rate ratio of 12.5 (He 500/BD 40, total gas flow rate 540 ml/min), emission peak of He at 667.5 nm became more intense followed by 706, 587 and 726.5 nm peaks. The emission peak at 655.8 nm (observed with pure He) was not visible. However, when the carrier to precursor gas ratio was increased to 50 (He 500/BD 10, total gas flow rate 510 ml/min), the emission peaks of He in He/BD plasma were observed to be similar to that obtained with pure He (without precursor gas) along with some additional peaks. Interestingly, the emission peak of He at 655.8 nm was also observed to reappear. The ratio of intensity of emission peaks of He 587/667.5 nm is nearly 1.0 for the He plasma. This ratio decreased to 0.42 for the He to BD gas ratio of 12.5 and increased to 1.25 for the He to BD gas ratio of 50.

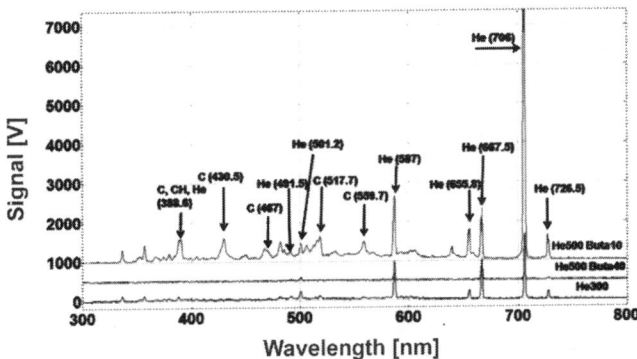

Figure 4.15 Effect of carrier to precursor gas ratio on ionization of He/BD plasma at discharge voltage of 3.96 kV and frequency of 21.8 ± 0.7 kHz

At (He) to (BD) gas flow rate ratio of 50, several additional peaks were observed; some of these were C line at 430.5 nm, 467 nm, 517.7 nm and 559.7 nm. The intensity of the peak at 388.6 nm also appeared to have increased presumably due to the presence of C, CH lines in the same region. These results indicated that the fragmentation of 1,3-butadine could be controlled by changing the carrier to precursor gas ratio. In the presence of cellulosic textile, no significant change in the number or position of OES peaks was observed.

In order to understand the fragmentation of BD, plasma exhaust gases were captured in n-hexane and analysed using GC-MS. The analysis of plasma exhaust by GC-MS gave seven (8) major components other than the n-hexane (solvent) peak as shown in Fig. 4.16. The major components corresponded to molecular mass of 96 a.m.u. (5 different peaks), 108 a.m.u. (1 peak) and 110 a.m.u. (2 peaks). The appearance of different GC peaks with the same mass may be arising due to small changes in the chemical structure. Peak corresponding to 1,3-butadiene (54 a.m.u.) was not observed, except for the dimer of it at 108 a.m.u, even at high flow rate.

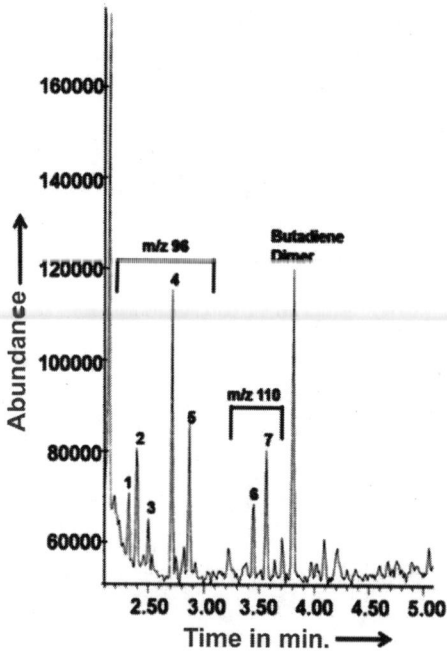

Figure 4.16 GC peaks obtained from plasma exhaust trapped in n-hexane at He 2500 ml/min and BD 500 ml/min

Interestingly, the major components of plasma did not change when the textile substrate was introduced in He/BD plasma. Higher oligomers of

1,3-butadiene were not observed in the plasma exhaust. This may be due to the fact that either they are being formed in low concentration or are formed and depositing on the substrate due to their higher mass.

After He/BD plasma treatment, the water absorbency time was found to increase from less than a second to more than an hour and water contact angle from 0° in untreated fabric to 142° in plasma-treated and soap-washed sample. Water contact angle images are shown in Fig. 4.17.

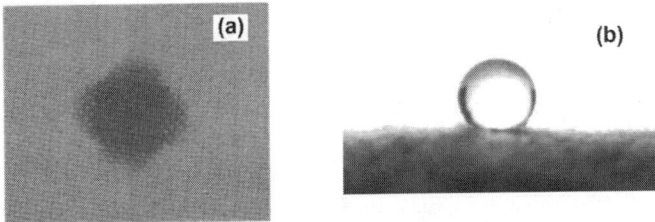

Figure 4.17 Water contact angle of (a) untreated, (b) plasma-treated soap-washed samples

(b) *Surface analysis*

X-ray photoelectron spectroscopy was used to analyze surface chemistry of the untreated and He/BD plasma-treated samples in terms of atomic percentage, atomic ratios and the presence of different chemical bonds. Figure 4.18 shows XPS survey spectra of the untreated and plasma-treated

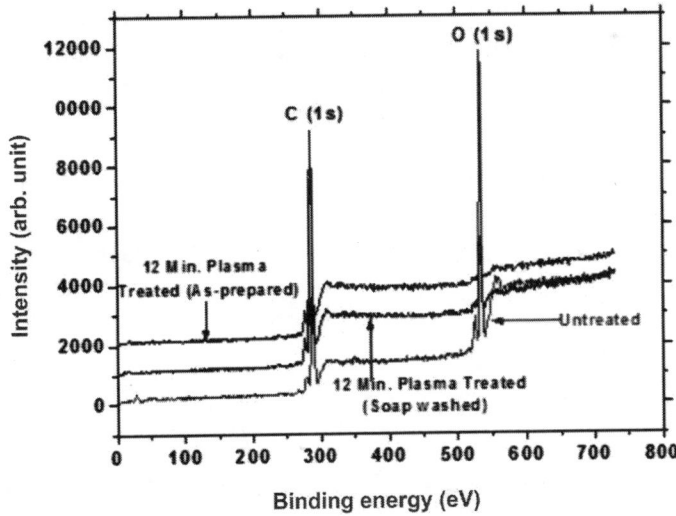

Figure 4.18 XPS survey spectra of untreated and 12 min plasma treated (as-prepared and soap-washed) cellulosic textiles

samples before and after soap washing over the binding energy range of 0–730 eV. Cellulosic textile was found to be composed of carbon (C), oxygen (O) and hydrogen (H), as expected. As XPS cannot detect H atom, therefore, the untreated sample shows only C (1s) and O (1s) peaks to be present at the binding energy of 285.3 eV and 533.3 eV, respectively. Similarly, in the plasma-treated cellulosic samples, peaks corresponding to C (1s) and O (1s) were also observed. This is because 1,3-butadiene is also composed of only C and H atoms.

It was found that in the untreated sample the carbon atomic percentage was 57.06%. With increase in plasma reaction time, a continuous increase in the carbon percentage was observed. In 12 min plasma-treated and soap-washed sample, carbon percentage was as high as 90.9%. In the untreated sample, the surface oxygen percentage was 42.9%. This value was found to reduce to 9.1% in 12 min plasma-treated and soap-washed sample. These results were an indication of reaction between the different fragment of 1,3-butadiene and cellulosic textile inside the plasma zone. These changes in the surface atomic composition of the substrate explained the hydrophobic behaviour of cellulose on He/BD plasma treatment. The atomic percentages of 12 min plasma-treated as prepared sample were only a little different than the washed sample. This indicated the stability of the treatment on washing.

High resolution C (1s) spectra of the untreated and He/BD plasma-treated samples were used for deconvolution to find the type of chemical bonds present in different cellulosic samples. Surface charging in the different cellulosic samples was corrected by considering the C(1s) peak to be located at 285 eV. This correction resulted in consistent values for all other peaks in all samples, such as $-C-C-/-CH_x$ at 284.8 eV, COII at 286 eV and $-C-O-C-$ at 287.1 eV, and $-O-C-O-$ at 288.1 eV as shown in Fig. 4.19.

It was found that He/BD plasma treatment resulted in increase of $-C-C-/-CH_x$ percentage. With increase in plasma treatment time from 1.5 min to 12 min, there was a gradual increase of $-C-C-/-CH_x$, bonds from 73.8% to 91.2% and a corresponding decrease of oxygen-containing species. The $-C-OH$ bonds decreased significantly from 20.3% for the untreated sample to 16.6% and 2.2%, respectively, for the 1.5 min and 12 min plasma-treated and soap-washed samples. However, ether, i.e. $-C-O-C-$ and $-O-C-O-$ bonds were found to remain nearly same for all the samples.

The increase in the $-C-C-/-CH_x$ bond and a similar decrease in the $-C-OH$ bond are responsible for the developed hydrophobic functionality in the otherwise hydrophilic cellulosic textile. These observations indicated that hydrophobic species containing $-CH_x$ groups such as CH, CH_2 and $-CH_3$, which are likely to be formed from butadiene in plasma, might have chemically attached to the surface of cellulosic textile possibly replacing

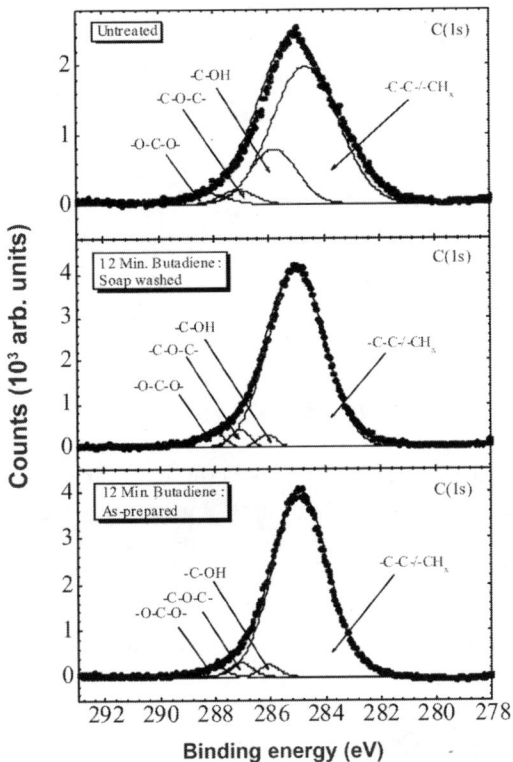

Figure 4.19 XPS deconvoluted core level of C (1s)-spectra of the different cellulosic samples

hydrophilic –C–OH groups. There may be two phenomena towards the enhancement of –C–C–/–CH$_x$ percentage and reduction of –C–OH percentage. One is XPS signal was collected from the thinly deposited/reacted layer of butadiene fragments, which might have masked the cellulose signals. Second may be from the actual reduction of –C–OH sites due to reaction with fragment of butadiene. The modification was firmly attached to the textile, i.e. it is durable to washing, which indicates that the butadiene fragments are chemically attached to cellulose. Therefore, it was proposed that the attachment site is likely to be –C–OH as the maximum change –C–OH percentage was observed compared to –C–O–C– and –O–C–O– percentages.

A small change of chemical bond percentages between 12 min as-prepared and soap-washed samples was probably because of the removal of physically deposited condensates of butadiene from the surface. However, the degree of change in these values remained largely unaffected showing durable nature of the treatment.

Surface chemistry of the untreated and He/BD plasma-treated samples was also analyzed using secondary ion mass spectrometer. Figure 4.20 shows the negative ion mass spectra of the untreated and 12 min plasma-reacted soap-washed samples. The negative mass spectrum of the untreated sample showed the presence of major species at mass values of 13 a.m.u. for CH^-, of 16 a.m.u. for O^-, and 17 a.m.u. for OH^-. In addition to these, there were also peaks for C^- at 12 a.m.u., CH_2^- at 14 a.m.u., $C-C^-$ at 24 a.m.u. and $CH-C^-$ at 25 a.m.u. These imply that during SIMS analysis, the groups present in cellulose such as $-CH_2OH$, $-CHOH$, $-CHOH-CHOH-$ etc., were broken down into the smaller fragments such as: C^-, CH^-, CH_2^-, O^-, HO^-, $C-C^-$ and $C-CH^-$.

The negative ion mass spectra of 12 min. plasma-treated soap-washed samples were similar to the untreated cellulosic substrate, i.e. no additional mass peaks were observed. However, the intensities of certain signals were enhanced and certain signals were suppressed, significantly. In the untreated sample, the intensities of major species at mass values of 16 a.m.u. for O^- and 17 a.m.u. for HO^- were significantly more compared to the CH^- at 13 a.m.u. (Fig. 4.20a). On the other hand, in the plasma-reacted samples, the abundance of CH^- was much more compared to O^- and HO^- (Fig. 4.20b). In the plasma-treated and soap-washed samples, the peak intensities of the other hydrocarbon molecules such as $C-C^-$ at 24 a.m.u. and $CH-C^-$ at 25 a.m.u. were also observed to increase significantly. The attached hydrophobic species generated from 1,3-butadiene plasma were likely to give enhanced signal at C^-, CH^-, CH_2^-, $C-C^-$ and $CH-C^-$ during the SIMS analysis. The XPS analysis also showed an increase of hydrophobic groups such as $-CH_x$ in the plasma-treated samples.

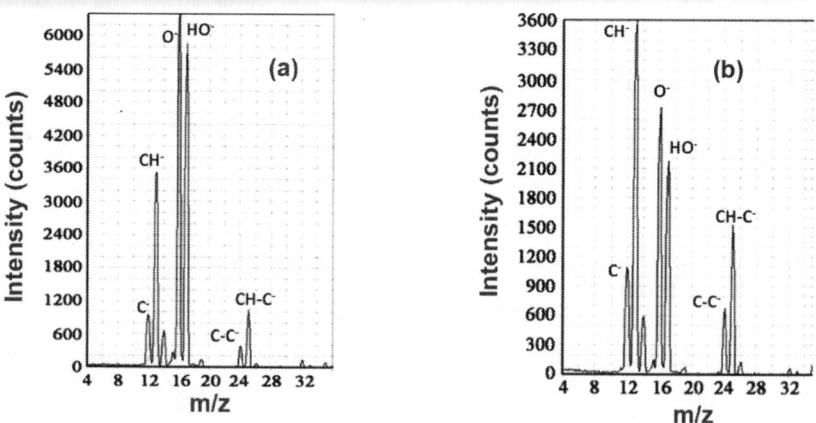

Figure 4.20 Negative ion mass spectra of cellulosic samples: (a) untreated, (b) 12 min. He/BD plasma treated soap washed

Based on the GC-MS analysis, the possible structures of the molecules that were formed during the plasma reaction of 1,3-butadiene were dimmer-based structures of 1,3-butadiene (108 and 110 a.m.u.) and species with seven carbon atoms (96 a.m.u.). SIMS and XPS analysis showed a significant increase in CH_x groups such as $-CH_3$ and $-CH_2$, which indicated that BD fragments, formed in the plasma had attached to the substrate. The possible sites (1, 2 and 3) at which the bonds of cellulose might have broken and reacted with various fragments of 1,3-butadiene are indicated in Fig. 4.21.

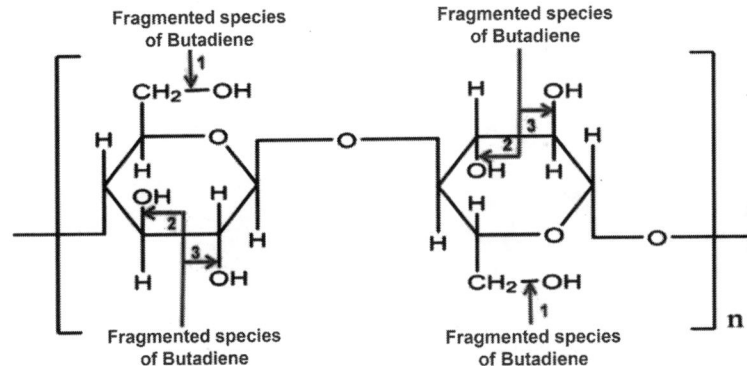

Figure 4.21 Schematic of cellulose depicting the possible places of plasma reaction of 1,3-butadiene with cellulosic textile

4.6   In-situ hydrophobic functionalization of textile using liquid precursor

In this section hydrophobization of cellulosic textile using the non-fluorocarbon liquid precursors such as dodecyl acrylate, styrene, acrylonitrile and hexamethyl disiloxane has been discussed.

4.6.1   Dodecylacrylate as a precursor (Panda et al. 2013)

(a) *Plasma generation and analysis*
In this study, the scoured fabric was pretreated with solution of precursor DA in ethanol by dip-pad method and dried. Thereafter, the pretreated fabric was treated with helium glow plasma for 0.5–2.5 minute. The effect of various plasma-treatment parameters such as power density, frequency, gas flow rate, concentration of precursor and treatment time was studied. Plasma-treated samples were thoroughly washed in hexane (which is a solvent for precursor and its polymer) and acetone to remove unreacted and loosely deposited materials. The dried samples were used for further evaluation.

An averaged profile of 8 scans was taken and shown in Fig. 4.22. This averaged current profile is very smooth and consists of a single peak along with a residual current peak, which are the typical features of a He glow discharge. This implies that He/DA discharge plasma is glow in nature. Similar current-voltage I-V profiles were obtained for other values of parameters in the range studied.

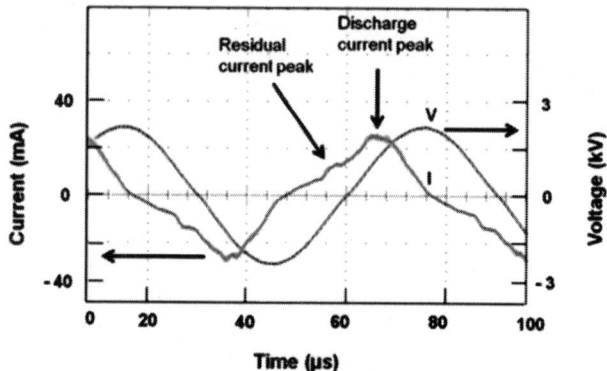

Figure 4.22 Averaged scan profile of discharge current and discharge voltage

Optical emission spectra of the discharges at the optimized set of parameters were taken for (1) pure helium plasma, (2) helium plasma in presence of liquid precursor DA placed on a glass slide and (3) helium plasma in the presence of precursor treated fabric. Overlay of these three spectra is shown in Fig. 4.23. Pure helium spectrum shows transition peaks of helium atom and low intense peaks at 314, 374, 380 and 391 nm, these are likely to be due to the presence of impurities (C/O/N) in the helium gas.

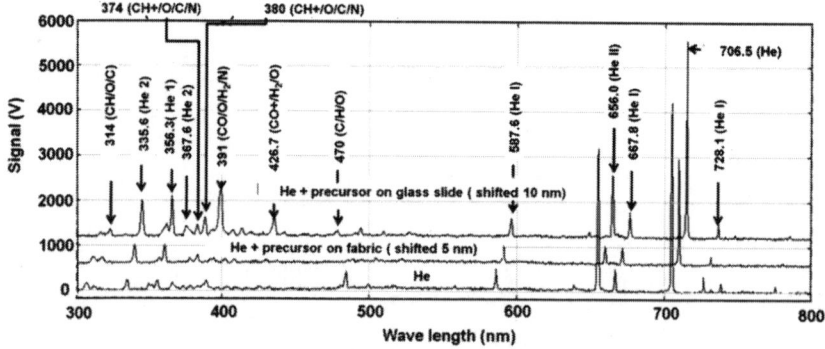

Figure 4.23 OES spectra of helium, helium with DA on a glass slide and helium with DA-treated viscose fabric

When liquid precursor was kept on a glass slide inside the helium plasma, intensity of peaks at 314, 374, 380 and 391 nm was significantly enhanced and two new peaks appeared at 426.7 and 470 nm. The enhancement of peaks intensities at 314, 374, 380 and 391 nm can be attributed to the presence of CH, CH^+ and CO fragments of the precursor. The new peak at 426.7 nm can be assigned to $CO+/H_2/O$ fragments and at 470 nm to C/H/O fragments of the precursor. No peaks could be detected for CH_2, which is a likely fragment of long hydrocarbon chain precursor. CH_2 being very unstable is likely to get converted to CH or CH^+. When precursor-treated fabric was introduced inside the He plasma, the intensity of all the peaks of He except those at 336 and 356 nm were found to reduce significantly. This indicated the interaction of He plasma with fabric and the precursor. However, the new peaks observed with pure precursor (spectrum 2) could not be detected possibly due to the low concentration of precursor in the fabric.

In the plasma zone, DA may undergo fragmentation and further reactions with the substrate. Plasma may act on precursor to create active species, which may combine to form different moieties. In order to understand the possible changes that DA molecules might be undergoing in the atmospheric plasma zone, the exhaust from the plasma zone was trapped in solvent (hexane) and analyzed by GCMS. Chromatographs of exhausts from treatment of viscose with DA at 11 kHz and 15 kHz frequencies are shown in Figs. 4.24(a) and 4.24(b), respectively. In chromatograph of the sample collected at discharge frequency 11 kHz, three major peaks at retention times of 6.0, 9.5 and 9.8 min were observed. Based on the analysis of their mass spectra, the peak at 9.828 was assigned to DA, 9.5 to lauryl alcohol (186 a.m.u.) and 6.0 to $C_{15}H_{32}$ (212 a.m.u.), a hydrocarbon fragment. At 15 kHz, an additional peak at 19.8 min was also observed. This corresponds to a molecule of 284 a.m.u. with a possible formula of $C_{18}H_{36}O_2$. The mass spectra for the chromatograph peaks at 6.0 and 19.8 mine are shown in Figs. 4.25(a) and (b), respectively. The structures of the various molecules collected in exhaust are shown in Table 4.4.

GCMS results indicated that at low frequency of 11 kHz, the monomer did not undergo significant fragmentation. This was evident from the fact that small carbon species were not visible in OES and also only a simple fragment of DA, such as lauryl alcohol, was present in significant concentration. Additionally, a small amount of $C_{15}H_{32}$ hydrocarbon component, which is a combination of two fragments of DA, was also observed in GCMS. However, at higher frequency of 15 kHz, the DA fragmentation was significantly higher as various OES peaks related to C were found to be enhanced. Also, it appeared that these fragments were able to recombine with other DA molecules to produce larger molecule of 284 a.m.u. in a significant amount as shown in chromatograph (Fig. 4.24b).

Table 4.4 Chemical structures of various species collected from the exhaust of plasma reaction chamber.

Retention time (min)	Mass (a.m.u.)	Chemical formula	Possible structure
9.828	240	$C_{15}H_{28}O_2$	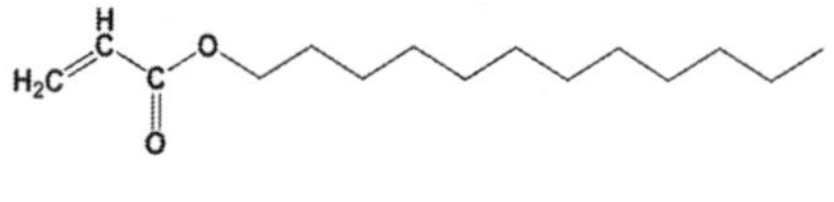
9.533	186	$C_{12}H_{26}O$	
6.085	212	$C_{15}H_{32}$	
19.857	284	$C_{18}H_{36}O_2$	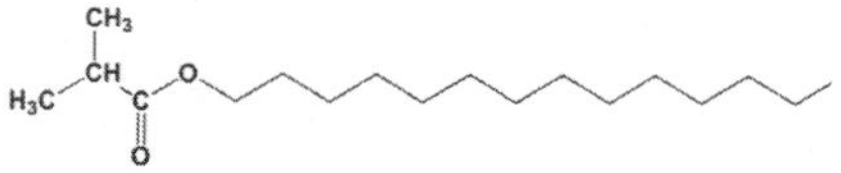

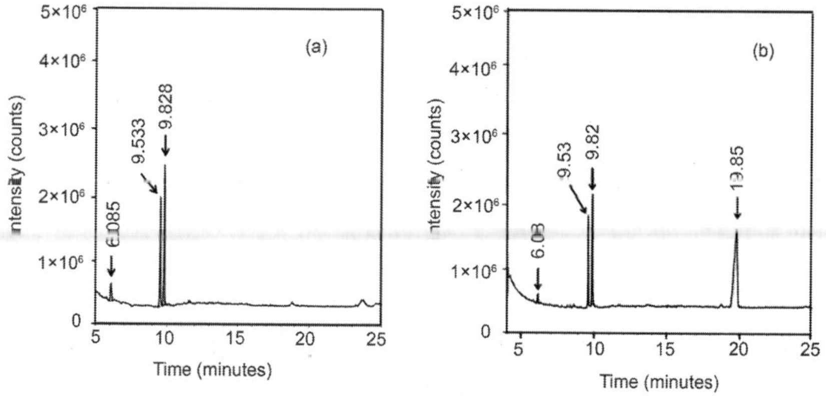

Figure 4.24 Gas chromatograph of plasma exhaust collected at (a) 11 kHz and (b) 15 kHz

Interestingly, the functionalization of substrate was observed at both the frequencies 11 and 15 kHz. This implies that DA, being an unsaturated monomer, might be undergoing oligomerization or reaction at double bond with the activated substrate. However, functionalization of the substrate is significantly higher at 15 kHz, which possibly indicates that in addition to reaction at the double bond, the fragment at 284 a.m.u. might also be playing a significant role in hydrophobic functionalization of the substrate. It may also be noted that oligomers or the fragments as formed above were chemically attached to the cellulosic substrate, which is indicated by the excellent durability to solvent wash.

(b) *Effect of parameters on added functionality*
Effect of precursor concentration, discharge frequency, power density and helium flow rate on hydrophobic functionalization was studied in separate experiments keeping other parameters fixed. The concentration of DA solution was varied from 0.05 to 0.25 M, frequency from 11 kHz to 21 kHz, helium flow rate from 0.05 SLPM to 1 SLPM, power density from 0.30 to 3.33 W/cm^3 and treatment time 0.5 to 2.5 minute.

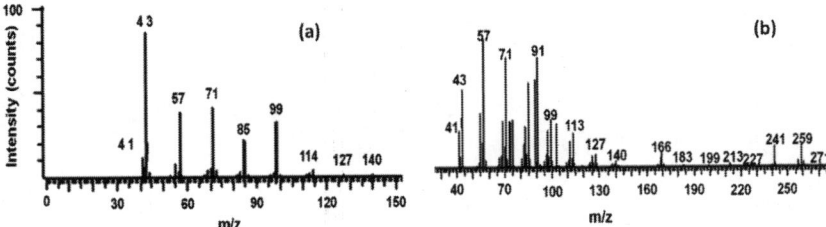

Figure 4.25 Mass spectra of GC peak observed at (a) 6.085 min and (b) 19.857 min

From the result it was found that water absorbency time and the water contact angle were nearly same at >60 min and 136–142°, respectively, for concentration ranging from 0.05 to 0.25 M. Only in the case of 0.01 M concentration, the water absorbency time was significantly less at 40–50 min. However, there was no significant variation in water contact angle. Based on these results, the concentration of 0.05 M, which is equivalent to an add-on concentration of 2.4 wt% of precursor on the weight of fabric, was found to be sufficient to provide a uniform treatment of the surface of the fabric. Images of water contact angle on untreated and treated–solvent–washed fabric are shown in Fig. 4.26.

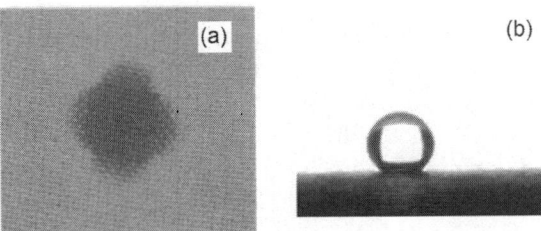

Figure 4.26 Images of water droplet on (a) untreated (b) solvent and soap-washed fabrics

It was found that at discharge frequency of 15 and 17 kHz, the water absorbency time was more than 60 min. However, at other frequencies it was significantly lower in the range of 15–35 min. Though there was no significant variation in water contact angle, the water absorbency time clearly indicated that the treatment was better in the frequency range of 15–17 kHz. To elucidate the reason behind it, the OES of the plasma was recorded during the treatment at different frequencies. The relative intensity of important peaks was plotted by normalizing them against the intensity of helium peak at 706.5 nm. It was found that intensity of peaks was maximum at 15 kHz and the discharge frequency of 15 kHz was found to be effective for functionalization.

Water absorbency time and water contact angle of the fabric did not change significantly with the change in helium gas flow rate. For both solvent-washed and soap-washed samples, the water absorbency time of more than 1 h was observed. Similarly water contact angle remained unaltered and was similar to the best cases reported above.

Discharge power density, which is a measure of plasma energy is an important parameter and is likely to influence functionalization of the substrate. Power density and treatment time, together, determine the total energy given to the system for carrying out a reaction. The power density, in turn, depends on the discharge voltage and the gap between the electrodes. These three parameters were varied to change the total energy supplied to the system. It was found that higher plasma energy was not helping the functionalization of the substrate. An optimum level of plasma energy, which was necessary to impart required level of functionalization, could be achieved by altering either the power density or treatment time. However, at energy levels higher than the optimum values, the functionalization was adversely affected. From the above results, it was inferred that there appeared to be time–power density equivalence similar to the time–temperature equivalence observed in chemical reactions.

(c) Surface analysis

The ATR-FTIR spectra of untreated, as-plasma-treated, plasma-treated solvent (hexane-acetone) washed viscose fabrics are shown in Fig. 4.27. In the spectra of as-plasma-treated fabric, two distinct peaks at 2818 cm^{-1} and 2844 cm^{-1} were observed for the symmetric and asymmetric stretching of CH$_2$. A new peak at 1731 cm^{-1} was also observed for C=O stretching of ester groups. Appearance of these peaks showed the presence of either DA or its fragments in the sample. On solvent washing, these peaks remained with only a small reduction in their relative peak intensities. This confirmed the reaction of DA or its fragments with cellulose.

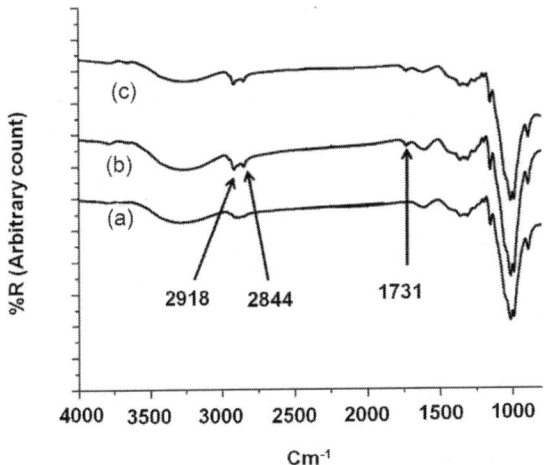

Figure 4.27 ATR-FTIR spectra of (a) untreated (b) as-plasma-treated
(c) plasma-treated solvent-washed viscose fabrics

Raman spectra of untreated, as-plasma-treated and plasma-treated-solvent-washed viscose fabrics were compared. A small hump was observed at 1738 cm^{-1} in spectra of as-plasma-treated and plasma-treated and solvent-washed samples. To elucidate the chemical differences between the three samples, Raman spectroscopy was carried out in a short range of 1600–1800 cm^{-1}. An overlay of these spectra is shown in Fig. 4.28. It was observed that the peak at 1738 cm^{-1} was present in both as-plasma-treated and solvent-washed

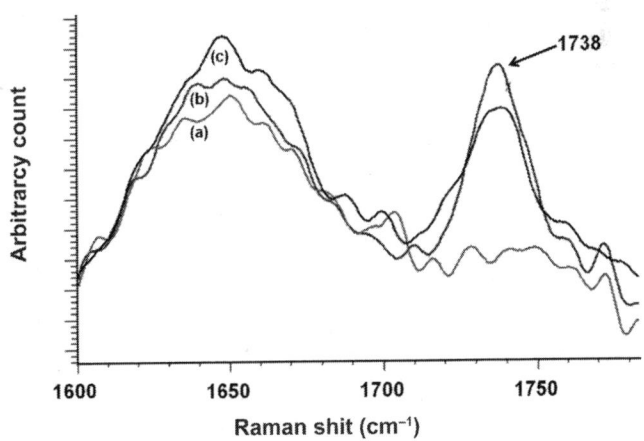

Figure 4.28 Raman spectra in range of 1600–1800 cm^{-1} of (a) untreated,
(b) as-plasma-treated and (c) plasma-treated, solvent-washed viscose fabrics

samples. However, this peak was absent in untreated sample. Presence of peak at 1738 cm^{-1}, which corresponds to C=O stretching of ester groups, indicated the incorporation of DA or its fragments on the surface of the substrate. Also, since intensity of the peak did not change with solvent washing, it appeared that these moieties were chemically attached to the substrate.

Negative secondary ion mass spectra (SIMS) of untreated and plasma-treated and solvent-washed viscose fabrics are shown in Fig. 4.29. Both the spectra showed peaks at m/z 12 for C$^-$, 14 for CH$_2^-$, 13 for CH$^-$, 16 for O$^-$, 17 for OH$^-$ and 25 for C$_2$H$^-$. The comparison of the two spectra clearly showed difference between the relative intensities of O$^-$ and OH$^-$ peaks. In untreated sample, OH$^-$ peak was higher compared to O$^-$, which was likely due to the presence of primary and secondary alcohol groups in cellulose. On plasma treatment, this ratio got changed in favour of O$^-$ presumably due to the fact that DA moieties, which were attached to the substrate, contained ester groups. Further, the intensity of CH$^-$ peak increased substantially compared to both O$^-$ and OH$^-$ peaks in treated–washed sample. Enhanced intensity of hydrophobic ions and decrease in intensity of OH$^-$ ions indicated the interaction of precursor fragments with the substrate.

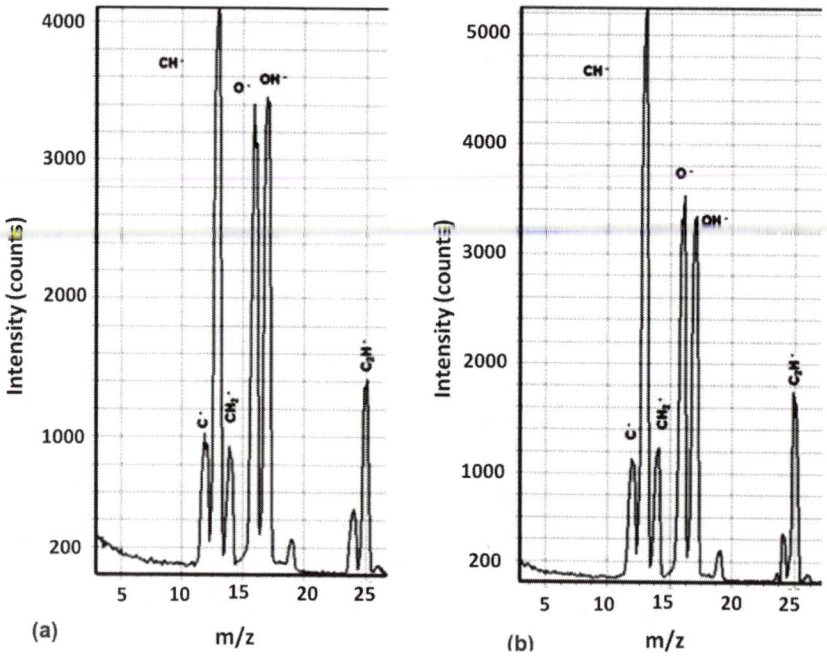

Figure 4.29 SIMS spectra of (a) untreated viscose fabric
(b) plasma-treated solvent-washed viscose fabric

4.6.2   Styrene as a precursor (Parida et al., 2012)

(a) *Plasma treatment and analysis*

In this study, the vapours of styrene were mixed with another stream of He and carried to the plasma zone for in-situ reaction with the textile substrate. The schematic diagram of the set-up is shown in Fig. 4.30. The plasma was stabilized and the fabric was plasma treated in the presence of styrene vapor in helium (He) plasma for 5 minutes.

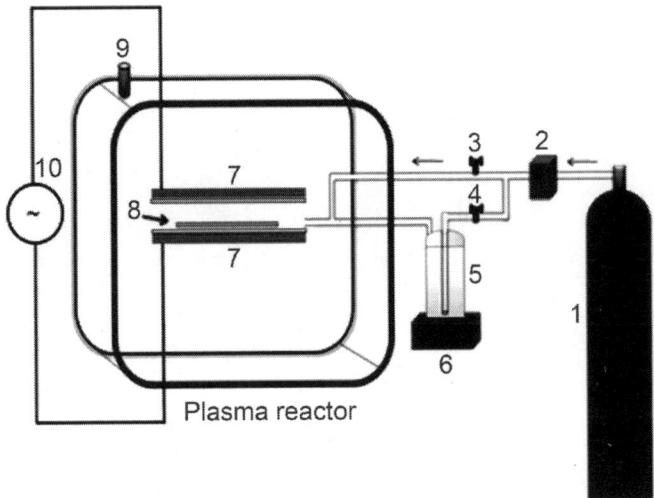

Figure 4.30 Schematic of atmospheric pressure plasma reactor. (1) Gas cylinder, (2) Mass flow controller, (3 & 4) Valves, (5) Bubbler, (6) Heater, (7) Electrodes, (8) Sample, (9) Exhaust, (10) Power supply

Figure 4.31 shows the current-voltage (I-V) wave-form of styrene/He plasma glow plasma. Styrene/He plasma current-voltage profile was smooth and free of micro-discharges indicated by the absence of short or long spikes in the current wave-form.

GC-MS spectra of the plasma exhaust gases trapped in n-hexane showed only two compounds – one the monomer, i.e. styrene, and the other saturated derivative of styrene. Interestingly, neither benzene radical nor its other fragments could be trapped in n-hexane. From GC-MS, it was indicated that styrene may also be undergoing reactions such as hydrogenation and polymerization. However, oligomers of styrene could not be found in GC-MS possibly because of the high tendency for large molecules to condense inside the plasma zone.

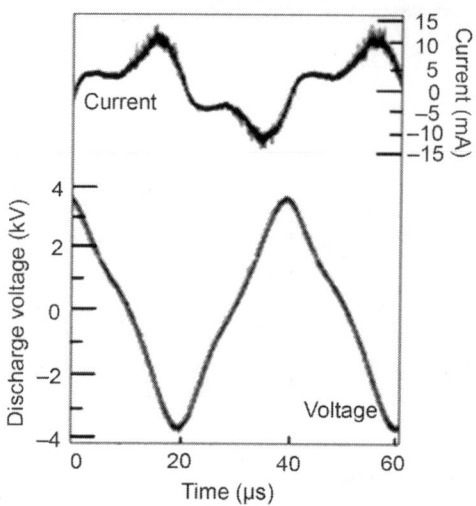

Figure 4.31 Current-voltage (I-V) wave-form of styrene/He plasma glow plasma. Styrene/He plasma current-voltage profile was smooth and free of micro-discharges indicated by the absence of short or long spikes in the current wave-form

(b) *Effect of different plasma parameters*

The effect of plasma discharge voltage and frequency on the reaction was studied in the range of 1.55–6.0 kV and 17.9–21.0 kHz, respectively, for treatment time of at 5 min. It was found that water-absorption time increased quickly with increase in plasma voltage up to 1.73 kV, and thereafter, it decreased with further increase in plasma voltage. This phenomenon was observed at each plasma treatment frequency. A maximum water drop disappearance time of 60 minutes with contact angle of 133° could be obtained after the plasma treatment. Effect of frequency was a bit different. Among the 6 different frequencies tried, three from 17.9 to 18.94 kHz gave similar effect, showing higher water drop disappearance time than the other three from 20.2 to 21 kHz, which formed a group giving much lower hydrophobic effect

Emission spectra of styrene/He plasma acquired in the wavelength range of 300–800 nm at different process voltage and frequencies are shown in Fig. 4.32. The OES spectrum of He plasma had characteristic peaks at 706 nm. On addition of styrene vapors, the plasma characteristics changed, and several new peaks corresponding to CH, CH_2, C_3H_3, C_6H_5 and H_2 appeared. Changes in spectral intensity obtained from OES of different species generated at different plasma process parameters were analyzed. It was observed that intensity of peak at 332 nm related to C_3H_3 increased with increasing plasma frequency, whereas intensity of peaks at 483.1 nm and 504.8 nm related to

C_6H_5 decreased. It was inferred that these changes might be due to the fact that at higher frequency, the styrene molecule and its major fragment C_6H_5 were getting trapped in the plasma zone for a longer duration of time, leading to higher fragmentation of styrene and benzene radicals by ring-breaking mechanism. Higher fragmentation of benzene ring is also supported from the fact that the intensity of CH (374 nm, 395 nm) peaks increased with increase in frequency.

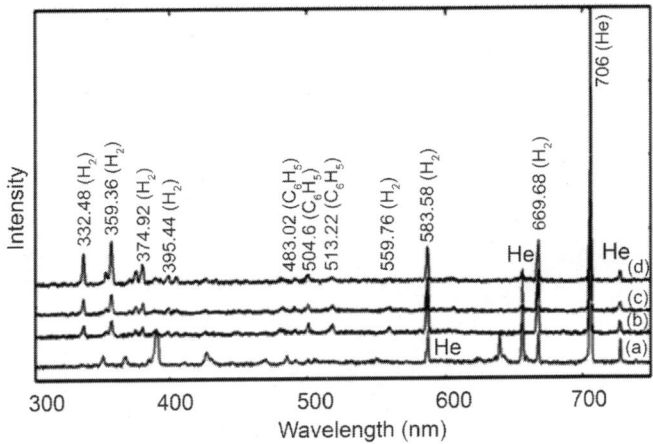

Figure 4.32 OES spectra of (a) He plasma and (b-d) styrene/He plasma at frequency of (b) 17.9 kHz, (c)18.94 kHz, (d) 21.0 kHz at discharge voltage 1.73 kV

(c) Surface analysis
Raman spectra of control cotton, styrene/He plasma-treated cotton, and plasma-treated washed cotton samples are shown in Fig. 4.33. The chemical changes were determined by comparing the normalized intensity of various emissions (peaks). The treated sample showed several new peaks notably at 1602, 1030, 1000 and 618 cm⁻¹ which are characteristic peaks for poly (styrene). In addition, a few weak peaks were observed at 306, 329 (shoulder), 1036, 1061, 1412 and 1629 cm⁻¹, which appeared to be due to the chemical modification of the cellulose by other fragments of styrene. In Raman spectrum of the treated and solvent-washed sample all the peaks corresponding to poly (styrene) were found to be missing, indicating that modification of cellulose with compounds similar to poly (styrene) was loosely bound to the surface and could easily get removed. However, some other modifications represented by appearance of other peaks such as 306, 329, 1061 and 1629 cm⁻¹ remained and their relative intensity increased compared to the as-treated sample. Peak at 1629 cm⁻¹ could be assigned to –C=C– stretching of aromatic moieties, 306 and 329 cm⁻¹ could be assigned to cellulose ring deformation, which appeared

to have shifted from 350 and 380 cm^{-1} peaks on substitution by fragments from styrene. This was accompanied by significant reduction in intensities of original ring deformation peaks of cellulose at 350 and 380 cm^{-1}.

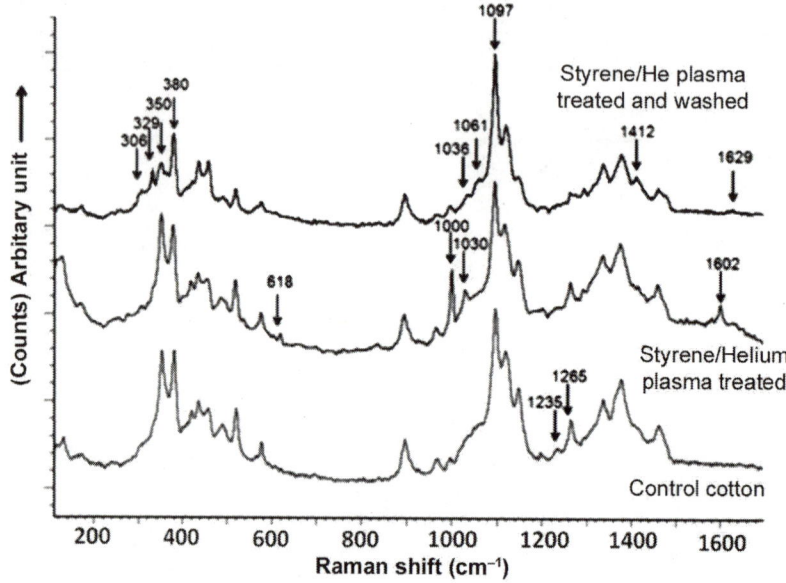

Figure 4.33 Raman spectra of control cotton, Styrene/He plasma-treated and styrene/He plasma-treated and washed samples

The 1036 cm^{-1} and 1061 cm^{-1} may be assigned to vibration of monosubstituted benzene ring, which were very weak and shifted compared to strong peaks at 1000, 1030 cm^{-1} of poly(styrene). The band at 1412 cm^{-1} was assigned to –CH$_2$ bending, which might have become prominent due to the attachment of styrene molecule (–CH$_2$–CH$_2$–Ar) or its fragment (–CH$_2$–Ar). It was interesting to observe that a strong band due to –CH$_2$OH groups present in control cotton at 1265 cm^{-1} and weak band at 1235 cm^{-1}, significantly reduced after the styrene/He plasma treatment. The appearance of the peaks corresponding to monosubstituted aromatic structure, which are different from that of poly(styrene) and reduction of the peaks corresponding to –CH$_2$OH suggested that the reaction of hydrophobic styrene fragments had taken place on the cotton surface preferably at the primary alcohol sites.

4.6.3   Acrylonitrile as a precursor

(a) Plasma generation and analysis
Another liquid precursor, acrylonitrile, was also introduced to the plasma chamber by vaporizing the liquid by bubbling He through the heated precursor

liquid as shown schematically in the previous section. After many trials glow plasma was possible to establish at a discharge voltage of 5 kV, frequency 17 kHz and discharge gap of 2.1 mm. Hydrophobic functionalization of cellulosic textile by plasma polymerization was carried out at this condition. Helium (He) plasma that gives bluish purple colour turned into bluish white colour when acrylonitrile vapor was present along with He, indicating that acrylonitrile was participating as an active species in generating plasma.

The current and voltage profile of the discharge is shown in Fig. 4.34. The measured current-voltage (I-V) waveform of the He/acrylonitrile plasma indicated the formation of glow plasma. After plasma treatment, the substrate turned hydrophobic. As result of this, a water droplet took 417 s to get fully absorbed by the fabric. Upon soap washing, the water absorbency time reduced significantly. However, the measured time of 65 s was still significantly higher than the < 1 s measured for the untreated sample. Also, the soap-washed sample gave a high water contact angle of 145° compared to the ~0° for the untreated sample.

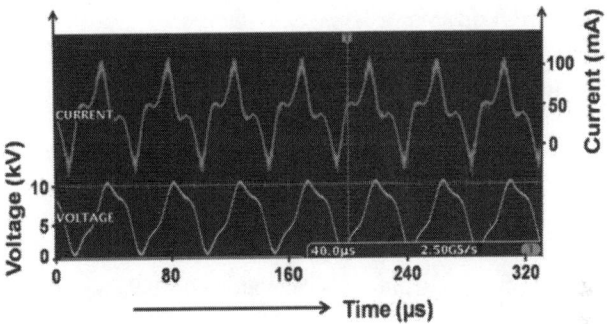

Figure 4.34 I –V characteristics of He/acrylonitrile plasma

Figure 4.35 shows the OES spectra of the He/acrylonitrile plasma in the presence and absence of cellulosic textile. When the cellulosic textile was introduced in the plasma zone, there was more fragmentation of acrylonitrile. As a result of this, a few additional peaks appeared such as at 450.1 nm for N & CN, at 482.2 nm and 491.9 nm for N, and 559.7 nm for C & CN. There was also a shift of C & N peak at 515.8 by 2 nm to 517.8 nm. It was observed that the ratio of atomic lines at 667.5 nm for combined effect of He, C & N and at 706 nm for the HeeHe was 0.95 at the frequency of 21.1 ± 1 kHz, which increased to 1.23 at the frequency of 17.2 ± 0.7; however, when cellulosic textile was introduced at the same frequency of 17.2 ± 0.7, this ratio further increased to 1.33. This indicated that acrylonitrile moieties were interacting with cellulosic textile in the plasma zone.

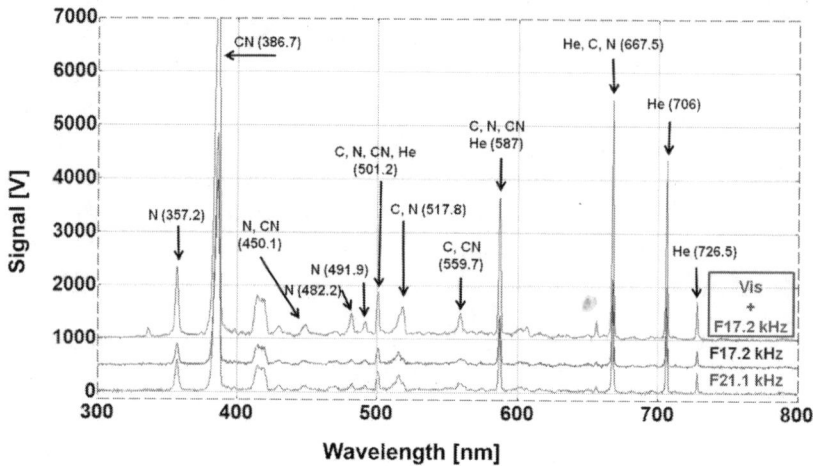

Figure 4.35 Effect of the presence of cellulosic textile on the OES spectrum of He/acrylonitrile plasma at discharge voltage of 5.63 kV

He/acrylonitrile plasma was also studied using gas chromatography-mass spectrometry (GC-MS) technique to understand the types of fragments that were produced from acrylonitrile inside the plasma zone. The plasma exhaust was passed (bubbled) through three different solvents, such as acetone, isopropyl alcohol and tetrahydrofuran (THF), to capture fragments of acrylonitrile. These liquids were then run on GC-MS. In acetone, three major peaks were detected at 73 a.m.u., 79 a.m.u. and 93 a.m.u. When isopropyl alcohol was used, similar results were observed. However, in tetrahydrofuran, two new species at 44 a.m.u. and 86 a.m.u. along with 93 a.m.u. molecule were detected. The solid deposits of condensed product that were obtained from the di-electric plates of the plasma reactor were also collected and dissolved in tetrahydrofuran (THF). The GC-MS of this solution gave molecules with mass unit of 86 a.m.u., 87 a.m.u. and 93 a.m.u. The possible molecular structures of all the masses are shown in Fig. 4.36.

It was found that three major kinds of species were present in the plasma zone. One was the acrylonitrile itself, which may react directly to the radicals present on the substrate. Second was formation of $(CH-CN)_n$ type oligomers, and lastly, the hydrogenated forms of acrylonitile fragments, i.e. amines. In GC-MS of the plasma gases, $(CH-CN)n$ was present in major fraction, while the amines were comparatively in lower fraction. On the other hand, the GC of deposits on the dielectric plates, which were principally a combination of the later two, had amines as the major fraction. From these results it was apparent that part of the acrylonitrile was fragmenting into smaller species such as CH, CH_2, CN, $CH-CN$, etc., which were combining to give oligomeric molecules larger than acrylonitrile itself. Interestingly, there was no evidence of acrylonitrile getting polymerized directly into oligomers inside the plasma zone.

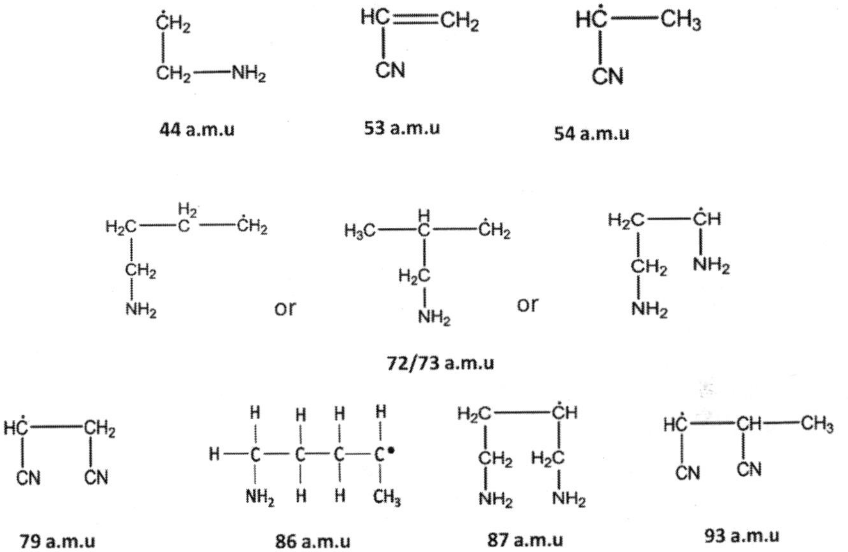

Figure 4.36 Possible structures of the molecules formed during the plasma polymerization of acrylonitrile in He

(b) *Surface analysis*

Figures 4.37(a) and (b) show the negative secondary ion mass spectra of the untreated and acrylonitrile plasma-treated and soap-washed samples. Negative ion mass spectrum of the untreated sample showed the presence of major fragments of cellulose molecule at different masses such as 12 a.m.u. for C^-, 13 a.m.u. for CH^-, 14 a.m.u. for CH_2^-, 16 a.m.u. for O^-, 17 a.m.u. for OH^-, 24 a.m.u. for $C-C^-$, and 25 a.m.u. for $CH-C^-$. Also a very small peak at 26 a.m.u. was also present likely be due to $CH-CH^-$ from cellulose. In the negative mass spectrum of the plasma-treated and soap-washed sample, mass peaks at 26 a.m.u. for $CH-CH^-$ or $-CN^-$, 28 a.m.u. for $CH_2-CH_2^-$, 35 a.m.u. for $-C=C-C^-$, and 37 a.m.u. for $N=C-C^-$ or $-C=CH-C^-$ were also detected in addition to the mass peaks seen in the untreated sample. However, the plasma-treated samples, the peak at 14 a.m.u. might be for CH_2 or N. The peak at 26 a.m.u., which was present as a very weak peak in the untreated sample, got intensified significantly possibly due to the attachment of hydrocarbon and cyanide fragment of acrylonitrile to the cellulose substrate during the plasma reaction. At the same time, the intensity of $-OH$ peak reduced significantly in the plasma-treated sample compared to the untreated sample. The increase in the intensity of hydrocarbon peak and nitrogen-containing peaks is an indication of plasma reaction of acrylonitrile fragments with the cellulosic substrate. It was found that the ratio of O–/C⁻ was 6.73 in the untreated sample, which decreased to 4.48 in the plasma-treated sample. Similar results were also observed for OH–/CH⁻ ratio. Because of these, the ratio of total oxygen-containing molecules to total hydrocarbon-containing molecules decreased significantly from 1.84 for the untreated sample to 0.481 for the plasma polymerized sample. This was the result of the attachment of hydrocarbon molecules from acrylonitrile fragments to the cellulose molecules at the expense of oxygen-containing groups present in native cellulose.

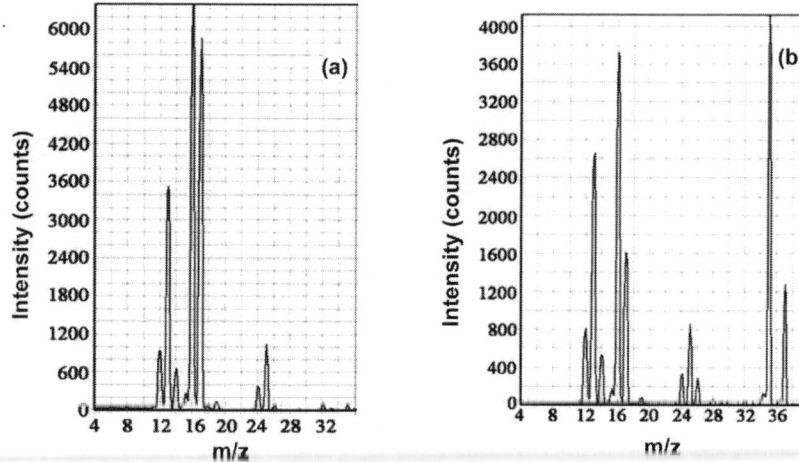

Figure 4.37 Negative ion mass spectra of (a) untreated and (b) plasma-treated and soap-washed fabric samples

Surface chemistry of the untreated and He/acrylonitrile plasma-treated samples in terms of atomic percentage, atomic ratios, and the presence of different chemical bonds was studied using X-ray photoelectron spectroscopy. Figure 4.38 shows XPS survey spectra of the untreated and plasma-treated and soap-washed samples over the binding energy range of 0–700 eV. In the survey spectrum of the untreated sample only C (1s) and O (1s) peaks were present at binding energy of 285.3 eV and 532.1 eV. On the other hand in the plasma-treated and soap-washed sample, there was a presence of N (1s) peak at 399.1 eV along with C (1s) and O (1s) peaks. The N peak was an indication that acrylonitrile fragments had reacted with cellulosic textile by forming covalent bonds. It was observed that the C percentage was 63.3% in the untreated samples and it was increased to 61% in the plasma-treated sample. Similarly, the O percentage reduced from 36.7% to 28.7% in the untreated to plasma-treated and soap-washed samples, respectively. The 10.3% atomic nitrogen was also detected in the plasma-treated sample due reaction of acrylonitrile fragments with the cellulose. The decrease occurred in atomic ratio of O/C from 0.578 for the untreated sample to 0.47 for the plasma-treated and soap-washed sample.

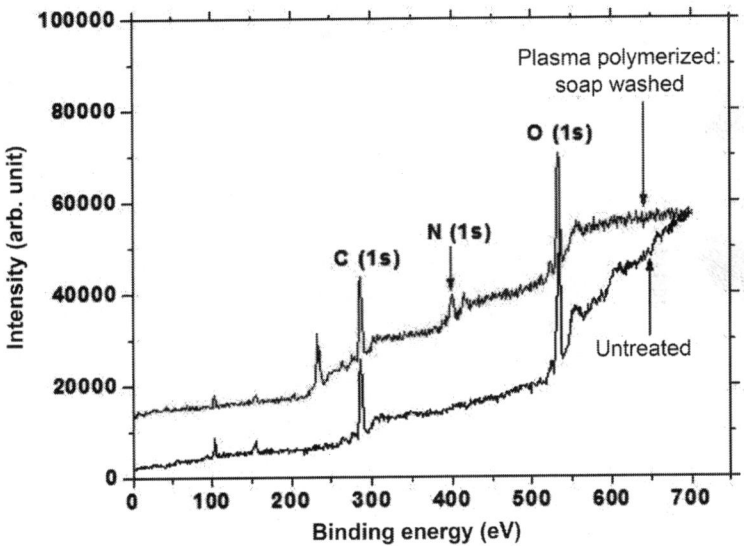

Figure 4.38 XPS survey spectra of the untreated and He/acrylonitrile plasma-treated and soap-washed samples

4.6.4   Hexamethylenedisiloxane as a precursor

(a) *Glow plasma generation and analysis*
The set-up of reactor used in this study was same as discussed in previous section. Different experiments were carried out by varying the plasma process parameters to generate the He/HMDSO glow plasma at atmospheric pressure. Glow nature of the plasma was conformed from the current-voltage (I-V) profile of the discharge.

(b) *Effect of different parameters*
Cotton fabric was treated for 5 min in He/HMDSO glow plasma and dried. The water absorbency time of more than 1 h and water contact angle of 150° was observed in plasma-treated and soap-washed fabric. The image of water drop shape is given in Fig. 4.39.

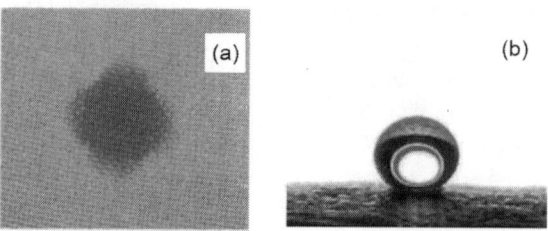

Figure 4.39 Water drop shape on (a) untreated
(b) HMDSO plasma-treated cotton fabric

Water drop disappearance time was found to increase with increase in voltage at a particular frequency. After washing, it reduced significantly at voltage below 4.2 kV but fabric treated at 4.2 kV showed water absorbency time greater than 1 h even after washing, which indicated that plasma reaction took place at this particular voltage probably due to better ionization and generation of more reactive species. Below this voltage, the extent of reaction was very less and possibly deposition was dominant. Low frequency in combination with high voltage was suitable for durable treatment. The water absorbency more than 1 h at less than 400 ml/min helium flow was observed.

(c) *Surface analysis*

In EDX spectra, shown in Fig. 4.40, Si peak was found in HMDSO plasma-treated sample even after soap washing. Si atom peak was also found in the acetone-washed sample indicating the reaction between fragments of HMDSO and viscose substrate. Amount of oxygen decreased in HMDSO plasma-treated viscose sample as compare to the control sample. This is probably due to removal of hydroxyl group and attachment of $Si - (CH_3)_3$ groups in place of hydroxyl group.

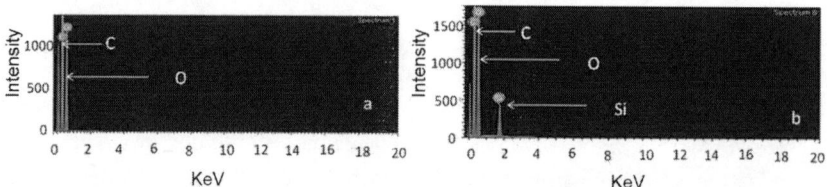

Figure 4.40 EDX spectra of (a) untreated (b) plasma-treated cotton fabric

Si peak in HMDSO plasma-treated cotton fabric was also confirmed in negative ion spectra obtained from secondary ion mass spectrometry. ATR-FTIR spectra of HMDSO as plasma-treated and acetone-washed HMDSO plasma-treated sample were taken and compared with the ATR FTIR of control viscose and pure HMDSO.

It could be seen from Fig. 4.41 that a few new peaks appeared in HMDSO plasma-treated sample as well as in acetone-washed sample at wave numbers 1250–1260 cm^{-1}, 840 cm^{-1} and 790–800 cm^{-1}. These peak position are characteristics of Si–(CH$_3$)$_3$, Si–CH$_3$, Si–CH$_2$, respectively. Ratio of peak intensity at new positions with respect to peak at 2360 cm^{-1} got reduced in washed sample, which indicated that superficially deposited layer of Si compound were removed during washing.

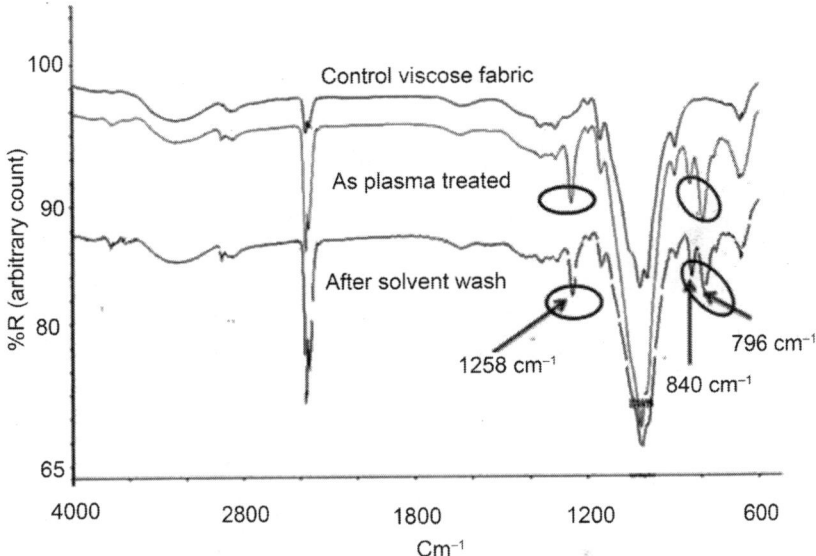

Figure 4.41 ATR-FTIR of control, as HMDSO plasma-treated and solvent-washed cotton fabric

The fragments of HMDSO sample collected from the deposits on glass slide after plasma treatment and residues at the time of solvent washing of HMDSO plasma-treated cotton fabric were analyzed by GCMS. It was found that spectral lines of fragments corresponding to m/z 147, 133,131, 73, which were present in the sample collected from the glass slide, were absent in the sample collected from treated fabric. This indicated that fragments corresponding absent peak position are mainly reacting with substrates and giving durable hydrophobicity to the cotton fabric.

(d) *Possible mechanism of reaction*
From GCMS analysis, possible fragments that might have reacted with substrate were at m/z 147 [(CH$_3$)$_3$–Si–O–Si (CH$_3$)$_2$], 131 [(CH$_3$)$_3$–Si–O–Si–CH$_2$$^-$], 133 [CH$_3$(CH$_3$)$_2$–Si–O Si(H)–CH$_3$], 73 [(CH$_3$)$_3$–Si–], 43 [(CH$_3$)–Si–].

From ATR-FTIR analysis, new peak at 1250–1260 cm−1 corresponded to $[(CH_3)_3–Si–]$, peak at around 840 cm^{-1} corresponded to $(CH_3)–Si$, peak at 790 cm^{-1} to 800 cm^{-1} corresponded to $(CH_2)_n–Si$ were found in all three: as-plasma-treated, soap-washed and solvent-washed samples of cotton substrates. Based on the results of characterization and their analysis, the possible reaction mechanism was proposed as given in Fig. 4.42.

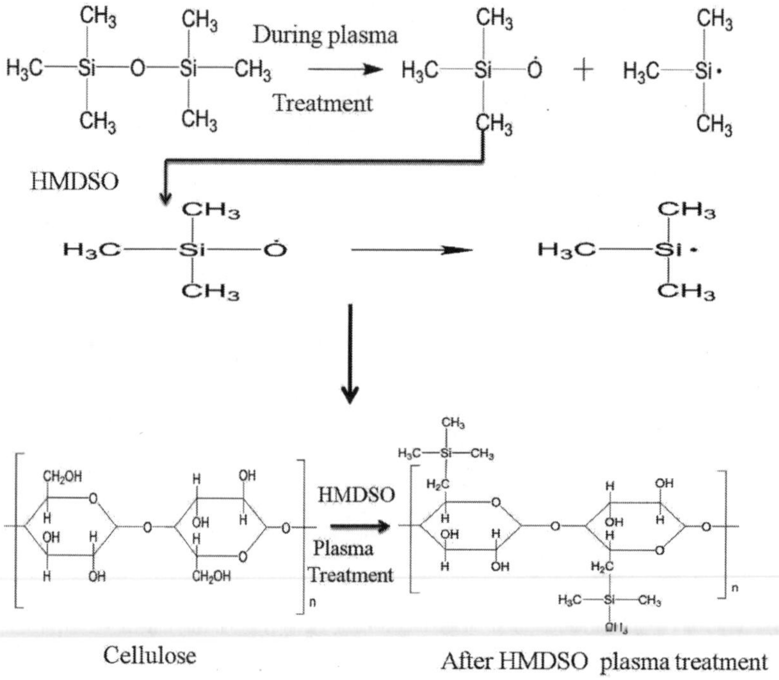

Figure 4.42 Chemical modification of cellulose after HMDSO plasma treatment

4.7   Summary

Different types of precursor such as gaseous, liquids having low boiling point, and liquids having relatively higher boiling point have been chemically attached in-situ on textile substrates using glow plasma at atmospheric pressure to produce hydrophobic effect. The precursors or its fragments were found to react with the substrate under suitable conditions and resulted in durable functionality. Attempts were made to understand the mechanism of functionalization involving different precursor molecules. The suitable combination of different plasma parameters such as discharge voltage, electrode gap, discharge frequency, carrier gas flow rate, carrier gas to precursor ratio and treatment time was found to be essential in controlling reaction chemistry inside plasma zone. A variety of non-fluoro compounds could be successfully used for the hydrophobic functionalization of textile substrates to avoid the hazardous effect of fluorocarbon compounds on the environment.

4.8   References

Abidi N. and Hequet E. (2005). J. App. Poly. Sci. **98**, pp. 896–902.

Bansode A.S., Babrekar H.A., Bhoraskar S.V., and Mathe V.L. (2012). Arch. Appl. Sci. Res. **4**(6), pp. 2356–2360.

Caschera D., Cortese B., Mezzi A., Brucale M., Ingo G.M., Gigli G., and Padeletti G. (2013). Langmuir **29**, pp. 2775–2783.

Chaivan P., Pasaja N., Boonyawan D., Suanpoot P., and Vilaithong T. (2005). Surf. Coat. Technol. **193**, pp. 356–360.

Cabrales L. and Abidi N. (2012). App. Surf. Sci. **258**, pp. 4636–4641.

Davis R., El-Shafei A., and Hauser P. (2011). Surf. Coat. Technol. **205**, pp. 4791–4797.

Farsari E., Kostopoulou M., Amanatides E., Mataras D., and Rapakoulias D.E. (2011). J. Phys. D: Appl. Phys. **44**, p. 194007.

Hochart F., De Jaeger R., Levalois-Gru"tzmacher J. (2003). Surf. Coat. Technol. **165**, pp. 201–210.

Iriyama Y.U., Yasuda T., and Cho D.L. (1990). J. App. Poly. Sci. **39**, pp. 249–264.

Jahagirdar C.J. and Tiwari L.B. (2007). Pramana – J. Phys. **68**(4), pp. 623–630.

Jafari R., Asadollahi S., and Farzaneh M. (2013). Plasma Chem. Plasma Process. **33**, pp. 177–200.

Ji Y., Hong Y., Lee S., and Kim S. (2008). Surf. Coat. Technol. **202**(22–23), pp. 5663–5667.

Kale K.H. and Palaskar S.S. (2012a). J. Text.Inst. **103**(10), pp. 1088–1098.

Kale K.H. and Palaskar S.S. (2012b). J. App. Poly. Sci. **125**, pp. 3996–4006.

Khoddami A., Avincb O., and Mallakpourc S. (2010). Progress in Organic Coatings **67**, pp. 311–316.

Leroux F., Campagne C., Perwuelz A., and Gengembre L. (2008). App. Surf. Sci. **254**, pp. 3902–3908.

Lin L.H., Wang C.C., Chen C.W., and Chen K.M. (2006). Surf. Coat. Technol. **201**, pp. 674–678.

Lee C.M., Pai Y.H., Zen J.M., and Shieu F.S. (2009). Materials Chemistry and Physics **114**, pp. 151–155.

Nattinen K., Nikkola J., Minkkinen H., Heikkila P., Lavonen J., and Tuominen M. (2011). J. Coat. Technol. Res. **8**(2), pp. 237–245.

Parida D., Jassal M., and Agrawal A.K. (2012). Plasma Chem Plasma Process. **32**, pp. 1259–1274.

Panda P.K., Jassal M., and Agrawal A.K. (2013). Surf. Coat. Technol. **225**, pp. 97–105.

Palaskar S., Kale K.H., Nadiger G.S., and Desai A.N. (2011). J. App. Poly. Sci. **122**, 1092–1100.

Roth J.R., Rahel J., Die X., and Sherman D.M. (2005). J. Phys. D: Appl. Phys. **38**, pp. 555–567.

Samanta K.K., Joshi A.G., Jassal M., and Agrawal A.K. (2012). Surf. Coat. Technol. **213**, pp. 65–76.

Tsoi W.I., Kan C., and Yuen C.M. (2011). Bio resources **6**(3), pp. 3424–3439.

Teshima K., Sugimura H., Inoue Y., Takai O., and Takano A. (2005). App. Surf. Sci. **244**, pp. 619–622.

Vohrer U., Muller M., and Oehr C. (1998). Surf. Coat. Technol. **98**, pp. 1128–1131.

Zhang J., France P., Radomyselskiy A., Datta S., Zhao J., and Ooij W.V. (2003). J. App. Poly. Sci. **88**, pp. 1473–1481.

5

In-situ plasma reactions for hydrophilic functionalization of textile substrates

Kartick K. Samanta, Manjeet Jassal and Ashwini K. Agrawal

Abstract: Wet chemical processing of textile is important for its value addition in terms of colouration and functionalization. However, the process is water, energy, time, and chemical intensive in addition to creating noticeable water pollution. Hydrophilic functionalization of polymeric/textile substrates is carried out for applications in textile, sport, packaging, and coating industry. Plasma, an ionized gas, can be used for nano-scale surface engineering of textile substrates in dry state. It also helps in reducing the processing time, the consumption of chemical and the production cost, while addressing the issue of environment pollution. Low temperature plasma either at low pressure or atmospheric pressure has been utilized for surface activation of textile substrates for achieving improvement in water absorption, colouration, adhesion strength, antistatic, and dust repellent properties. Cellulosic textile (cotton), ligno-cellulosic textile (jute, flax, sisal etc), protein fibres (wool and silk), and synthetic textile (nylon, polyester, polypropylene etc) have been modified using either inert gaseous plasma (He, Ar, N_2) or reactive plasma using various precursors. Improvement in hydrophilic property is due to removal of impurities like, wax, fatty acid layer, residual size material etc., formation of polar groups and increase in surface roughness/ area. It has been found that formation of nano-sized channels may also help in improving wicking tendency of the substrate to both polar and non-polar liquids. The chapter also discusses the role of plasma parameters such as the discharge voltage, frequency, treatment time, and gas flow rate on fragmentation of precursors and functionalization of substrates.

Keywords: Plasma, Hydrophilic, Textile, Surface functionalization, SIMS

5.1   Introduction

Plasma is a partially ionized gas composed of many type of species, such as positive and negative ions, electrons, neutrals, excited molecules, photons and UV light (Karmakar, 1999). The main attraction of plasma in industrial application is to avoid the chemical effluents. Other advantages are low cost of operation, rapid processing, high efficiency and dry process. Now-a-days, plasma is being used for a number of industrial applications ranging from arc welding, metal hardening, metal coating, nuclear fusion, synthesis of nano materials, creation of nano-structures, surface cleaning, and functional polymeric coating. Among the various types of plasmas, only the cold plasmas, also known as low temperature or non-thermal plasma with bulk temperature close to ambient temperature, can be used for nano-scale surface engineering of heat-sensitive polymeric and textile substrates. An alternating current (AC) with high frequency power supply helps to dissociate various gaseous molecules into a collection of ions, electrons, charge-neutral gaseous molecules and other species. It is, hence, an energetic chemical environment, where the generation of plasma species opens up diverse chemical reactions resulting in various end applications. Plasma processing of textiles brings physico-chemical modifications on the top surface of the polymeric substrates without altering the core of the fibre. Different value-added new functionalities can be imparted by modifying the fibre surface in dry state, while avoiding the usage of water as a processing medium. This leads to saving of large quantities of water, chemicals and energy; hence reduction of production of large quantity of effluent is possible (Karmakar, 1999; Muthukumar, 2004; Banchero, 2008). During the last four decades, use of polymers in medical, packaging, sports, automobile, composite and textile sector has been growing at a phenomenal rate due to its low cost, ease of availability, light weight, and flexibility in moulding and making complex design. In many of these applications, functional hydrophilic polymeric surfaces play an important role. Among the available technologies of hydrophilic surface modification of polymer, plasma technology is gaining interest in the research community for similar modification due to its advantage of dry and faster processing technology. Plasma treatment of hydrophobic polymers, such as polypropylene (PP), polyamide (PA) and polyester (PET), is known to enhance the surface hydrophilicity due to the generation of hydrophilic functional groups. Protein fibres such as wool and silk though they have high moisture regain, do not absorb or spread liquid (water) faster due to the presence of fatty hydrophobic layer on surface. Generally, these protein fibres are plasma treated in the presence of helium (He), argon, oxygen, air or mixture of gaseous in an open or closed reactor. The radicals generated on the plasma-treated substrate then react with the oxygen present in air inside the plasma reactor or with oxygen present during exposing the sample in the open atmosphere. The presence of oxygen, either during the reaction or afterwards, plays an important role in formation of hydrophilic groups. The improvement in hydrophilic property leads to the improvement in dyeing, printing, water and oil absorbency and adhesion strength. It could also be used for imparting antistatic properties. However, it is also reported that the plasma-induced imparted hydrophilic functionality is not durable particularly in the thermoplastic polymers, such as polypropylene (PP), polyamide (PA), polyester (PET) i.e., it decays due to the ageing effect associated with time, temperature (heat treatment) and number of washing cycles. This phenomenon is reported to be associated with the mobility of the generated polar molecules from the surface of the polymer to bulk of the materials.

This chapter discusses the generation of atmospheric pressure plasma in the presence of various gases such as He, He/O_2, He/N_2, He/air, etc., and in-situ plasma reactions with the different textile substrates, such as polyamide, polyester, wool, cotton, etc., to impart antistatic functionality and improvement in dyeing, printing, water and oil absorbency.

5.2   Plasma processing of textiles

It is well known that during the wet chemical processing of textiles, industries consume large quantity of water and generate huge effluent. The cost of the final products also increases due to multiple numbers of drying of wet textiles. Of late, due to increase in environmental awareness and effluent norms, textile industries are slowly moving towards implementation of environment-friendly low water-based technologies such as digital printing, spray, foam finishing, super critical fluids and solvent processing. They are also using eco-friendly chemicals and agents such as natural dyes, enzymes and plant and animal extracts for textile processing and finishing. In this context, cold plasma seems to be a promising cost-effective environment-friendly technology for wet chemical processing of textiles. The plasma treatment of textiles modifies only the surface of the material without altering its bulk properties to increase the uptake of dyes and finishes, and imparts a unique functionality. Different value-added functionalities, such as water, stain and oil repellency, hydrophilicity, antimicrobial, flame retardant, UV protective, dirt-repellent, and antistatic properties, electromagnetic radiation (EMF) protection, and improvements in dyeing, printing, bio-compatibility and adhesion can be achieved by modifying the fibre surface at nano-meter level (Samanta, 2006; Vaideki, 2009; Vaideki, 2008; Jahagirdar, 2007; Bhat, 1999; Shin, 2006; Samanta, 2009; Qiuran, 2009; Krump, 2005; Nobuyuki, 2008; Park, 2008; Zhang, 2009; Naebe, 2010; Leroux, 2009). Low pressure plasma has already been extensively studied for such applications but the process technology has not been commercialized in textiles due to its techno-economical limitations. Atmospheric pressure cold plasma can overcome the limitations of low pressure plasma technology and is being explored for similar applications in textile and allied industries. Therefore, in this chapter an emphasis has been given on atmospheric pressure plasma reaction for hydrophilic finishing of various textile substrates.

5.3   Mechanism of hydrophilic finishing

Plasma chemistry is a quite complex process and involves a large number of elementary reactions. These are mainly homogenous and heterogeneous reactions. Homogeneous reactions occur between species in the gaseous phases as a result of inelastic collisions between electrons and heavy species. On the other hand, heterogeneous reactions occur between the plasma species and the solid surface immersed in it. Some homogeneous reactions are excitation of atoms or molecules, de-excitation from electronically excited state to ground state, ionization of neutral species through electron detachment to form positively charged particles, dissociation of a molecule, dissociative attachment, dissociative ionization, recombination of ions, electron–ion recombination, ion–ion recombination and Penning dissociation or ionization when neutrals collide with energetic meta-stable species (Chen, 1995). In plasma, there is a possibility of various types of constructive and destructive reactions that may take place among the plasma species themselves or with substrate. Different types of plasma surface reactions that are responsible for hydrophilic surface modification of textile are discussed below.

(i) *Oxidation:* In treatment by oxygen-containing plasma, surface excitation leads to absorption of oxygen and formation of polar surface groups such as ketone, hydroxyl, ether, peroxide and carboxylic acid that are much more hydrophilic (wettable) than the untreated surface. On exposure to air discharge, articles like polyethylene (PE) film, teflon, and polypropylene nonwoven fabrics show significant increase in surface wettability. Corona discharge treatment is routinely applied to PE film to make it printable. Such kind of plasma treatment also leads to improvement in adhesion properties (Karmakar, 1999).

(ii) *Peroxide formation:* When a typical polymer/textile substrate is exposed to argon (Ar) plasma followed by exposure to air, a high proportion of reactive sites are converted to peroxide form. Since peroxides are known to act as initiators for vinyl polymerization, a vinyl monomer under suitable conditions will react with the peroxide groups to produce a graft copolymer with the substrate. Since peroxide groups are indefinitely stable under ambient conditions, they can be stored as reactive sites to be used later on. Such surfaces also show improved interfacial strength.

(iii) *Radical formation:* Carbon-free radicals are formed when the energetic ions from the plasma break the organic bonds at the surface of the substrate with the evolution of gaseous products, e.g. hydrogen from hydrocarbons. Other processes besides radical formation are degradation, cross-linking, solid-state polymerization and etching action.

(iv) *Polymerization:* Initiation of polymerization produces a thin uniform polymeric film on the electrodes or substrates. Ionization of organic monomers in the vapour phase by bombardment of the electrons under discharge conditions lead to rapid polymerization resulting in formation of thin pin hole-free films with good stability, insulating properties and uniform thickness. Grafting of block co-polymers may also be achieved through plasma polymerization. The free radicals formed on the surface of the fibre by exposure to plasma can be used directly to initiate polymerization with a new polymer bonded firmly to the surface by carbon linkage. This makes coating possible by continuous movement of textiles, paper, etc., through flat plate electrodes.

5.4   Generation and characterization of plasma

Samanta et al. studied the generation of plasma in an indigenously developed atmospheric pressure plasma reactor as shown in Fig. 5.1 (Samanta, 2010). Cold plasma was generated in the mixture of He, He/oxygen, He/air and He/nitrogen gases. Both the plasma electrodes were covered with glass or teflon dielectric sheets and had a mechanism to cool the electrodes by passing cold water, as required.

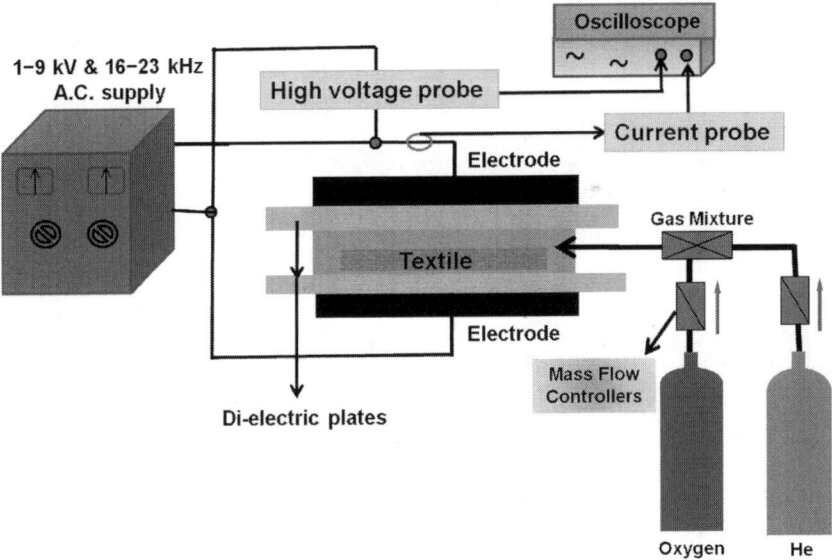

Figure 5.1 Schematic of atmospheric pressure plasma processing of textile in a continuous manner (Samanta, 2010)

Online analysis of the effect of different plasma parameters, such as discharge voltage and frequency on the quality of generated plasma was studied. The electrical properties of plasma was determined by studying the current-voltage (I-V) waveform using digital oscilloscope, model Tektronix DPO 3012 attached with P6015A high voltage and TCP 0030 current probes. Optical properties of the plasma were studied by measuring the light emitted by the excited atoms and molecules over the wavelength of 200–1100 nm using Optical Emission Spectrometer (OES), Model-Mikropack made PlasCalc 2000. It is known that helium (He) gas is easy to ionize to form glow plasma. However, other gases such as oxygen, air and nitrogen are difficult to ionize at atmospheric conditions and to form glow plasma. A mixture of small amount of these gases with He could help to generate glow plasma without significantly deteriorating the plasma quality.

Depending upon the ionization pattern of a gaseous molecule and the excited energy states of the atoms or molecules in the plasma zone, a particular colour is produced. For example, when He gas is ionized, it gives light bluish purple colour. Visually it forms glow plasma, which is free from any micro-discharge. When 2% (v/v) oxygen was introduced along with He, it produced milky white colour. Similar to He plasma, He/oxygen plasma also produced uniform glow plasma and was free from micro-discharges. When 2.5% (v/v) air was introduced along with He, it produced pinkish purple colour, whereas when the same amount of nitrogen was introduced with He, strong bluish purple glow was observed. The different colours of glow produced in the above gaseous mixtures indicate that both components of the gas mixtures are ionized to form glow plasma.

It is known that plasma with micro-discharges or filaments give several spikes in the current waveform. The presence of spikes increases peak to peak current value, significantly. However, if the plasma is glow in nature, the current and voltage waveform would be smooth and free from any short or long spike. Figure 5.2 shows the current-voltage (I-V) waveform of He plasma, when it visually appeared to be uniform glow. It can be seen that current profile is smooth and free from any short or long spikes. Similar result was also observed in He/nitrogen plasma as shown in Fig. 5.3. The spike-free smooth profile of current waveform is an indication of glow-type plasma. In He/air plasma, two short spikes are visible in each wave cycle of current wave. However, the amplitudes of these spikes are comparable to the peak current value. As a result of this, peak-to-peak current does not increase significantly, unlike the case of the micro-discharge plasma. Also, the numbers of spikes are few. This indicated that the nature of generated plasma in the mixture of He/air was still a glow. Similar result was also observed in the case of He/oxygen plasma.

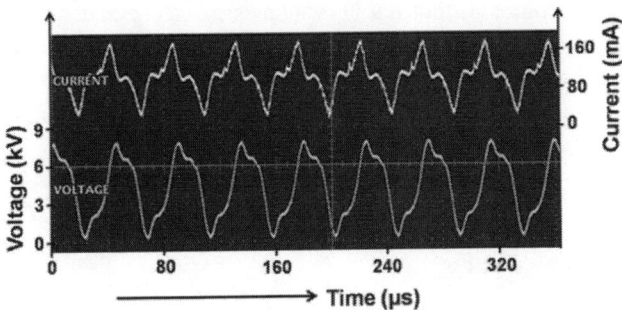

Figure 5.2 Current-voltage (I-V) waveform of He plasma (Samanta, 2010)

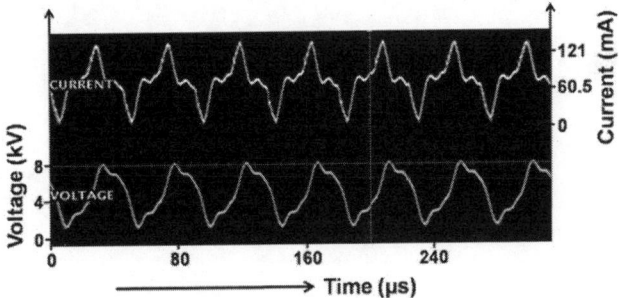

Figure 5.3 Current-voltage (I-V) waveform of He/nitrogen plasma (Samanta, 2010)

Figure 5.4 shows the OES spectrum of He plasma. Upon ionization, He mainly emits photons at five different wavelengths such as at 706 nm, 667.5 nm, 587 nm, 655.8 nm and 726.5 nm in the presence of pure helium and mixture of gases, such as He/O$_2$, He/N$_2$ and He/air gases. However, in the pure He, the He 706 nm peak appeared at 705.1 nm. Among the various atomic line of He at different wavelengths, He peak at 705.1 nm is the most prominent. Several other peaks in between the wavelengths of 330 nm 430 nm were also visible in the He spectrum, which were mainly due to the atomic lines of nitrogen. Similar results were also observed in He/air and He/N$_2$ plasmas. However, in these plasmas, the intensity of atomic lines of helium was much lower compared to nitrogen atomic lines or lines for pure He gas. The presence of trace amount of nitrogen in He gas produces these peaks at lower wavelengths viz. mostly in UV region. Figure 5.5 shows the intensity of He 705.1 nm line with plasma discharge time. It can be seen that intensity of He 705.1 nm line remains constant with plasma discharge time in real time. Since the emission intensity of a species is directly related to its concentration, this observation indicates that stable glow plasma has been established in pure helium gas. The presence of any filaments or micro-discharges would have resulted in fluctuation of the intensity of He lines.

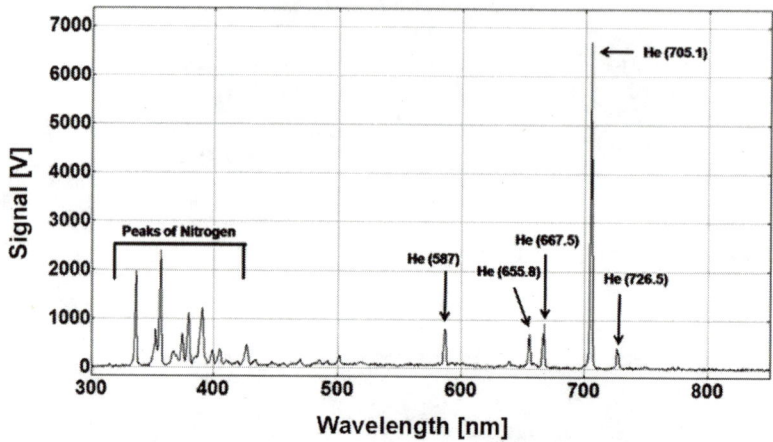

Figure 5.4 OES spectrum of He plasma at plasma discharge voltage of 3.85 kV and frequency of 22.1 ± 0.2 kHz (Samanta, 2010)

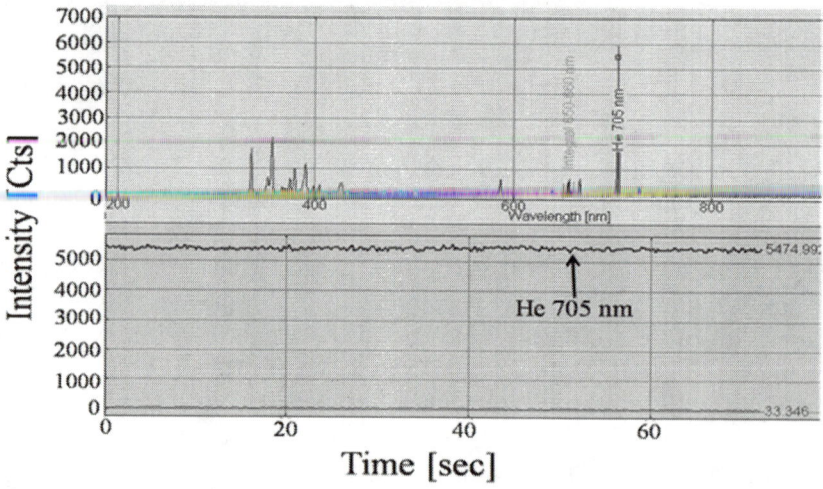

Figure 5.5 Intensity of He atomic line at 705.1 nm with plasma discharge (Samanta, 2010)

Figure 5.6 shows the effect of plasma discharge voltage on emission intensity of He atomic line at 705.1 nm. It can be seen from the figure that the intensity of emission increases substantially by 48% with an increase of plasma discharge voltage from 2.31 kV to 3.81 kV. However, the discharge frequency has little effect on the ionization of He gas.

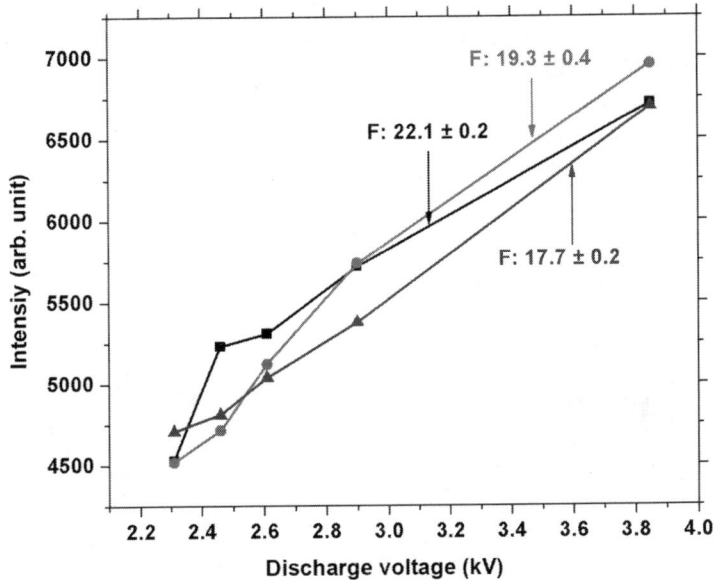

Figure 5.6 Effect of discharge voltage on emission intensity of He atomic line 705.1 nm (Samanta, 2010)

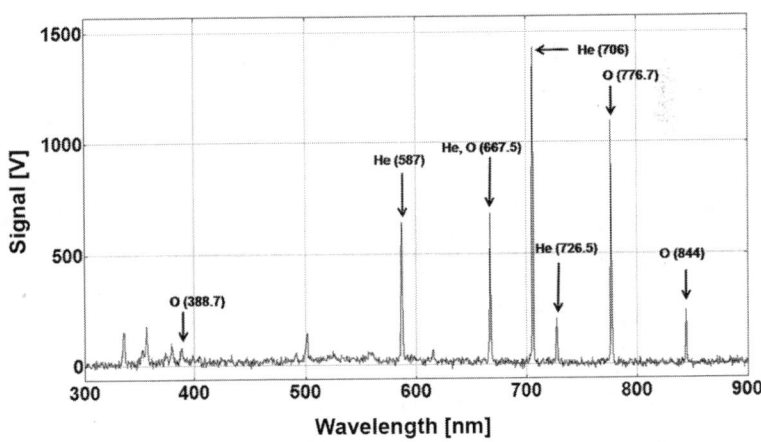

Figure 5.7 OES spectrum of He/oxygen plasma at a voltage of 5.9 kV and frequency of 21.2 ± 0.8 kHz (Samanta, 2010)

Figure 5.7 shows the OES spectrum of He/oxygen plasma. Upon ionization, oxygen emits atomic line at 776.7 nm and 844 nm. The main peak for oxygen was at 776.7 nm (Samanta, 2010). Emission line of He was similar to that indicated for He plasma, except the major helium atomic line appeared at 706 nm instead of 705.1 nm. It was interesting to note that the helium peak at 655.8 nm did not appear in He/O$_2$ plasma. The intensity of He 706 nm line and oxygen 776.7 nm lines remained constant during the plasma discharge with time as shown in Fig. 5.8. This indicated that the concentration of He and oxygen species remained constant with plasma discharge time. As mentioned earlier, this was a characteristic of stable glow plasma. It can also be seen that the peaks for the nitrogen atoms that appeared in the wavelength range of 330 nm 430 nm in He plasma were hardly visible in the case of He/oxygen plasma.

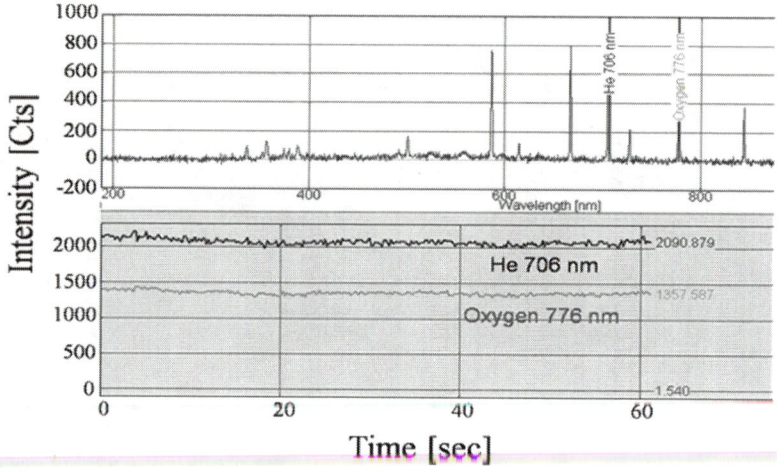

Figure 5.8 Intensity of He line at 706 nm and oxygen line at 776.7 nm with plasma discharge time (Samanta, 2010)

Similar to helium plasma, the effect of discharge voltage and frequency was studied by measuring the intensity of oxygen atomic line at 776.7 nm (Fig. 5.9). It was seen that with increasing discharge voltage, the intensity of oxygen emission increased linearly by 179%. However, unlike helium plasma, the discharge frequency had little effect on ionization.

5.5   Application of plasma in textile chemical processing

5.5.1   Improvement in desizing and scouring

Plasma may be used for removal of contaminants, finishing and sizing agents from the fabrics. Cotton warp yarns are required to be sized prior to weaving

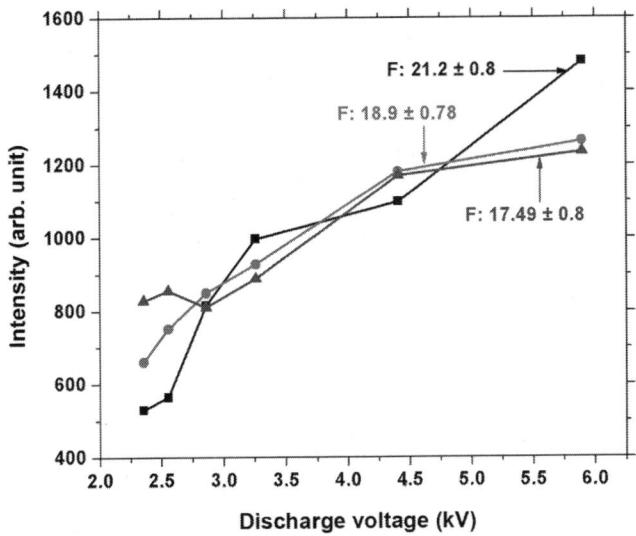

Figure 5.9 Effect of discharge voltage on optical emission of He/oxygen plasma at oxygen atomic line of 776.7 nm (Samanta, 2010)

to provide a protective coating for improving the yarn strength and reducing the yarn hairiness (Goto, 1992). Polyvinyl alcohol (PVA) is primarily used for sizing of synthetic yarns, as a secondary sizing agent to starch for cotton yarns. Sizing materials must be removed by desizing process prior to dyeing and finishing operation of woven textiles (Sparavigna, 2008). A complete removal of PVA size is difficult due to high energy and water consumption. Atmospheric plasma treatment has shown to greatly increase the solubility of PVA present on cotton in cold water (Cai, 2003). A percent desizing ratio of 99% was obtained with both air/He and air/O_2/He plasma treatments in an atmospheric pressure glow discharge (APGD) reactor, followed by cold and hot washings. Scanning electron microscope (SEM) images revealed that for both air/He and air/O_2/He plasma-treated fabrics, fibre surfaces were nearly as clean as the un-sized fabric, indicating almost complete removal of PVA size from the cotton fibre/fabric surface. Desizing of polyester fabric sized with polyvinyl alcohol could also be removed by plasma treatment (Riccobono, 1973). Both air and oxygen plasma treatments allowed the scouring process to be eliminated (Tsriskina, 1991).

5.5.2   Improvement in adhesion strength

Man-made fibres generally have smooth surfaces and low surface energy resulting in poor adherence to matrix component. The synthetic fibres such as polyester, poly(propylene), poly(ethylene) etc. usually do not have any functional group to form covalent bonds in fibre–matrix interface. The most common treatment is to modify the fibre surface by removing the superficial layer, changing the topography and chemical nature of the surface. Plasma treatment can improve fibre-matrix adhesion strength largely by introducing polar or excited groups that can form strong covalent bonds between the fibre and the matrix (Kalia, 2009). In addition, roughening of fibre surface can also increase mechanical interlocking between the fibre and matrix. Plasma polymerization is being used as a promising tool to modify the surface of the fibre to improve the adhesion strength of the fibre-reinforced composites (Seki, 2009).

Poly(propylene) is a good industrial fibre; however, its inherent hydrophobicity hinder its end-use in different applications. The improved wettability and adhesion properties are advantageous for its use in different applications. Using O_2 plasma treatment, an improvement in wettability was observed. Similar effects were also observed in air and NH_3 plasma treatment. Similar to PP, improvement in wettability and adhesion in the poly(ethylene) (PE), poly(ethylene terephthalate) (PET), and poly(tetrafluorethylene) (PTFE) were also noted upon plasma treatment with air, O_2 and NH_3 gases (Deshmukh, 2003; Morent, 2007). Inducing permanent changes in the wettability of fluoropolymer and polyolefin surfaces pose a challenge due to their chemical inertness and tendency to undergo hydrophobic recovery upon surface oxidation. Selective modification of the surface is desirable in order to introduce functional groups and wettability, while preserving the material's virgin bulk properties (Herbert, 2011). Air plasma treatment of PET films for different time duration produced improvement in adhesion and printability properties, which are correlated to surface energy and surface roughness (Deshmukh, 2003). To improve the interfacial strength between poly(ethylene) fibres and epoxy resins, which are cured by amino cross-linking, amino groups were introduced on the fibre surface to promote covalent bonding. Plasma treatment with NH_3 gas gives rise to N-functionalities such as amino ($-NH_2$), imino ($-CH=NH$), cyano ($-C\equiv N$), and oxygen-containing groups due to post-plasma atmospheric oxidation (Chappell, 1991).

Graft polymerization of vinyl monomers was carried out onto the natural and synthetic textile fibres by glow discharge plasma. After graft polymerization of 2-hydroxyethyl methacrylate, the breaking strength of yarns increased, which could be due to the binding effect of the grafted polymers (Zubaidi, 1996). Oxygen plasma treatments at low RF frequency in the presence of Ar gas at 0.1 mbar vacuum were used to modify the surface of jute fibres at different plasma powers (30, 60 and 90 W) for 15 minutes. Composite laminates having in-plane dimensions of 180 × 180 mm² were fabricated by embedding jute fabrics between HDPE sheets in a hydraulic moulding press. It was observed that the inter-laminar shear strength (ILSS) increased by 32% and 47%, when the samples were plasma treated at powers of 30 W and 60 W, respectively. However, there was a decrease in ILSS for the 90 W plasma-treated sample. The decrease in mechanical strength at higher power may be due to greater etching of bulk of the fibre, which made the fibres weak (Seki, 2009). The effect of atmospheric air pressure plasma on the mechanical and interfacial properties of the different lingo-cellulosic fibres, such as abaca, flax, hemp and sisal has been studied. The apparent interfacial shear strength (IFSS) of the fibres (modified) to cellulose acetate butyrate (CAB) increased after short plasma treatment of 1 minute duration. This was due to the introduction of functional groups, cleaning of contaminants that hinder the adhesive process, and enhanced surface roughness, which favours mechanical interlocking between fibres and matrix (Jimenez, 2008).

5.5.3   Improvement in oil absorbency

The atmospheric pressure glow plasmas, obtained in He, argon (Ar), oxygen (O_2) and air gases by optimizing the various parameters such as voltage, frequency, distance between the electrodes, gas flow, and thickness of the dielectric barriers (1.7 mm) were used to improve the oil absorbency in the different textile substrates. The effect of plasma treatment on oil absorbency was measured for nylon, polyester and cotton woven textiles (Samanta, 2009). The oil absorbency of the fabrics was measured within 30 min of plasma treatment according to AATCC test 39-1971. A drop of 37 µl of mustard oil was placed at the centre of the marked area (3.79 cm^2) of the fabric held in a frame. For cotton, three oil drops were placed simultaneously for the test, because one drop of oil was found not to be sufficient to cover the specified area. The time taken for the oil to spread in the marked area was recorded. It was interesting to observe that there was significant improvement in oil absorbency (mustard oil) for all the plasma-treated fabrics. The time required for spreading of an oil droplet over 3.79 cm^2 area of nylon fabric decreased from 152 s for the untreated fabric to 52 s for the 60 s He-plasma treated, as shown in Table 5.1. Similar results were also observed for the nylon fabrics treated with the plasma generated in presence of Ar, O_2 and air gases. However, argon, oxygen and air plasma-treated samples took slightly more time compared to helium plasma-treated sample. The oil spreading time in the plasma-treated sample was reduced by approximately half to one third of the time measured in untreated sample.

Similar to nylon, oil-spreading time also decreased for polyester fabrics from 28.6 min in the untreated sample to 2.8 min in the 60 s plasma-treated sample (Table 5.1). The plasma-treated samples showed better rate of oil spreading, though the surface had turned more hydrophilic. It was interesting to observe that even for hydrophilic cotton fabric, oil (hydrophobic liquid) spreading time decreased from 59.5 s in the untreated sample to 30.4 s in the 60 s He plasma-treated sample. Similar results were also observed when the cotton fabric was plasma treated in the presence of air, argon (Ar) and oxygen (shown in Table 5.1).

Table 5.1 Oil absorbency time in the different textile substrates (Samanta, 2009)

Type of plasma used	Type of fabrics		
	Nylon (Time in s)	Cotton (Time in s)	Polyester (Time in min)
Untreated	152	59.5	28.6
60 s He-plasma treated	52	30.4	2.8
60 s Ar-plasma treated	64	29.7	
60 s O_2-plasma treated	72	29.7	
60 s air-plasma treated	75	33.3	

Mechanism of oil absorbency

The cotton substrates being made up of cellulosic polymer are highly hydrophilic in nature. Therefore, based on chemical nature, it is expected to show slow spread of oil. However, it was observed that untreated cotton could easily spread oil (Samanta, 2009). The results indicated that the spreading of fluids, whether water or oil, may be due to its physical structure. Cotton has a textured and convoluted structure, which may help in wicking of fluids. Therefore, SEM and AFM micrographs of the untreated and plasma-treated nylon and polyester were carried out to understand the mechanism of fluid wicking. Figures 5.10(a) and (b) show the surfaces of the untreated and

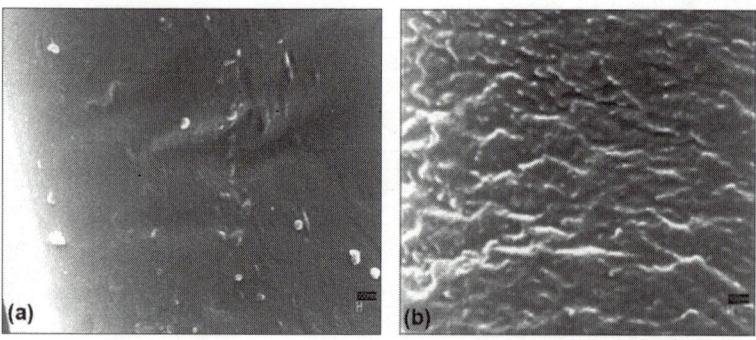

Figure 5.10 SEM micrographs of nylon filament (magnification: 35000×) [(a) untreated and (b) He plasma treated (60 s)]. The indicated bar is 100 nm (Samanta, 2009)

60 s He-plasma-treated nylon samples at 35,000× magnification to reveal the surface features at less than 100 nm scale. It was seen from the SEM micrographs that untreated nylon had smooth surface, whereas the 60 s He-plasma-treated nylon had surface that was full of new features. Similar surface features were also observed in polyester samples.

The new features, i.e. hills and valleys, were present in the plasma-treated samples of both polyester and nylon. However, in treated nylon sample, these features were more pronounced compared to the treated polyester. The observed variations in features may be a result of differences in the physical properties and crystallinity of the two samples. Figures 5.11(a) and (b) shows the surface features and profile of untreated and plasma-treated nylon samples over an area of 4 × 4 μm², respectively. At a magnification of 500 nm/div in the direction of Z-axis, the untreated surface appeared to have smooth surface. After the plasma treatment, vertical features could be easily seen. The vertical channel-like features had dimensions of <200 nm. The channels were formed by ions and electrons bombardment during the plasma treatment. This type of surface is similar to that observed in SEM micrographs. These nano-sized vertical channels were uniformly distributed over the entire surface of the plasma-treated samples.

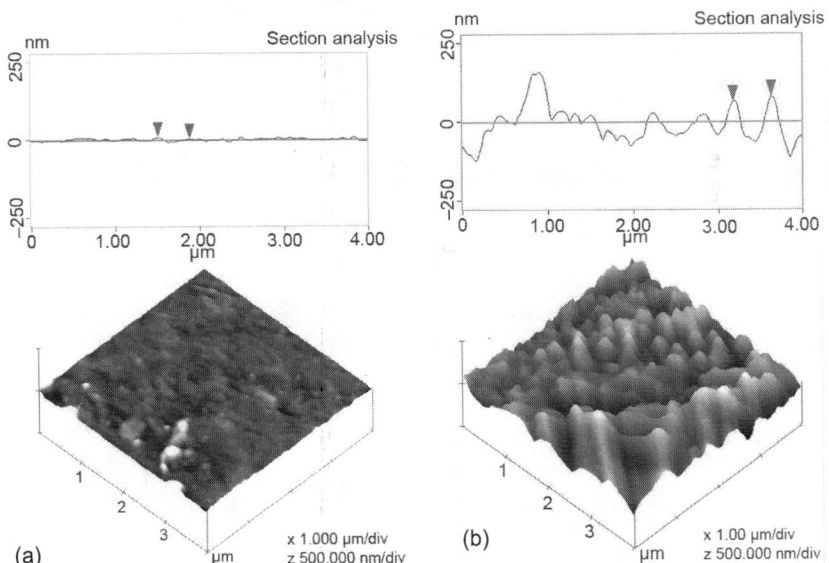

Figure 5.11 AFM micrographs of nylon filament and surface profile (a) untreated and (b) He plasma treated (60 s) (Samanta, 2009)

Figures 5.12(a) and (b) show the surface features and profiles in the untreated and 60 s He-plasma-treated polyester samples, respectively, at a magnification of 200 nm/division in the direction of Z-axis. It can be seen from Fig. 5.12(a) that it has only few broad curvature like features, which are inherent to the otherwise smooth surface. After plasma treatment, the surface developed horizontal channel-like features. These channels are ~100 nm in height and were separated from each other by about 350 nm in the horizontal direction. Similar to nylon, these nano-sized horizontal channels were also uniformly distributed over the entire surface of the plasma-treated sample.

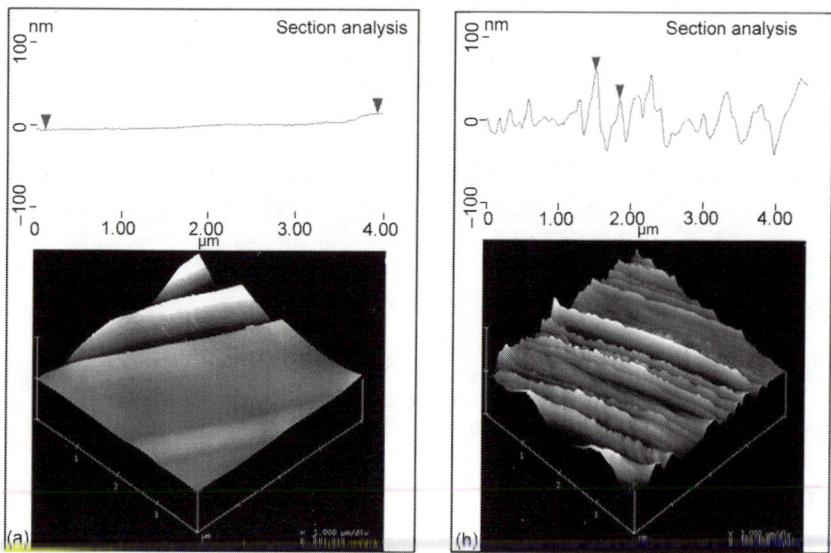

Figure 5.12 AFM micrographs of polyester filament and surface profile (a) untreated and (b) He plasma treated (60 s) (Samanta, 2009)

The formation of the horizontal and vertical channels, upon plasma treatment, is probably responsible for the increase in the spreading rate of the oil on the hydrophilic surface. This may be explained on the basis of Lucas-Washburn equation, given below, for horizontal wicking in a fabric (Chen, 2001).

$$L_2 = \gamma \, R \cos\theta \, t \, / \, 2 \, \eta \qquad \qquad ...(5.1)$$

where, γ is surface energy of the fluid, R the effective radius of capillaries in a fabric, θ the angle made by the fluid with the substrate, η the viscosity of the fluid and L the distance to which fluid is wicked in time t.

Since oil has higher viscosity and lower surface energy than water, it is likely to spread to a much smaller distance. Further, the angle of contact is likely to be higher for a fluid (in this case oil) with larger difference in surface energy in comparison to that of the textile surface. Therefore, oil should take longer time to spread over the same distance after the surface of the fabric has been converted to a more hydrophilic surface on plasma treatment. However, in the study, the oil took significantly lesser time to spread in the treated samples. This implied that spreading of the oil on the plasma-treated samples is primarily due to the increase in effective capillary radius of the treated fabric. The formations of nano-channels, on plasma treatment, are likely to increase effective capillary radius and would facilitate transport of fluid through enhanced capillary action. Increased oil absorption by textile substrates is likely to have direct application in cleaning of surfaces contaminated with oil such as metal industry, household cleansers and cosmetics.

5.5.4   Improvement in antistatic property

During the last decade, significant effort has been made to design suitable personal protective clothing to ensure worker's health and safety in hazardous work environments. Regarding this, protection from static electrical hazard is an important aspect because it causes accidental ignition and fire in paper, printing, oil, gas and plastic industries. Static charge build-up also adversely affects garment's fall by clinging to the wearer's body, and at the same time increases soil and lint pick-up by the fabric (Joshi, 1996; Bajaj, 1992). The term "static electricity" refers to the phenomena associated with the accumulation of electrical charge, for example, by contact and/or rubbing of the two objects. The hydrophobic character of the synthetic fabrics does not allow the formation of a conducting wet layer on the surface of the fibres, and hence, results in accumulation of static charge. The static charge problem can be reduced by applying antistatic agents on the fabrics. Many chemical formulation for antistatic finishing of textile has been reported by the various research groups, such as non-durable to durable-based chemical finishes, graft and co-polymerization with suitable monomer, UV excimer treatment, use of metal particle, metal coating on textile and use of conducting metallic/organic fibres in the yarn and fabric structure (Joshi, 1996; Bajaj, 1992; Cullough, 1989; Whitacre, 1964; Friedman, 1979; Latschar, 1969; Holme, 1998). Atmospheric pressure cold plasma (APCP) treatment on hydrophobic nylon and polyester fabrics to impart effective antistatic property without using any antistatic chemical is an environment-friendly approach of anti-static finishing (Samanta, 2010). The developed static charge and ½ decay time were measured using Static Honestmeter (Type H-0110-B, Shishido Electrostatic Ltd., Japan). Sample size was kept at 40×40 mm^2. Static charge was generated on the samples by applying corona discharge voltage of $+10$ kV at a discharge gap of 20 mm.

(a) *Static charge and half decay time*

The static charge is the charge developed on the surface of a substrate under the standard conditions of charging in a standard environment. Half decay time is the time that is required to dissipate half magnitude of the developed static charge. A substrate with antistatic effect should develop low amount of static charge and also should show shorter decay time. Figures 5.13 and 5.14 show the amount of static charge developed and the half decay time, respectively, for the plasma-treated and untreated nylon samples.

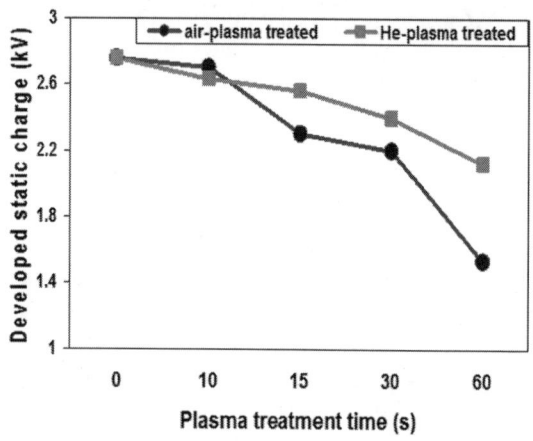

Figure 5.13 Effect of plasma treatment time on developed static charge in nylon fabric (Samanta, 2010)

It is evident from the above figures that both developed static charge and the half decay times decreased significantly for the plasma-treated samples (Samanta, 2010). The 60 s He-plasma-treated sample produced significantly lower static charge of 2.12 kV compared to 2.76 kV produced by the untreated samples, reduction of 23%. Interestingly, the ½ decay time decreased exponentially with plasma treatment time from 8.9 s for the untreated sample to only 1.1 s for the 60 s He-plasma-treated sample, amounting to a reduction of ~88%. It was also observed that with increasing plasma treatment time both developed static charge and ½ decay time decreased. Figures 5.13 and 5.14 also show the static charge properties for the air-plasma-treated samples. In this case, the developed static charge decreased even more significantly from 2.76 kV for the untreated nylon sample to 1.53 kV for the 60 s air-plasma-treated sample, resulting in a 45% reduction. Similar to He-plasma-treated sample, the ½ decay times were found to decrease exponentially with air-plasma treatment time. Compared to the untreated sample, 10 s air-plasma-treated sample took 2.8 s and the 60 s air-plasma-treated sample took only 0.63

s to dissipate half the amount of developed charge compared to 8.9 s observed in the untreated sample. These correspond to 68.5% and 93% reduction in ½ decay time for the 10 s and 60 s air-plasma-treated samples, respectively.

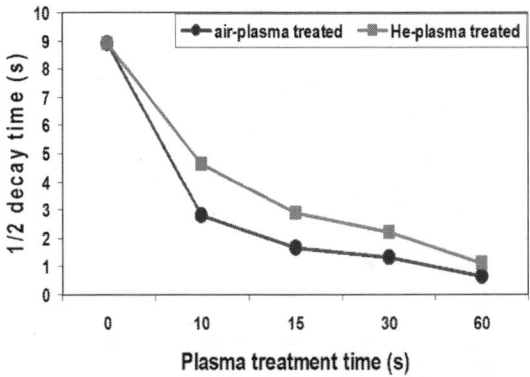

Figure 5.14 Effect of plasma treatment time on half charge-decay time in nylon fabric (Samanta, 2010)

Polyester fabrics, which are highly susceptible to static charge built-up, also showed significant improvement antistatic property. The developed static charge decreased from 1.53 kV in untreated sample to 1.42 kV in 60 s He-plasma-treated sample. The half decay time also reduced significantly from 107 s for the untreated sample to 19.8 s for the 60 s He-plasma-treated sample. The untreated polyester sample showed a much lower static charge built-up than untreated nylon sample, though polyester is known to be more susceptible to static charge built-up. This might be because of the inherent differences between the two fabrics with respect to construction, denier per filament and number of filaments per inch, etc.

It was evident from the above results that the plasma-treated samples showed significantly better antistatic properties than the untreated samples. Also, the air plasma treatment gave better results compared to He-plasma treatment. It was an important result as air is inexpensive and readily available in open atmosphere. Durability of the plasma treatment was evaluated by investigating the decay of antistatic properties of treated nylon fabrics with washing cycles. For the nylon sample treated with 60 s He-plasma, the developed static charge increased by 17% after one washing cycle. With the increasing number of washing cycles, the sample retained more charge. After five washing cycles, the sample produced static charge of 2.55 kV, which was still lower than that of the untreated sample (2.76 kV). Similar trend was also observed for the ½ decay time, where the treated nylon samples, even after 5 washing cycles, showed significantly lower ½ decay time of 4.7 s compared to 8.9 s for the untreated sample. It was found that the first washing treatment had the most deteriorating effect on the anti-static properties of the plasma-treated samples, and the further washings had only a marginal effect.

(b) *Mechanism of antistatic property development*
(i) Increase in hydrophilic molecules
In order to measure the improvement in hydrophilicity of the plasma-treated sample, surface energies of the nylon and polyester fabric were measured. The surface energy of nylon fabric was found to increase slowly from 49.5 to 63 mJ/m^2 as the treatment time was increased from 10 s to 60 s in He-plasma. Surface energy of the polyester fabric also increased significantly from 55 mJ/m^2 for the untreated sample to >71 mJ/m^2, the maximum measurable value, after 60 s of He-plasma treatment. The increase in the surface energy of both the nylon and polyester fabrics on plasma treatment indicated transformation of the hydrophobic surfaces to hydrophilic surfaces possibly due to the generation of hydrophilic groups such as ketone, hydroxyl, ether, peroxide and carboxylic acid. Interestingly, even though the plasma-treated polyester samples showed higher surface energy, they also showed significantly higher half decay time than treated nylon samples. This may be explained based on the differences in surface morphology of the samples. Due to the improvement in surface hydrophilicity in nylon and polyester samples after plasma treatment, a water droplet took significantly lower time to spread over a specified area (Section 5.5.6) due to the generation of hydrophilic groups and formation of nano-sized channel in the surface.

The ATR-FTIR spectra of the untreated and 60 s helium (He) and air plasma-treated samples were taken to determine the chemical changes upon plasma treatment. Plasma treatment is known to break the covalent bonds present on the surfaces of organic substrates and generate radicals. These are highly reactive sites and can combine with other radical species such as organic molecules, unsaturated monomers, or reactive gases such as oxygen to generate functional groups on the surface. Since the samples were either exposed to air soon after plasma treatment, or were treated in presence of air, it is expected that hydrophilic groups would be generated by reaction of active radicals with the surrounding oxygen molecules.

It can be seen from Table 5.2 that the value of normalized peak intensity for $-NH_2$ groups increased significantly for the 60 s plasma helium as well as air-plasma-treated samples. In the air-plasma-treated sample, in addition to the above, there was a small increase in 1633 cm^{-1} peak as well, which indicated the formation of additional $-CONH-$ groups due to the presence of oxygen in plasma zone. Possibly, the presence of O_2 in air-plasma promoted the formation of amide groups over the amine groups during the plasma treatment. On the other hand, in the plasma-treated polyester, very small increase (from 1.02 to1.07) in the value of normalized peak intensity for –C=O group was observed possible due to the formation of more –COOH groups in the presence of atmospheric oxygen. The peak intensity was higher for air-plasma-treated samples, where oxygen was readily available during the plasma treatment itself. The variation in the peak intensity as recorded by ATR-FTIR was small in all samples because plasma treatment can only alter the surface of the textile materials. It was also indicated that chemical changes that occurred in nylon were more significant than those in polyester fabrics upon plasma treatment. This may be one of the possible reason for better antistatic property in nylon than polyester sample. The formation of more hydrophilic groups could have helped in better transport of the static charge.

Table 5.2 ATR-FTIR data for the nylon fabrics (Samanta, 2009)

Parameters	Normalized peak intensity at *	
(Nylon sample)	3295 cm^{-1} (–NH$_2$)	1633 cm^{-1} (–CONH–)
Untreated	1.2615	3.153
60 s air-plasma treated	1.409	3.22
60 s He-plasma treated	1.4285	3

*Peak at 2928 cm^{-1} for C–H stretching was used for normalization

(ii) Increase in surface area
In order to understand the change in surface morphology of the nylon and polyester samples after plasma treatment, AFM scans of the samples were taken over an area of 4 × 4 μm^2 (Samanta, 2010). At a magnification of 500 nm/division in the direction of z-axis, the untreated sample had comparatively a feature-less smooth surface. However, after the 60 s He-plasma treatment, the surface developed channel-like features. Features with vertical dimensions of <200 nm were easily seen. These appeared to be nano-sized channels. These vertical channels were evenly distributed over the entire plasma-treated surface. Due to the formation of these channels in the plasma-treated samples, the surface area increased to 19.27 μm^2 from 16.28 μm^2 in the untreated sample, corresponds to 18.3% increase. Similarly to nylon, in polyester samples at a magnification of 200 nm/division in the direction of z-axis, the untreated sample showed few broad curvatures like features, which were inherent to the otherwise smooth surface. However, on He-plasma treatment, the surface developed horizontal channel-like features that were <100 nm in height and separated from each other by ~350 nm in the horizontal direction. As are a result of this, surface area increased to 17.18 μm^2 from 16.01 μm^2 in the

untreated sample. Similar to plasma-treated nylon samples, these nano-sized horizontal channels were also uniformly distributed over the entire treated surface. Because the dissipation of static charge occurs through the surface of the substrate, the increase in surface area helped in better dissipation of the developed static charge from the plasma-treated substrates. The charge leakage to the surroundings might also be increased from the sharp-pointed edges of the channels. It was also indicated that the better antistatic properties shown by the plasma-treated nylon samples compared to the polyester samples may be due to the higher increase in the surface area.

5.5.5   Improvement in dyeing and printing

As indicated above, the DBD plasma treatment of cotton, wool, silk and polypropylene change their surface from hydrophobic character to hydrophilic. It was observed that after plasma treatment, specific surface area increased significantly from 0.1 m^2/g to 0.35 m^2/g in cotton and wool, thereby increasing percentage of dye uptake (Hocker, 2002). Atmospheric pressure plasma treatment with He/Ar and acetone/Ar on Merino wool with 30 s treatment increased dyeing rate (Wakida, 1993). Air plasma and dichlorodifluoromethane (DCFM) plasma treatment on cotton fabrics caused surface modification, and also improved dyeability with reactive and natural dyes but slightly decreased dyeability with direct dye (Bhat, 2011). Nitrogen-containing plasma has been widely used to improve wettability, printability, bond forming tendency and biocompatibility of polymer surfaces (Labay, 2012). When nylon 6 fabrics were treated with low temperature plasma using three non-polymerizing gases such as oxygen, argon and tetrafluoromethane, there was a slight decrease in the air permeability of the treated fabrics, probably due to plasma changing the fabric surface. In synthetic fibres, plasma causes etching of the fibre and introduction of polar groups, which improves its dyeability with basic dyes. When different monomer coatings were applied to polyester, polyamide and polypropylene fabrics, these improved the affinity of these fibres to other classes of dyes (Vesel, 2009). The wettability of cotton and silk was found to increase a few fold by N_2 plasma pre-treatment. In polyester fabrics too, wettability increased significantly. Low temperature plasma treatment of grey, mercerized cotton and polyester/cotton-blended fabrics before dyeing has been reported to be effective for improvements in dyeing (Tsriskina, 1991).

Using dichloromethane with a RF plasma (10 Pa), for short treatment times (10–45 s) on cotton and polyester fabrics, the dyeability with reactive dyes could be enhanced without affecting other properties (Jahagirdar, 2004). An increase in the diffusion coefficient of acid dyes into wool has been reported using atmospheric plasma (Holme, 2000). Acetone/Ar plasma polymerization produces a hydrophilic polymer, which modifies the surface of the endocuticle or cell membrane complex causing accelerated dye diffusion. Silicon tetrachloride hydrophilic plasma coatings on PET fabric cause a roughened surface, improving the PET fabric dyeing properties (Kang, 2009). Plasma polymerization of acrylic-like coatings on polyester and polyamide fabrics was found to enhance wettability, dyeability using basic dyes, and improve their soil resistance (Cireli, 2007). A glow-discharge argon and argon–oxygen mixture $(Ar–O_2:10:1)$ was used to continuously modify polyester fabric (Ren, 2007). The spectral value (K/S) of the dyed polyester fabric increased by 50% and relative dye up-take also increased by 18% after the plasma treatment. The improvement of the dyeability was not at the expense of the dyeing fastness of the material. Dyeing of polyamide fabrics was carried out after surface modification by dielectric barrier discharge (DBD) plasma treatment at plasma dosage of 0.5–2.5 kW/min/m (Souto, 2012). The fibre nature changed from hydrophobic to hydrophilic, which is the key point for adsorption of aqueous dye solutions to achieve excellent dye uptake, high rate of dyeing and good uniformity, good fastness levels for obtaining darker shades using less concentration of dyestuffs at lower temperature and shorter dyeing times. The colour depth of the dyed fabric (polyester/cotton blend) increases with the acrylic-like film thickness. The surface modification of PP fabrics by acrylonitrile cold plasma to deposit poly-acrylonitrile like layers leads to improved water absorption and dyeing treatments (Sarmadi, 1993). This is due to the presence of nitrogen and carbon-based unsaturated linkages and the formation of second generation $=C=O$ groups.

Dyeing of plasma pre-treated cotton woven fabrics showed deeper and brighter colour shades than untreated samples (Nasadil, 2008). Plasma helps the cotton textiles to absorb more dye from dyeing bath, but has little effect on colour fastness of textile material (Malek, 2004). Plasma treatment on cotton in presence of air or oxygen gas increases both the rate of dyeing and the direct dye uptake of Chloramine Fast Red K, in the absence of electrolyte in the dye bath (Sung-Spitzl, 2003). This effect depends more on the oxygen component of air than the nitrogen component. Oxygen plasma treatment was more effective than air plasma treatment. The increase in dye uptake was attributed to: (i) the oxidative mechanism of attack on the cotton that modifies the surface properties, (ii) change of fabric surface area per unit volume caused by surface erosion, (iii) etching of fibre surface by plasma and removal of impurities like cotton wax, or any remaining size, etc., (iv) chemical changes in cotton fibre leading to formation of carbonyl and carboxyl groups on the fibre surface, and (v) possibility of formation of free radicals on the cellulosic chains of cotton (Guglani, 2002). Plasma treatment of cotton fabric in amine plasma (ethylenediamine or triethylenetetramine) resulted in improved colour yield (K/S) and better fastness properties, when the treated fabric was dyed with a reactive dye (Remazol Black B) (Özdogan, 2002).

Among the various methods of textile printing, inkjet printing is an important technique and was carried out on hard-to-print polypropylene after plasma pre-treatment (Nasadil, 2008). The results before and after a couple of washing cycles showed distinct difference between the plasma-treated and untreated samples. Plasma treatment improved inkjet printing on polypropylene partly due to hydrophilic modification of the substrate and partly due to improved sorption of inks on samples.

Effect of atmospheric pressure glow plasma treatment on dyeing of protein fibre (wool) was studied by Panda (2012) using cold brand dichlorotriazine-based reactive dye (C.I. Reactive Red 2) at lower temperature. The samples were plasma treated for 6 min in glow cold plasma generated in the presence of He gas at a discharge voltage of 3 kV and frequency of 20 kHz (Panda, 2012). Urea is reported to increase the fibre swelling and disaggregation of dye leading to improve dye penetration, whereas sodium bisulphite reacts with the disulphide residues in wool to generate highly nucleophilic cysteine thiol groups. Presence of these additives would however increase the chemical loading of the waste water and cause environment pollution. Any treatment, which could otherwise reduce the surface barrier of wool fibres and makes it reactive, will be beneficial for dyeing and effluent load. The improvement in hydrophilic property after plasma treatment and its stability to number of washing cycles and storage time in addition with anti-felting property were also studied. The dye bath liquor consists of 5 g/l or 10 g/l reactive dye, 10 g/l TRO, 0–300 g/l urea, and 0–10 g/l sodium bisulphite and pH was kept at either 5 or 7. Wool fabric was padded at the room temperature (30–35°C) and kept for 24 hours at 35°C and dyeing time varied from 10 min to 30 h. K/S values were measured before and after stripping unfixed dye. Concentration of dye in fibre (dye uptake) was determined by dissolving dyed fabric in 4% NaOH solution and measuring the optical density in UV–VIS spectrometer.

Plasma treatment resulted in significant improvement in the colour yield of reactive dye on wool. Both the dye pick-up and fixation on wool increased by almost twice based on K/S values after the plasma treatment. The liquid expression of wool was found to increase substantially from 60% to 70% in untreated wool to 105–115% after plasma treatment. This increase in expression led to a very high concentration of dye application on the surface of plasma-treated wool fibres observed from the K/S values. The increased absorption of dye liquor was feasible due to the removal of surface barrier (hydrophobic cuticle layer) and increased wettability of wool after the plasma treatment. Even after the removal of unfixed dye, the K/S values increased by almost 100% (range of 1.5–2.5 for the untreated wool to 3.7-4.6 for plasma-treated wool). The percentage fixation was in the range of $80 \pm 5\%$ for both untreated and treated wool. This translated to a net increase in the amount of dye fixed in the treated fibre as the dye uptake was significantly higher in this case. It was indicated that the generation of new amino groups during plasma treatment might have contributed towards the increased fixation of the dye on the treated fibre.

The effect of different additives in dyeing of wool was studied by varying urea concentration in the range of 0–300 g/l, which resulted in slight increase in the colour yield in both untreated and treated samples. Sodium metabisulphite and acidic pH did not appear to have any effect on the colour yield. In fact, the K/S value of plasma-treated wool at neutral pH without the addition of urea and sodium metabisulphite was almost double (4.26) as compared to that of the untreated wool-dyed under similar conditions (1.77) or in acidic pH in the presence of 300 g/l urea and 10 g/l sodium metabisulphite (2.55) (Table 5.3). Cold pad-batch dyeing of untreated and plasma-treated wool was further carried out using increased concentration of reactive dye (10 g/L) in the dye bath at neutral pH and without any additives. The pickup of dye on the treated wool increased by approximately double on increasing the dye concentration from 5 g/l to 10 g/L, whereas, in the case of untreated wool, the increase was lower. The resultant K/S of the stripped samples was still significantly higher in the treated samples, whereas untreated wool showed only a small increase.

Table 5.3 Effect of plasma treatment on colour yield of C. I. Reactive Red 2 dye on wool at varying concentrations of urea and sodium metabisulphite (Panda, 2012)

Sample	K/S Before stripping Urea (g/L)				K/S After stripping Urea (g/L)				% Fixation Urea (g/L)			
	0	100	200	300	0	100	200	300	0	100	200	300
Acid + sb UT	2.33	2.43	3.15	3.55	1.77	1.73	2.27	2.55	75	70	72	71
Acid + sb T	4.66	4.95	4.34	5.37	3.68	3.98	3.66	4.15	78	80	84	77
Acid UT	2.55	2.82	2.91	2.67	2.11	2.51	2.55	2.49	82	88	87	93
Acid T	4.83	5.03	5.42	5.54	4.02	4.32	4.50	4.57	80	86	82	82
Neu + sb UT	2.11	2.38	2.64	2.43	1.49	1.87	2.29	1.94	70	78	86	79
Neu + sb T	5.62	5.07	5.60	5.62	3.70	3.74	4.63	4.27	65	73	82	75
Neu UT	2.95	2.61	3.06	2.77	2.49	2.07	2.49	2.31	84	79	81	83
Neu T	4.94	5.88	5.77	5.58	4.26	4.35	4.19	4.31	86	73	72	77

[a] Acid: acidic pH 5 [b] Neu: neutral pH [c] sb: 10g/L sodium metabisulphite [d]T: plasma treated [e] UT: untreated [f]dye concentration: 5 g/L .

Dyeing kinetics

In order to elucidate the change in the kinetics of dyeing after plasma treatment, the dye kinetics of reactive dye on wool was studied in infinite bath at 40°C using exhaust method of dyeing. The dye uptake after different dyeing periods was determined by measuring the K/S values of the dyed fabric as well as by measuring the optical density of the dissolved dyed fabric. As observed in the cold-pad-batch experiments, the K/S values in the exhaust dyeing were also found to be almost double in the case of the treated samples compared to the untreated samples (Fig. 5.15). The increase in the exhausted and fixed dye on treated wool could be seen right from the initial dyeing period and the extent of the increase was maintained till the equilibrium. When the total concentration of dye in the treated sample was determined by optical density, it was found to be higher by 40–50% as compared to the untreated sample (Fig. 5.16). This difference in the K/S values and total dye uptake revealed an interesting behaviour of the treated samples. It indicated that a large amount of dye is being absorbed and reacted on the surface of the fibre after the plasma treatment, though only a part of it is diffusing inside the fibre.

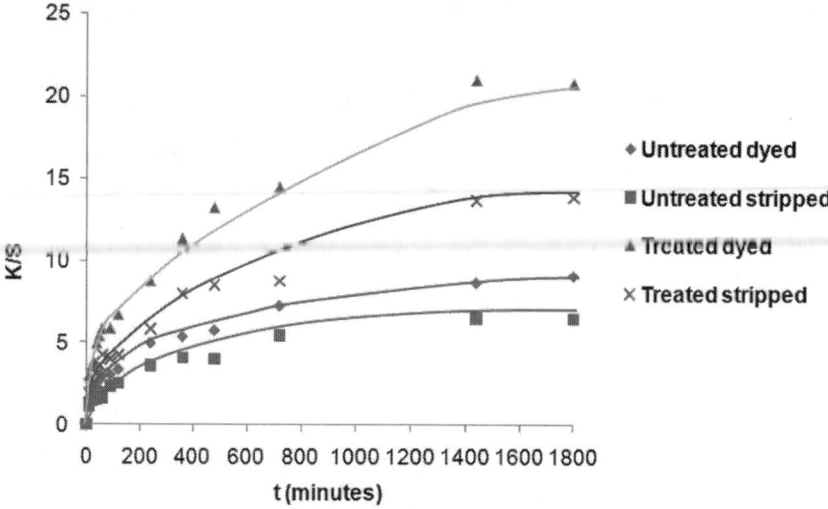

Figure 5.15 K/S values of plasma treated and untreated wool dyed in infinite bath of C.I. Reactive Red 2 for different dyeing periods (Panda, 2012)

Further, the concentration of dye in fibre (before stripping) was plotted against the square root of the dyeing time for plasma-treated and untreated wool fabric for initial dyeing period of 60 min as well as, for longer dyeing periods of up to equilibrium. Curves were linear, similar results are reported in literature (Bird, 1968). For initial periods of dyeing, the slope of the treated sample was 0.9×10^{-6} moles/g.s$^{0.5}$ and for untreated one was 0.5×10^{-6} moles/g.s$^{0.5}$. The ratio of the slope for the treated wool to that of the untreated wool was 1.8. For longer dyeing times, the slope of the line for the treated fabric was 1.0×10^{-6} moles/g.s$^{0.5}$ and for the untreated fabric was 0.8×10^{-6} moles/g.s$^{0.5}$, and ratio of treated to the untreated was 1.2. As the slope of line is proportional to the square root of the diffusion coefficient, it shows that the initial difference in dye uptake is due to significantly faster diffusion through the modified cuticular layer of the plasma-treated wool. For longer dyeing times, there was only a small difference in the average diffusion coefficient between the treated and untreated wool. The higher concentration of dye inside the treated fibres may also have been a result of higher concentration gradient of dye present between the surface and the inside of the fibre. The above results conform to the fact that the plasma treatment mainly changed the surface properties of the wool.

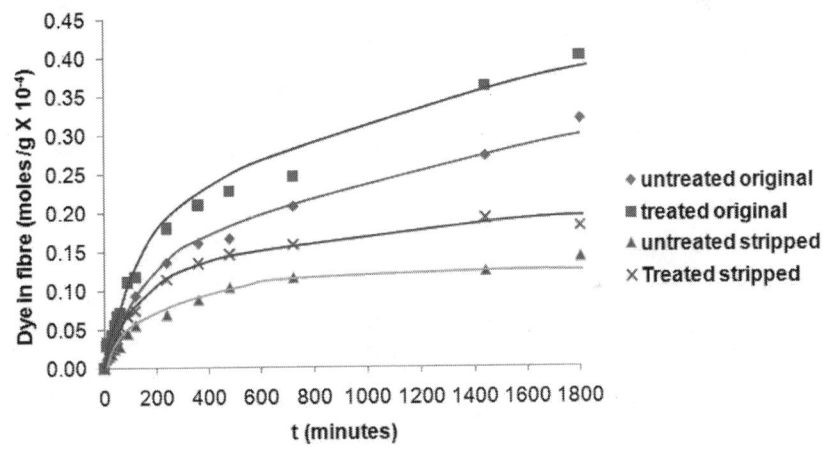

Figure 5.16 Amount of dye taken up by plasma treated and untreated wool dyed in infinite bath of C. I. Reactive Red 2 for different dyeing periods (Panda, 2012)

Mechanism of dyeing

Figures 5.17(a) and (b) show the secondary negative ion spectra (SIMS) of the untreated and plasma-treated wool. The intensity counts of the HO$^-$ and NH$^-$ groups increased, whereas intensity count of CH$^-$ group decreased after the plasma treatment (Panda, 2012). Figures 5.18(a) and (b) show the change in intensity of the selected ions with increasing depth of the samples. In the untreated sample, the intensity of CH$^-$ ion is very high compared to HO$^-$ and NH$_2^-$ ions. This is likely due to the presence of fatty acid layer on the cuticle surface. The relative ratios of the CH⁻ to HO⁻ and NH₂⁻ do not seem to change significantly with depth of up to 10 etched layers. However, in the plasma-treated samples, the intensity of CH⁻ ions has reduced significantly compared to the untreated wool. On the other hand, the intensities of HO⁻ and NH₂⁻ ions were increased significantly in the first few layers. With increasing number of depth, the intensity of HO⁻ and NH₂⁻ ions drop to the values similar to that of the untreated sample. However, CH⁻ concentration remained low throughout with a gradual increase in its concentration with increasing depth.

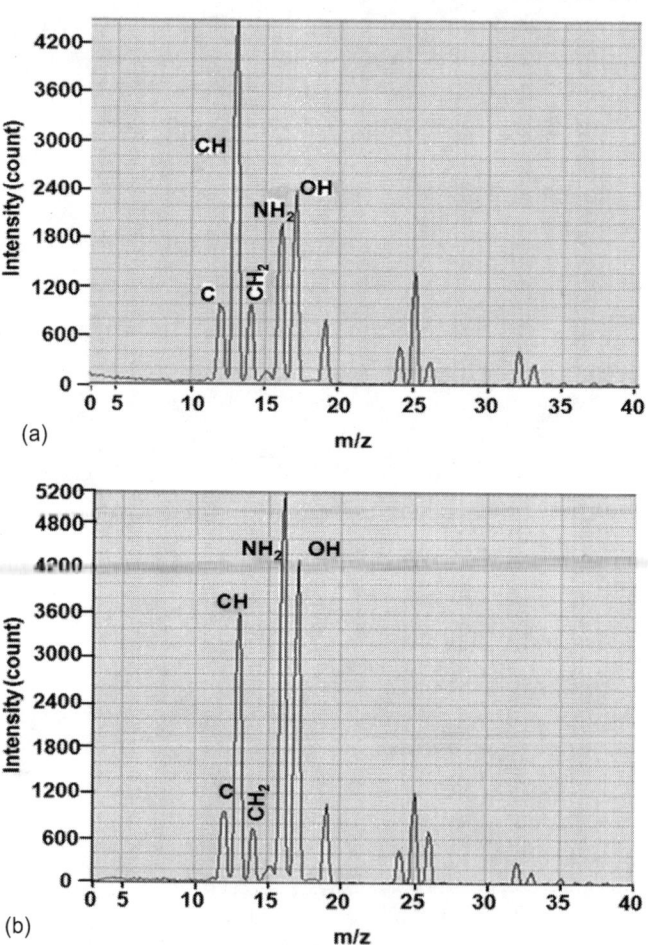

Figure 5.17 Negative ion mass spectra of the (a) untreated wool and (b) plasma-treated wool (Panda, 2012)

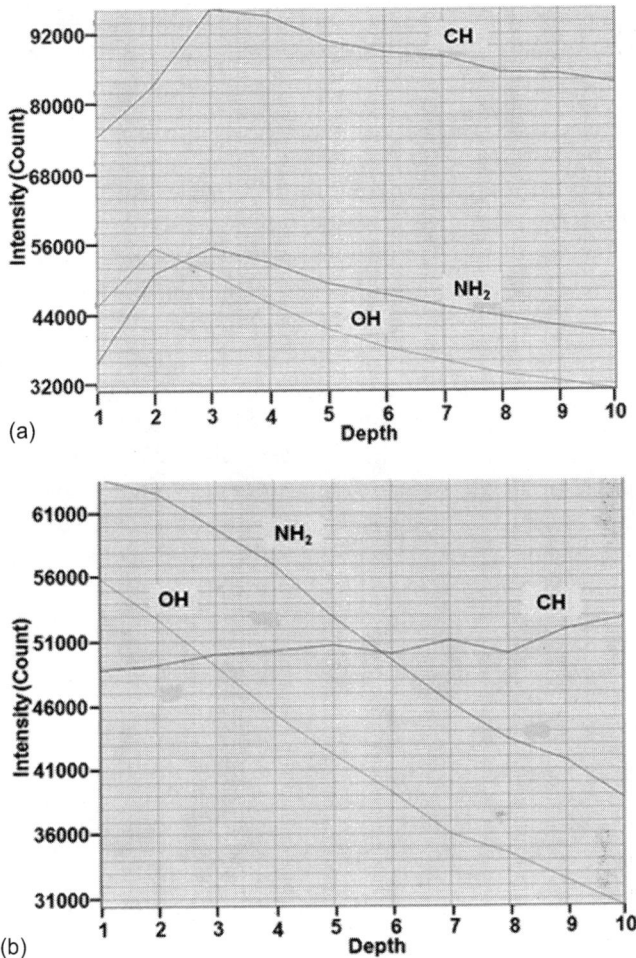

Figure 5.18 Depth profile of selected negative ions from mass spectra of (a) untreated wool (b) plasma-treated wool (Panda, 2012)

The depth profile showed that the effect of plasma treatment was mainly on the surface, where the fatty acid molecules seemed to be etched out and polar groups from peptide chains generated. The wettability of wool after plasma treatment was enhanced significantly as the water absorption time decreased from more than 1 h in the case of untreated wool to less than 1 s in the case of helium plasma-treated wool. There was no detectable effect of ageing observed up to 9 weeks time and 30 wash cycles on the wettability of the plasma-treated wool sample. This indicated that the effect of plasma treatment under the specified conditions was permanent. The change in fibre wettability was likely to be because of the removal of the covalently bound fatty acid layer from the surface of the wool fibres, exposure of the underlying hydrophilic protein material (Naebe, 2010), and generation of additional polar groups as discussed above.

Due to the reduced surface barrier in treated wool, higher amount of dye was able to diffuse quickly through the cuticular layer compared to the untreated wool. This was seen as an increase in the dye uptake by the treated fibre was about 40–50% (determined by dissolving the fibres). Since the bulk properties of wool did not change by plasma treatment, not all of this adsorbed dye was able to diffuse inside the fibre towards the core. The plasma treatment was also able to increase the concentration of polar groups near the surface, in particular NH_2 groups (SIMS analysis), which helped in fixing the absorbed reactive dye in the first few atomic layers of the treated wool fibres. This manifested as a much higher colour yield by almost 100% in terms of K/S values in the plasma-treated wool. Though most of dye molecules were present on the surface of the fibres, dye fastness properties did not get affected. Based on the dyeing studies, a possible mechanism of dyeing in plasma-treated wool was proposed, which is shown schematically in Fig. 5.19.

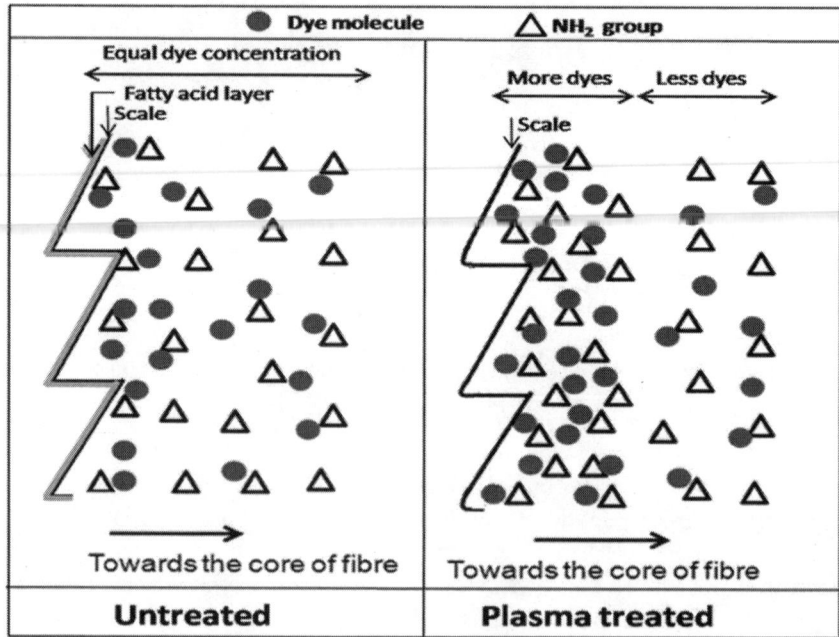

Figure 5.19 Schematic representation of dyeing mechanism in the untreated wool and plasma-treated wool (Panda, 2012)

5.5.6   Improvement in water absorbency and wicking

The effect of atmospheric pressure glow discharge cold plasma (APGD) treatment on the improvement in hydrophilicity in term of water absorbency time was measured in the nylon and polyester woven textile substrates without any detriment to the mechanical properties of the samples in a continuous manner (Samanta, 2006; Samanta, 2008; Samanta, 2009). It was inferred that nano-sized physical structures, formed due to plasma treatment, had profound effect on the spreading of different fluids on the surface of the fabric. The water absorbency or horizontal wicking was measured according to AATCC test 39-1971 within 30 min of the treatment. A 37 μl distilled water droplet was placed at the centre of the marked area (3.79 cm^2) of the fabric. The time for the water droplet to spread in the marked area was recorded. It was observed that a water droplet took 540 s to spread over an area of 3.79 cm^2 in the untreated nylon sample, whereas it took only 1.1 s in the 60 s He-plasma-treated nylon samples. In the untreated PET fabric, water took 700 s to spread over the same area, whereas it took only 6.7 s in the 60 s He-plasma-treated sample. In the untreated sample, absorption and spreading of water was very slow. This was because nylon and polyester are essentially hydrophobic in nature and they have low moisture regain percentage. However, after plasma treatment, the surface energy of the samples increased significantly, likely due to the generation of hydrophilic groups, which are expected to assist absorption and spreading of water.

Surface energy of the as-procured nylon fabric was below 43.4 milliJ/m^2 and after 60 s plasma treatment, it was found to increase to 63 milliJ/m^2 in the 60 s He, Ar, or air-plasma-treated samples. As mention above (antistatic finishing section), in the polyester sample, the surface energy increased from 36 milliJ/m^2 in the untreated sample to reach the maximum measurable value of 71 milliJ/m^2 in the plasma-treated sample. This increased in surface energy was due to formation more hydrophilic groups and change in surface morphology. The faster spreading of water in the plasma-treated sample was possible partially due to the presence of these nano-sized channels and capillaries in addition of generation of hydrophilic groups.

The objective of increasing hydrophilic features and reducing chemical waste from existing pre-treatment processes for cotton fabrics was achieved by using corona discharge in an air atmosphere (Hegemann, 2006). It effectively increased hydrophilicity without affecting the integrity of the fibre or yarn. The treatment affected chemical and physical changes in the waxy cuticle layer of the cotton without damaging the cellulose backbone. Treatment of polyester fibres by glow discharge in air or oxygen causes a partial degradation of the fibre surface with an increase in the capillary sorption of iodine or cations in aqueous solution. Polyester has the hydrophobic surface, because it is made up of ether oxygen (C–O–C) linkages, while the hydrophilic ester oxygen (C=O) linkages are facing towards the core of the fibre. When it was plasma treated either the ester oxygen (C=O) came closer to the surface as a result of etching or some new C=O bonds were formed due to the oxygen ions present in the plasma chamber (Karmakar, 1999). Wettability of polyester can be achieved by treatment with corona discharge, where the fabric could be processed without using a wetting agent. With respect to the soiling of fabrics, treatment of polyester with air plasma considerably decreased the soiling. During plasma treatment, fabrics attained negative charge. The soils are also generally negatively charged; hence, they repel each other. Plasma etching increases the hydrophilic nature and that also decreases soiling (Yasuda, 1982). Etching by air plasma was found to cause greater weight loss of nylon, cotton, silk or wool fibres than etching by nitrogen or carbon tetrafluoride plasma treatment. It was observed that degree of surface modification was lower for plasma-etched nylon or polyester fibres than for cotton/wool fibres. Polyester/cotton (P/C)-blended woven fabric was treated with atmospheric pressure plasma to improve its hydrophilic properties (Kale, 2010). A dielectric barrier discharge (DBD) plasma generated from He/O$_2$ gas was used in such treatment. It was found that plasma process parameters played a critical role in deciding the efficiency of the treatment, and at optimum conditions, hydrophilicity of P/C-treated samples measured by wicking behaviour was found to be higher than the untreated samples.

In another study, atmospheric pressure plasma treatment with He/Ar or acetone/argon on wool and PET fabrics was carried out. Wettability was found to increase significantly with increasing plasma treatment time. Plasma treatment by He/Ar was found to be more effective than acetone/argon (Wakida 1993). Atmospheric pressure plasma treatment with helium (He) and monomer, acrylic acid, on PET produced acrylic acid polymers having a high value of surface energy with hydrophilic character (Topala, 2008). Low pressure plasma treatments using oxygen-containing gaseous mixtures (Ar/O$_2$ and He/O$_2$) on polyester (PET) fabric, improved wettability significantly due to formation of polar groups on the surface; hydrophilicity of loose-structured fabrics improved remarkably as compared to tightly woven fabrics (Hossain, 2006). Atmospheric air plasma treatment of poly(ethylene terephthalate) fibre and polyamide 66 fabrics decreased the wetting time for polyamide fibre and poly(ethylene terephthalate) as roughness in etched surface caused increase in wettability. In addition, soil-release behaviour of the treated fabrics was also improved (Nourbakhsh, 2012). Atmospheric pressure dielectric barrier discharge (DBD) in air at ambient temperature on polyester fabric produced distinct variation on textile surface, with increased hydrophilicity of fabric (Píchal, 2009). Upon 300–1000 W atmospheric pressure air plasma treatment on PET woven fabric, surface energy increased, water contact angle decreased and water capillarity improved. Weight of water absorbed by the capillary increased from 12 to 200 mg. Water contact angle on plasma-treated PET decreased from 80° to 40°, indicating an increase in the surface energy of PET fibres due to the change in chemical nature of the fibre surface (Guo, 2009). Atmospheric-pressure plasma treatment of polypropylene nonwovens with nitrogen and dry air resulted in surface activation and permanent hydrophilization of light-weight nonwovens. A homogeneous surface roughening was obtained due to the plasma treatment and the treatment had no wash-out effect on hydrophilicity for the aged samples (Cernák, 2002). Similarly, with acetylene mixed with ammonia plasma (C_2H_2/NH_3) treatment on textile, a slight increase of wettability was also seen (Hegemann, 2007).

Polyamide is widely used as a semi-crystalline engineering thermoplastic polymer as well as excellent fibre. Samanta (2010) studied plasma treatment of 100% polyamide 6 (Nylon 6) woven textile in continuous manner. The samples were treated with atmospheric plasmas of helium (He) and helium and oxygen (He/O_2) gases at a frequency of 21.2 kHz and with discharge gap of 2.1 mm. For He plasma, the output voltage of 2.7 kV and He gas flow rate of 3 l/min were used (Samanta, 2010). Slightly more discharge voltage and gas flow rate were kept for He/O_2 plasma (O_2 – 350 ml/min) to generate similar type of plasma. The samples were exposed to air after the plasma treatment to allow the formation of various oxidative hydrophilic groups. The plasma treatment time was varied from 10 s to 60 s by adjusting the linear speed of the fabric in the range of 13.5–81.0 cm/min. Water absorbency time was measured according to AATCC test Method 79-2007 and contact angle was by Sessile drop method. The lower water absorbency time or contact angle value is an indication of better hydrophilic property. Table 5.4 shows the water absorbency time and contact angle value of the untreated and plasma-treated nylon samples. In the untreated nylon sample a water droplet of 37 μl did not get absorbed even after 60 min (maximum measured time) as shown in Fig. 5.20(a). This is because the untreated nylon was hydrophobic in nature. After 10 s of He or He/O_2 plasma treatment, there was no improvement in the water absorbency time. However, in the 15 s plasma-treated samples, a significant improvement in the hydrophilicity was observed. As a result of this, the water absorbency time reduced from >60 min for the untreated sample to 9 min 08 s for the He/O_2 and 35 min 20 s for the He plasma-treated samples.

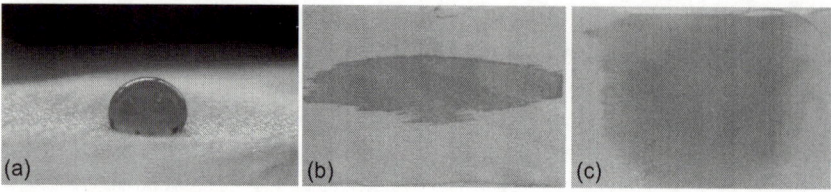

Figure 5.20 Colour water droplet on nylon: (a) untreated, (b) 60 s He plasma treated, and (c) 60 s He/O$_2$ plasma treated (Samanta, 2010)

Table 5.4 Water absorbency time of untreated and plasma-treated nylon samples (Samanta, 2010)

Plasma treatment time in (s)	He plasma	He/O$_2$ plasma
0 (Untreated)	> 60 min	> 60 min
10	>60 min	> 60 min
15	35 min 20 s	9 min 8 s
30	209 s	5.2
60	4.9 s	3.3 s

The water contact angle in the 15 s plasma-treated sample was measured to be 0° compared to 130° in the untreated sample. With increasing plasma treatment time to 30 s, water absorbency time further decreased to 209 s and 5.2 s in the He and He/O$_2$ plasma-treated samples, respectively. After the 60 s of plasma treatment, hydrophobic nylon was turned into highly hydrophilic nylon with a water absorbency time of only 3.3 s and 4.9 s in the He/O$_2$ and He plasma-treated samples, respectively, as shown in Fig. 5.20(b) and (c). In the plasma-treated hydrophilic nylon samples, the contact angle was measured to be 0° as water get absorbed by the fabric very fast. Interestingly, the He/O$_2$ plasma treatment was observed to impart more hydrophilicity compared to the He plasma treatment. This may be due to the formation of more oxygen-containing hydrophilic groups in the He/O$_2$ plasma-treated samples.

Surface chemical characterization

In order to elucidate the possible reason to improvement in hydrophilicity in the nylon fabric, the samples was analyzed using XPS and secondary ion mass spectrometer (Samanta, 2010). Figure 5.21 shows XPS survey spectra of the untreated and 60 s He and He/O$_2$ plasma-treated (as-prepared) samples over the binding energy range of 0–700 eV. Polyamide-6 (nylon-6) macromolecule has carbon (C), hydrogen (H), oxygen (O) and nitrogen (N) atoms. As XPS can not detect H atom, therefore in the survey spectra of different nylon samples C(1s), O(1s) and N(1s) atomic peaks only were observed with a binding energy of 285.2, 532.7 and 399.7 eV, respectively.

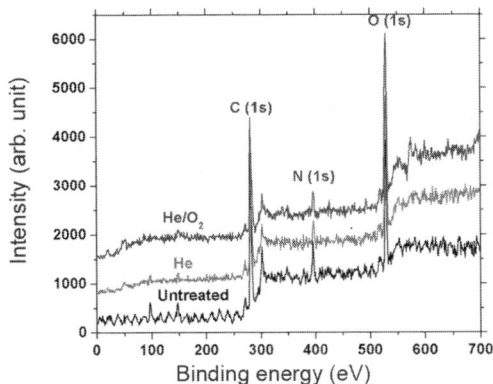

Figure 5.21 XPS spectra of the untreated, He and He/O$_2$ plasma-treated nylon samples (Samanta, 2010)

Table 5.5 shows the atomic percentage in the different nylon samples. In the untreated sample, surface oxygen percentage is 18.6 and it increased to 21.8% for the 60 s He-plasma-treated and 29.0% for the He/O$_2$ plasma-treated sample. This indicated that after the plasma, surface oxygen percentage increased and it was more pronounced in He/O$_2$ plasma-treated sample. In the plasma-treated samples, the surface nitrogen percentage was a little lower compared to untreated sample.

Table 5.5 Surface chemical composition of untreated and plasma-treated nylon samples (Samanta, 2010)

Different Sample	Atomic concentration (%)			Atomic ratio	
	C	O	N	O/C	O/N
Untreated	71.9	18.6	9.5	0.26	1.9
60s He plasma treated: As-prepared	71.1	21.8	7.1	0.31	3.1
60s He/O$_2$ plasma treated: As-prepared	63.5	29.0	7.5	0.46	3.9

The O/C ratio, hydrophilic to hydrophobic molecules, in the untreated sample was 0.26, and there was 19% and 77% increase in the He and He/O$_2$ plasma-treated samples, respectively. On the other hand the O/N ratio was 1.9 in the untreated sample and the ratio increased to 3.1 (63%) and 3.9 (105%) in He and He/O$_2$ plasma-treated samples, respectively. The increase in oxygen percentage could have helped in formation of different polar surface groups, such as ketone, hydroxyl, ether, peroxide and carboxylic acid that are responsible towards the improvement in hydrophilicity in the otherwise hydrophobic nylon textile.

Surface chemical composition, chemical mapping and depth profiling of the nylon samples were also carried out using Secondary Ion Mass Spectrometer (Model: MiniSIMS, from Millbrook Company, UK) using gallium (Ga) liquid metal ion source (LMIS). Negative ion mass spectra were used to determine the different molecules present on the surface, while the uniformity of distribution of different molecules on the surface was studied by chemical mapping. Figure 5.22 shows the negative ion mass spectra of the different nylon samples. Negative ion mass spectrum of the untreated sample showed the presence of major fragments of nylon at different mass values, such as 13 a.m.u. for CH^-, 16 a.m.u. for O^-, and 17 a.m.u. for OH^-. In addition to these, C^- at 12 a.m.u., CH_2^- or N^- at 14 a.m.u., $C-C^-$ at 24 a.m.u. and $CH-C^-$ at 25 a.m.u. and $CH-CH^-$ at 26 a.m.u. were also observed. The presence of these masses in the mass spectrum indicated that during the SIMS analysis major bonds of nylon, such as $-CH_2-CH_2-$, $-CO-NH-$ etc., were broken down into the smaller molecules and atoms, such as C^-, CH^-, CH_2^-, N^-, O^-, HO^-, $C-C^-$ and $C-CH^-$. In the plasma-treated samples also similar peaks were detected. However, the intensity of peak for C^- at 12 a.m.u. was similar to HO^- peak at 17 a.m.u. in the untreated sample. On the other hand in He and He/O_2 plasma-treated samples, the intensity of HO^- peak at 17 a.m.u. was more than C^- at 12 a.m.u. Table 5.6 shows the negative mass spectra data of the different nylon samples. It was observed that in untreated sample, the O^-/CH^- ratio was 2.4 and it increased to 2.7 in the He plasma-treated sample and 3.0 in the He/O_2 plasma-treated sample. Similar trend was observed for O^-/C^- ratio. The O^-/C^- and O^-/CH^- ratios in the plasma-treated samples were more compared to the untreated sample. In the untreated sample, the O^-/N^-

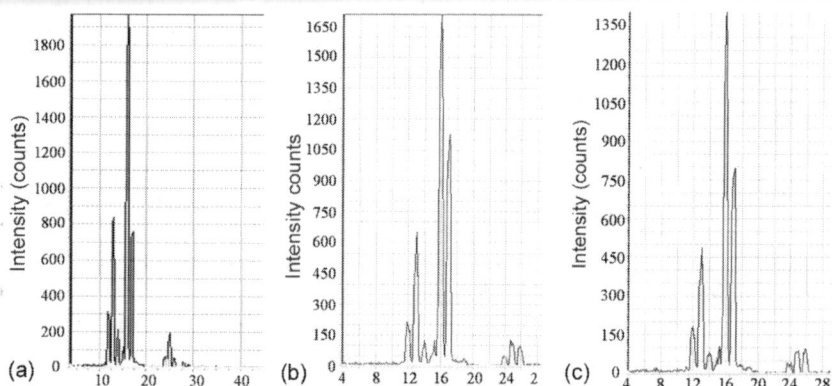

Figure 5.22 Negative mass spectra of: (a) untreated, (b) 60 s He plasma treated and (c) 60 s He/O_2 plasma-treated nylon samples (Samanta, 2010)

ratio was 9.2 and it increased to 14.0 and 16.9 in the He and He/O$_2$ plasma-treated samples, respectively. Similar changes in O/N were also observed in XPS atomic ratio analysis. The presence of more oxygen-containing groups, incorporated during plasma treatment, must be contributing towards the improvement in hydrophilicity in the plasma-treated samples.

Table 5.6 Relative abundance of different species in negative ion mass spectra of the different nylon samples (Samanta, 2010)

Different sample	O$^-$/C$^-$	O$^-$/CH$^-$	O$^-$/N$^-$
Untreated	6.3	2.4	9.2
60 s He plasma treated: As-prepared	8.1	2.7	14.0
60 s He/O$_2$ plasma treated: As-prepared	7.9	3.0	16.9

From the negative mass spectra of the untreated and 60 s plasma-treated sample, the mass of the CH$^-$ (13 a.m.u.) and O$^-$ (16 a.m.u.) were selected for depth profiling to study the change in surface hydrophilic (O$^-$) and hydrophobic molecules (CH$^-$) with number of atomic layers (depth). It was seen (Fig. 5.23a) that in the untreated nylon sample, the abundance of O$^-$ on the surface is more compared to CH$^-$ and the same trend was observed with increasing the number of depths. However, in the 60 s He plasma-treated sample, the surface oxygen abundance was observed to be much more compared to CH$^-$ as shown in Fig. 5.23(b). However, the intensity of O$^-$ atom decreased significantly after the first layer and in all the successive atomic layers it remained below the intensity of CH$^-$ molecule. Similar results were also observed for the He/O$_2$ plasma-treated sample.

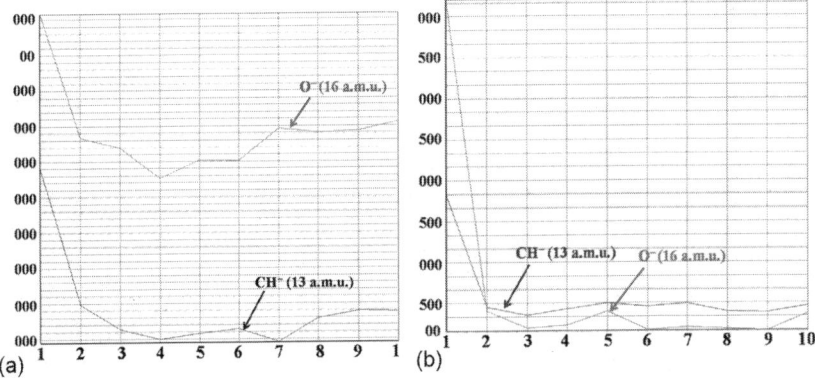

Figure 5.23 Depth profile of the different nylon samples: (a) untreated and (b) 60 s He plasma treated (Samanta, 2010)

Similar to nylon, a comparison was made by treating polyester fabric with He plasma, He/O$_2$ plasma and maleic anhydride (MA)/He plasma (Parida, 2010). Plasma-processing parameters, such as type of gas, voltage, frequency, and time of treatment, were varied to study their effect on improvement in hydrophilicity. To ascertain the change in fabric properties after plasma treatment, water drop disappearance time of samples were determined. Water drop disappearance time was determined for all treated samples and tabulated (Table 5.7). Table 5.7 shows the effect of discharge voltage in water absorbency time at three different frequencies level such as 19 kHz, 20 kHz and 22 kHz. The plasma treatment time was varied from 0.5 min to 5 min.

Table 5.7 Water drop absorbency time (in seconds) in the He plasma-treated PET fabric (Parida, 2010)

Treatment time (min)	V = 1.55			V = 1.73			V = 2.36		
	F = 19	F = 20	F = 22	F = 19	F = 20	F = 22	F = 19	F = 20	F = 20
Untreated	2100	2100	2100	2100	2100	2100	2100	2100	2100
0.5	519	478	542	445	529	542	478	489	522
1	261	345	512	409	506	494	393	423	442
2	163	315	321	254	341	329	312	310	334
3	112	192	271	201	224	320	221	182	234
5	105	109	205	218	248	213	145	123	143

F = frequency in kHz and V = voltage in kV

It could be seen that a water drop took 2100 s to get fully absorbed by the fabric. This is because polyester is hydrophobic polymer with low moisture regain percentage of 0.4%. However, when the sample was plasma treated for 30 s, water absorbency time decreased to 519 s (V = 1.55, F = 19) and further to 105 s in the 5 min plasma-treated sample. With increasing plasma discharge frequency, improvement in water absorbency was little lower. Effect of voltage had a marginal role in water absorbency. PET samples treated in He gas at 19 kHz and 1.55 kV for 5 min showed the best water absorption time of 105 s. Sample showed better hydrophilic property at low frequency level. Similar to the He/O$_2$ plasma treatment of nylon, mixture of He/O$_2$ gases were used to further improve the surface hydrophilicity of the polyester sample. Samples were treated in He/O$_2$ plasma in which flow rate of He and O$_2$ was 1.5 L/min and 350 ml/min, respectively. He/O$_2$ plasma-treated samples showed better absorbency time compared to the He plasma-treated sample (Table 5.8). A lowest measured water absorbency time 69 s was obtained at 23.5 kHz and 2.36 kV for the 3 min plasma-treated sample. With increasing plasma treatment time, water absorbency time reduced.

Table 5.8 Water drop absorbency time (in seconds) in the He/O$_2$ plasma-treated PET fabric (Parida, 2010)

Treatment time (min)	Gas = He/O$_2$			Voltage, V = 2.36	
	Frequency (F) in kHz				
	19	20	20.5	22	23.5
Untreated	2100	2100	2100	2100	2100
0.5	252	321	351	169	206
1	64	271	327	222	117
2	119	166	187	170	86
3	85	80	146	125	69
5	85	82	99	146	72

With the aim to improve the durability of the imparted hydrophilic functionality in the polyester (PET) sample, the PET fabric was first padded with maleic anhydride solution, then plasma treated with He plasma. In this experiment a minimum water drop disappearance time of 99 s was obtained for the plasma-treated sample at 19 kHz and 2.36 kV for 5 minutes (Table 5.9). As discussed in He and He/O$_2$ plasma-treated samples, with increasing plasma treatment time, water absorbency time decreased and samples showed better hydrophilic property at lower frequency level compared to higher frequency level. The durability of the treatment in these samples is discussed in the next section.

Table 5.9 Water drop absorbency time (in seconds) after He plasma treatment of MA-padded PET fabric (Parida, 2010)

Treatment time (min)	F = 21.5			V = 2.36	
	V = 1.55	V = 1.73	V = 2.36	F = 19	F = 24
Untreated	2100	2100	2100	2100	2100
0.5	423	222	312	327	365
1	314	176	212	212	286
2	243	161	152	226	224
3	223	130	146	112	212
5	212	110	118	99	146

5.6   Stability of hydrophilic finish

As discussed above, plasma treatment can increase the hydrophilicity of the hydrophobic polymer or textile substrates. However, the improved hydrophilicity is not durable/stable. It is known that the imparted hydrophilic property decreases with storage time and temperature (heat treatment). This is caused by re-distribution of the surface-oxygen containing polar groups or migration of such groups from surface to bulk of the material. This phenomenon is also known as hydrophobic recovery. The hydrophobic recovery of nylon to storage time at ambient temperature and to high temperature was investigated. The hydrophobic recovery of plasma-treated samples was monitored by water absorption time measurement over several weeks. To test the hydrophobic recovery in the plasma-treated sample, the samples were stored for 175 days. Similarly, to study the effect of heat treatment, the samples were heat treated at 70°C, 90°C and 110°C for different periods of time.

(a) Nylon fabrics

Effect of storage time

In this study, nylon woven fabric plasma treated for a period of 15 s, 30 s and 60 s in He or He/O$_2$ were selected to investigate the hydrophobic recovery by Samanta (2010). Figure 5.24 shows the hydrophobic recovery in the 60 s plasma-treated samples. It was seen that the hydrophobic recovery in He plasma-treated sample was very slow during the first few days; thereafter, a gradual increase in hydrophobicity was observed up to about 100 days. After that the recovery was very rapid. On the other hand, the 60 s He/O$_2$ plasma-treated sample remained stable up to nearly two-and-half (2.5) months, after that a very slow hydrophobic recovery was observed. After approximately 3 months (95 days), the hydrophobic recovery was only 3% and 22.3% in the He/O$_2$ and He plasma-treated samples, respectively; and after around 6 months it was 13.7% and 77.3% for these respective samples. Clearly, the imparted hydrophilic functionality in the He/O$_2$ plasma-treated samples was

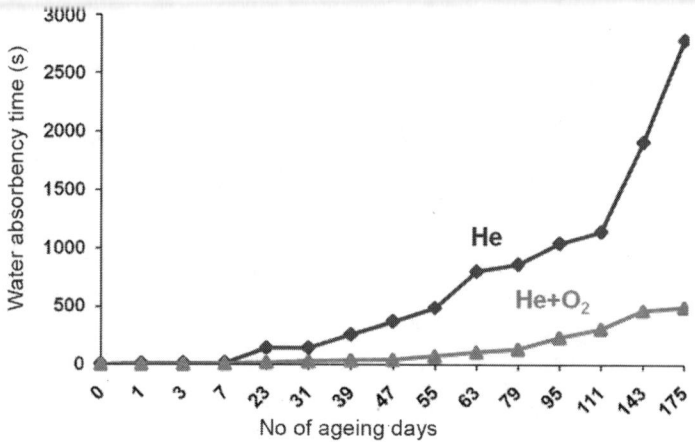

Figure 5.24 Hydrophobic recovery of the 60 s plasma-treated nylon samples (Samanta, 2010)

much more stable with time compared to He plasma-treated samples. This meant that incorporation of oxygen during the plasma treatment could help to generate and attach more number of oxygen-containing hydrophilic groups on the surface of the hydrophobic nylon textile. That helped not only in the improvement of hydrophilicity but also to retain the imparted hydrophilicity.

Contrary to 60 s plasma-treated samples, in the 30 s plasma-treated sample, the rate of hydrophobic recovery in both He and He/O_2 plasma-treated samples was observed to be uniform through the ageing days. However, as expected the extent of hydrophobic recovery was much higher in the 30 s plasma-treated samples compared to the 60 s plasma-treated samples. Further, it must be noted that the He/O_2 plasma-treated sample showed lower recovery than the He plasma-treated sample. It was seen that after 167 days, there was almost 100% hydrophobic recovery in the He plasma-treated sample, whereas it was only 67% for He/O_2 plasma-treated sample. The highest recovery in contact angle was 103° after 71 days in the He plasma and 90° in the He/O_2 plasma-treated samples. The 15 s He plasma-treated sample showed almost 100% hydrophobic recovery within 15 days of storage, while the He/O_2 plasma-treated sample showed only 55% recovery. The highest recovery percentage of 82% was measured in the He/O_2 plasma-treated sample after 23 days of storage. After aging for 103 days, the water contact angle of 112° was observed in both He and He/O_2 plasma-treated samples. From these experiments, it was evident that a longer plasma treatment leads to a more stable (durable) hydrophilic modification. After storage for 55 days (ageing to time), the surface oxygen in the He/O_2 plasma-treated sample reduced to 25.5% from 29%, measured in as-plasma-treated sample. However, in the He plasma-treated sample, the oxygen percentage increased a little. However, in the aged plasma-treated samples, surface oxygen percentage was still more than the untreated samples. This indicated the imparted hydrophilic finish was quite durable.

Effect of heat treatment

The decay of improved hydrophilic functionality was measured after subjecting the 60 s plasma-treated samples to heat treatment at different temperature, such as 70°C, 90°C and 110°C for long durations (Samanta, 2010). Figure 5.25 shows the hydrophobic recovery behaviour of the 60 s plasma-treated samples subjected to heat treatment at 70°C. Up to 7 hours of heat treatment hardly any loss was observed in the imparted hydrophilicity. After 7 hours, a slow recovery in hydrophobicity was observed. Interestingly, even after 196 hours of treatment at 70°C, the hydrophobic recovery was only 16% and 30% in the He/O_2 and He plasma-treated samples, respectively. This indicated that re-distribution of the surface-oxygen-containing groups or migration from surface to bulk of such groups was quite slow at 70°C; therefore, the hydrophobic recovery percentage was low in the both the samples. This was because nylon (polyamide 6) has a glass transition temperature (Tg) of 45–50°C and melting temperature of 215°C. As the experimental temperature was only a marginal higher (20°C more) than the glass transition temperature of the polymer, the molecular segmental mobility was quite low; hence, the hydrophilic molecules neither migrated from surface to bulk of the material nor redistributed among themselves.

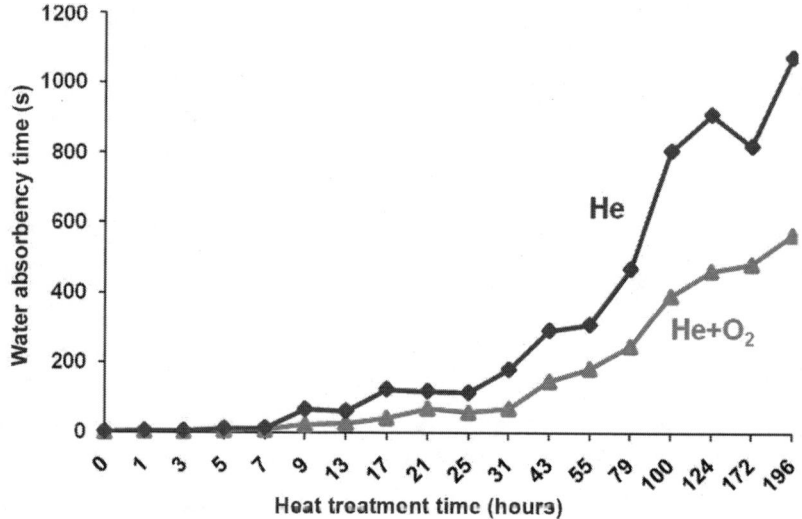

Figure 5.25 Water absorbent time of 60 s plasma-treated nylon upon heat treatment at 70°C (Samanta, 2010)

Similar result was also observed when the 60 s plasma-treated samples were subject to heat treatrment at 90°C. It was observed that heat treatment at 90°C for 196 hours, led to 18.1% and 34% recovery in hydrophobicity in the He/O_2 and He plasma-treated samples, respectively. On subjecting to heat-treatment for 43 hours at 110°C, the He/O_2 plasma and He plasma-treated samples showed a maximum hydrophobic recovery of 22.8% and 39%, respectively (Fig. 5.26). Thereafter, further increase in heat treatment duration did not show any substantial recovery in hydrophobicity. At 110°C temperature, the sample showed the maximum hydrophobic recovery within 43 hours compared to the 196 hours for the samples heat treated at 70°C and 90°C. This may be attributed to the higher mobility of the polymer molecules at 110°C, which might have helped in reorientation of the hydrophilic functional groups on the surface.

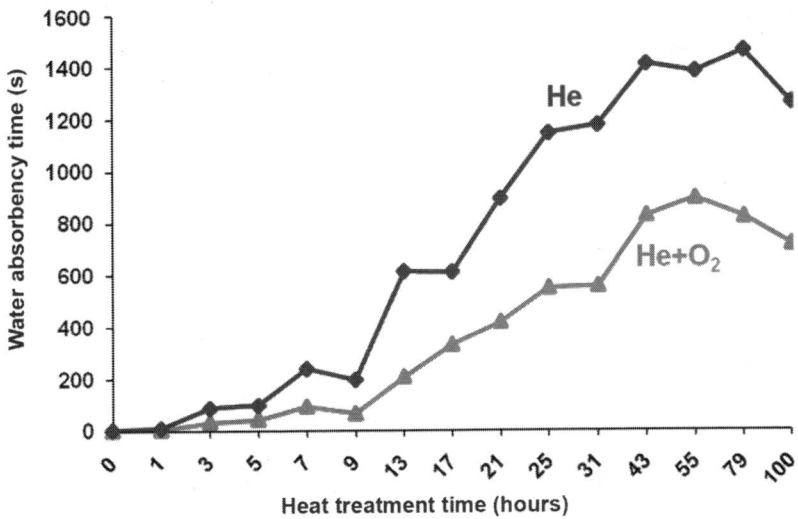

Figure 5.26 Water-absorbency time of 60 s plasma treated nylon upon heat treatment at 110 °C (Samanta, 2010)

(b) Polyester fabrics

Effect of heat treatment

As discussed earlier, the imparted hydrophilic functionality was not stable in the thermoplastic polymer/fibres, such as polyamide (nylon) and polyester (PET) due to the mobility of the polar molecules even at the room temperature. Therefore, similar to nylon, change in hydrophilicity with heat treatment for the polyester fabric was evaluated by keeping the sample at 60°C and 90°C for different periods of time. The hydrophobic recover in the plasma-treated samples were measured in similar manner as discussed for nylon. From Figure 5.27, it could be seen that due to thermal ageing, the water-drop disappearance time in the plasma-treated samples increased in all cases. In case of He plasma-treated sample, the rise in drop disappearance time was significantly faster as compared to He/O_2 plasma and maleic anhydride (MA) plasma-treated PET fabrics (Parida, 2010). This change in hydrophilicity could be explained by the mechanism of segmental mobility of polymer molecular chains. At high temperature, segmental mobility of molecular chains is higher; therefore, molecular chain segments can diffuse into the bulk of the polymer with time. Also the functional groups formed during the plasma treatment can move from surface to bulk of the polymer. This process of movement results in loss of developed surface hydrophilic functionality. It could be seen that He/O_2 plasma-treated samples showed better result, when the two samples were plasma treated at different voltage and frequency level. However, the MA padded followed by He plasma-treated sample showed much more stability of the imparted hydrophic functionality compared to He or He/O$_2$ plasma-treated samples. This is important, because though MA-treated samples showed less improvement in hydrophilic property compared to the He and He/O$_2$ plasma-treated samples, however, these samples could retain majority percentage of the imparted hydrophilic functionality when subjected to heat treatment at 60°C.

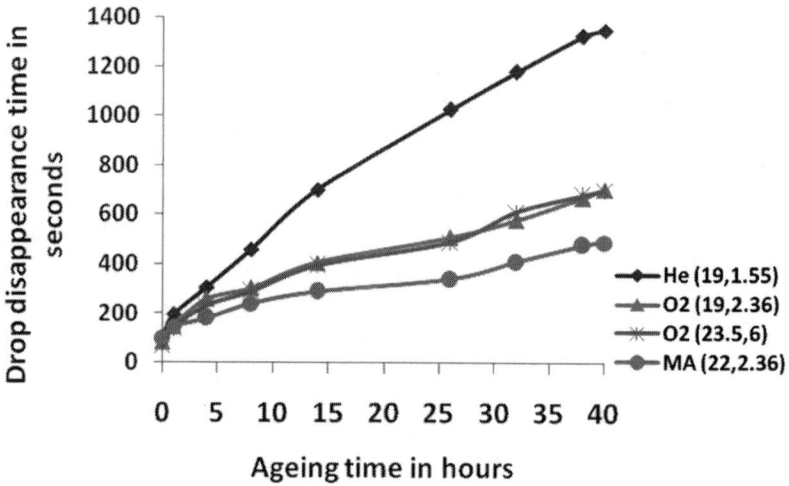

Figure 5.27 Change in drop disappearance time with ageing time at 60°C (values in the bracket are the plasma discharge frequency in kHz and voltage in kV) (Parida, 2010)

Similar trend in hydrophobic recovery was also observed when the samples were subjected to heat treatment at 90°C. However, the degree of hydrophobic recovery, as expected, was much more at 90°C compared to 60°C. The attachment of maleic anhydride groups (MA) during plasma treatment significantly slows down the reversal of hydrophilization of PET fabric. This may be attributed to the presence of bulky hydrophilic groups (possibly hydrolyzed MA) on the surface of PET fibre that is difficult to diffuse into the fibre bulk and thereby helps in retaining hydrophilicity for a longer duration during the ageing process.

Effect of washing

As discussed above, the imparted hydrophilic functionality is not stable to storage time and temperature in the thermoplastic textile, such as polyamide (nylon) and polyester (PET) due to their segmental mobility of the polymer chain molecules (Parida, 2010). Therefore, the hydrophobic to washing cycles was also studied. It can be seen from Fig. 5.28 that water-drop disappearance time of surface functionalized PET samples increased after successive washing cycles, which indicated destruction of functional groups due to mechanical abrasion and/or surface detachment during the washing of sample. The PET fabric modified by MA gave better results as compared to He and He/O$_2$ plasma-treated samples; similar phenomenon was also observed for hydrophobic recovery related to temperature effect. These results indicate that some amount of MA might have reacted with polyester fabric during the plasma treatment.

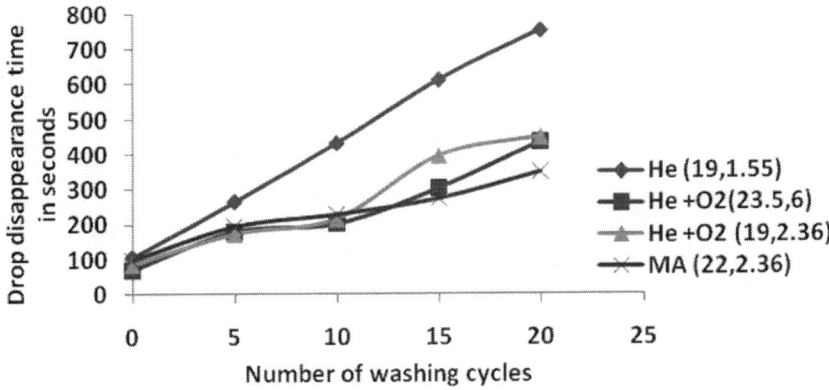

Figure 5.28 Comparison of water drop disappearance time of plasma-treated samples after different wash cycles (values in the bracket are the plasma discharge frequency in kHz and voltage in kV) (Parida, 2010)

In order to understand the better durability of imparted hydrophilicity in the maleic anhydride (MA) treated PET fabric, the concentrations –COOH groups were estimated for all samples. It was thought that maleic anhydride will react on PET surface and –COOH groups will be generated on the surface. Drop disappearance test result after plasma treatment of MA-padded fabric showed improvement in hydrophilicity due to formation of hydrophilic groups on the surface. Thermal ageing showed diffusion of this group was difficult as compared to other groups generated during plasma treatment. It was thought that groups generated after plasma treatment of MA-padded fabric might be –COOH group due to opening of maleic anhydride ring. Therefore, surface concentration of –COOH group after plasma treatment was determined. The polyester (PET) fabric of equal area and weight were taken, then fabric samples were dyed with TBO. The absorbed dye on the fabric was desorbed with 50% glacial acetic acid, and optical density (OD) of desorbed dye solution was measured at wavelength of 623 nm. Using the calibration curve of OD vs. dye concentration, the amount of dye present on fabric sample was calculated to estimate the number of –COOH groups present on fabric surface. To study the effect of plasma treatment time as well as hydrolysis time, MA-padded PET fabric samples were treated for different time and hydrolyzed for different duration to generate –COOH groups. It was seen (Fig. 5.29) that after plasma treatment of MA-padded PET fabric, surface concentration of –COOH groups increased significantly compared to all other cases. This clearly showed the role of MA in generating enough –COOH groups on PET surface. Interestingly, samples treated with He/O$_2$ plasma did not show higher –COOH groups than He plasma-treated samples, while the former samples showed better hydrophilicity than the later. This implied that the hydrophilic groups generated in He/O$_2$ plasma-treated samples were possibly of other types such as –OH than –COOH.

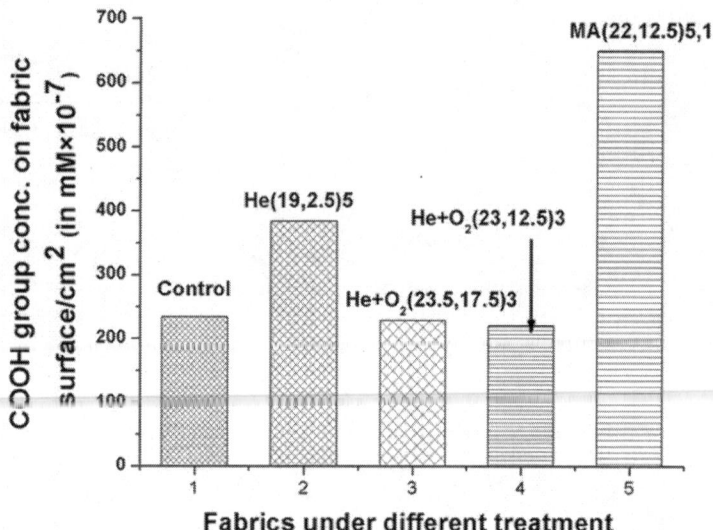

Figure 5.29 Surface concentration of –COOH groups on plasma-treated PET fabric (Parida, 2010)

It was seen that –COOH group concentration on surface increased with plasma treatment time, which indicated that by providing more time, probability of reaction between PET and maleic anhydride increased. Similarly, with increase in hydrolysis time, the more MA groups could convert to give higher concentration of –COOH groups. Increase in –COOH group concentration with hydrolysis (boiling in water) up to 5 min also indicated that functionalities developed by reacting MA on PET surface was wash stable to boiling water.

5.7   Breaking strength

As reported earlier, the nylon samples were subjected to atmospheric pressure glow plasma treatment, to evaluate the effect of glow plasma treatment on nylon a comparison in breaking strength between the untreated, 60 He and He/O$_2$ treated samples were drawn (Samanta, 2010). It was seen that untreated sample has breaking force of 181 gf; however, when the sample was subjected to 60 s He plasma treatment, the breaking force was 182.5 gf, the variation was within the CV% of the measurement. This indicated that there was no change in the breaking strength of the plasma-treated sample compared to untreated sample. Similar results were also observed for the He/O$_2$ plasma-treated sample (178.2 gf). As the samples had undergone glow plasma treatment, severe surface etching was not observed in SEM. Similar to breaking strength, there was also no change in the breaking extension of the plasma-treated samples compared to untreated sample. The results indicated that atmospheric pressure glow-type cold plasma can modify the surface chemistry of nylon fabric significantly to impart excellent hydrophilic functionality without adversely affecting the mechanical properties of the textile. Similar effect of atmospheric pressure plasma treatment for 5 min on breaking load of PET fabric was measured (Parida, 2010). Though plasma treatment is a surface phenomenon; however, a long duration of treatment time may change the mechanical properties adversely, if the sample is not treated in glow plasma. Therefore, the strength of warp and weft yarns of plasma-treated fabrics was determined. It was observed that the He, He/O$_2$, and MA padded followed by He plasma-treated samples were very stable to the plasma treatment of 5 minutes. The strength test results showed that He/O$_2$ plasma-treated samples showed the highest drop in tensile strength for warp yarns. Similar result was also observed for weft yarns. The drop in tensile strength was 8.2% and 9.0% for warp and weft direction, respectively for these samples. However, the decrease in tensile strength in He plasma treated and MA padded followed by He plasma-treated samples were slight less compared to He/O plasma-treated samples.

5.8   Summary

Wet chemical processing of textile is important as it impart highest aesthetic and functional values. However, in wet chemical processing of textile large amount of water is used as a processing medium, which is finally discharged as an effluent, contaminated with unused dyes, pigments, acid, alkali and others chemicals. Plasma, an ionized gas, can be used for nano-scale surface engineering of textile in dry state. It also helps in reducing the processing time, consumption of chemical and production cost. Mostly, low temperature plasma either at low pressure or atmospheric pressure has been utilized for activation of textile substrates for improvement in their hydrophilic properties, colouration, adhesion strength and antistatic properties. Glow plasmas produced using different precursors show different type of colours based on the emitted wavelength by the excited photons. From the OES plasma monitoring spectra, it was also observed that the intensity (concentration) of a plasma species remains constant with plasma discharge time (real time in-situ analysis). Among the various plasma process parameters, effect of voltage was found to have more profound effect on fragmentation of a precursor than the effect of frequency.

It was possible to completely remove polyvinyl alcohol (PVA) size from the surface of the cotton fibre by air/He and air/O_2/He plasma treatment. Low pressure argon and oxygen plasma treatments can be used to modify the surface of jute fibres. Similar results have also been reported when the other natural fibres such as abaca, flax, hemp and sisal or synthetic fibre such as polypropylene were plasma treated in the presence of reactive and non-reactive plasma. It was observed that plasma treatment in the presence of pure helium (He), argon (Ar), air, nitrogen and oxygen or mixture of He with these gases could improve liquid absorbency of natural and synthetic textiles. It was observed that a water droplet of 37 µl, which took 540 s to spread over an area of 3.79 cm^2 in the untreated nylon sample, took only 1.1 s in the 60 s He-plasma treated nylon samples. On the other hand, water droplet, which took 700 s to spread over the same area in the untreated PET fabric, took only 6.7 s in the 60 s He-plasma-treated sample. The water contact angle reduced to 0° compared to the untreated sample, where it was 130°. Plasma treatment not only improves the water absorbency, but also improves the oil absorbency. The time required for spreading of an oil droplet over a fixed area in nylon fabric decreased from 152 s for the untreated fabric to 52 s for the 60 s He-plasma treated. Similar results were also observed for the cotton and polyester samples. This improvement of both hydrophilic liquid (water) and hydrophobic liquid (oil) absorbency in the plasma-treated textiles was mainly due to the formation of nano-sized channels. Due to the improvement in hydrophilic characteristic in the nylon and polyester samples along with increase in their surface area (7.3–18.3%) on plasma treatment, the samples showed good antistatic property. It was observed that the 60 s He-plasma-treated sample produced less static charge of 2.12 kV compared to 2.76 kV produced by the untreated samples, a net reduction of 23%. Interestingly, the ½ decay time also decreased exponentially with plasma treatment time from 8.9 s for the untreated sample to only 1.1 s for the 60 s He-plasma-treated sample, amounting to a reduction of ~88%.

Improvement in textile colouration of cotton, wool, silk, nylon, and polyester has been reported using various dye classes such as reactive, acid and disperse by modifying the surface of the textile using plasma. Both simple gases (He, Ar, N_2, air, O_2) and polymerizable reactive monomers have been used for hydrophilic finishing of textile. The effect of plasma treatment on dyeing of wool, sodium metabisulphite and acidic pH did not appear to have any effect on the colour yield of reactive dye on wool. The K/S value of the plasma treated wool at neutral pH, without addition of urea and sodium metabisulphite, was almost double (4.26) as compared to that of the untreated wool dyed under similar conditions. It was interesting to observe that dye concentration on the surface of the fibre was more compared to the bulk of the fibre as plasma treatment could only modify the surface of the fibre.

After plasma treatment, the improvement in hydrophilic functionality was mainly due to formation of more hydrophilic groups. This was confirmed by ATR-FTIR, XPS and SIMS analysis in terms of hydrophilic to hydrophobic molecular ratio or atomic percentage. In the various textile substrates, with increasing plasma treatment time, hydrophilicity was found to improve linearly. In the polyester sample, this improvement was better at lower frequency level (19 Hz) as compared to higher frequency level (22 Hz). The hydrophobic recovery due to the re-distribution of surface-oxygen-containing polar groups and/or migration of such groups from the surface to bulk of the material with time, temperature and washing cycles was studied for nylon and polyester samples. The 60 s He/O_2 plasma-treated sample remained stable with hydrophilic property up to 75 days, after that a very slow hydrophobic recovery was observed. After 95 days the hydrophobic recovery was only 3% and 22.3% in the He/O_2 and He plasma-treated samples, respectively. Similar results were also observed even after 6 months. Slow hydrophobic recovery was also observed when the samples were subjected to heat treatment at 70°C, 90°C and 110°C. It was seen that plasma treatment in the presence of He/O_2 could impart better hydrophilic functionality compared to He or MA padded followed by He plasma-treated samples. However, the MA padded followed by He plasma-treated sample showed better hydrophilic durability as compared to He or He/O_2 plasma-treated samples. Plasma modifies the surface of textile at nano-meter level; hence, bulk physical and chemical properties remain unaltered. Therefore, plasma could be used for surface engineering of textile based on the end application requirement without altering the bulk properties. Plasma processing of textile is carried out in dry state; hence, adoption of such technology would help to develop superior quality products at lower cost, while addressing the environmental issues.

5.9   References

Bajaj P. and Sengupta A. K. (1992). 'Protective clothing', *Textile progress,* **22**(2–4), pp. 1–110.

Banchero M., Sicardi S., Ferri A., and Manna L. (2008). 'Supercritical Dyeing of Textiles - From the Laboratory Apparatus to the Pilot Plant', *Textile Res. J.,* **78**(3), pp. 217–223.

Bhat N.V., Netravali A.N., Gore A.V., Sathianarayanan M.P., Arolkar G.A., and Deshmukh R.R. (2011). 'Surface modification of cotton fabrics using plasma technology', *Textile Res. J.,* **81**(10), pp. 1014–1026.

Bhat N.V. and Benjamin Y.N. (1999). 'Surface Resistivity Behavior of Plasma Treated and Plasma Grafted Cotton and Polyester Fabrics', *Textile Res. J.,* **69**(1), pp. 38–42.

Bird C.L. (1968). 'The theory and practice of wool dyeing', *Society of Dyers and Colourists* (3rd Eds), Bradford Yorkshire England, 15.

Cai Z., Qui Y., Zhang C., Hwang Y.J., and McCord M. (2003). 'Effect of Atmospheric Plasma Treatment on Desizing of PVA on Cotton' *Textile Res. J.,* **73**(8), pp. 670–674.

Cernák M., Šimor M., Ráhel J., Kovácik D., Záhoranová A., and Mazúr M. (2002). 'Atmospheric-pressure plasma treatment of nonwovens using surface dielectric barrier discharges' *12th TANDEC International Nonwovens Conference, Knoxville, Tennessee, USA,* pp. 1–15.

Chappell P.J.C., Brown J.R., George G.A. and Willis H.A. (1991). 'Surface modification of extended chain polyethylene fibres to improve adhesion to epoxy and unsaturated polyester resins' *Surface and Interface Analysis,* **17**(3), pp. 143–150.

Chen X., Kornev K.G., Kamath Y.K., and Neimark A. (2001). 'The Wicking Kinetics of Liquid Droplets into Yarns', *Textile Res. J.,* **71**(10), pp. 862–869.

Chen F.F. (1995). 'Industrial Applications of Low Temperature Plasma physics', *Physics of Plasmas,* **2**(6), pp. 2164–2175.

Cireli A., Kutlu B. and Mutlu M. (2007). 'Surface modification of polyester and polyamide fabrics by low frequency plasma polymerization of acrylic acid' *J. App. Poly. Sci,* **104**(4), pp. 2318–2322.

Cullough Mc, Francis P., Hall J., and David M. (1989). U. S. Patent 4869951.

Deshmukh R.R. and Bhat N.V. (2003). 'The mechanism of adhesion and printability of plasma processed PET films, *Materials Research Innovations,* **7**(5), pp. 283–290.

Friedman L.A., Faulkner J.D., and King A.D. Jr (1979). U. S. Patent 4152288.

Goto T., Wakita T., Nakanishi T., and Ohta Y. (1992). 'Application of low temperature plasma treatment to the scouring of grey cotton fabric', *Journal of the Society of Fibre Science and Technology Japan,* **48**(3), pp. 133–137.

Guglani R. (2002). 'Recent developments in textile dyeing techniques', *Fibre2Fashion, www.fibre2fashion.com/industry-article/pdffiles/12/1171.pdf.*

Guo L., Campagne C., Perwuelz A., and Leroux L. (2009). 'Zeta Potential and Surface Physico-chemical Properties of Atmospheric Air-plasma-treated Polyester Fabrics', *Textile Res. J.,* **79**(15), pp. 1371–1377.

Hegemann D. (2006). 'Plasma polymerization and its application in textiles', *Indian J Fibre Text Res.,* **31**(1), pp. 99–115.

Hegemann D., Hossain M. M., and Balazs D. J. (2007). Nanostructured plasma coatings to obtain multifunctional textile surfaces. Prog Organic Coat **58**, 237–240.

Herbert P.A.F., O'Neill L., Jaroszy'nska-Woli'nska J., Stallard C., Ramamoorthy A., and Dowling D.P. (2011). 'A Comparison between Gas & Atomised Liquid Precursor States in the Deposition of Functional Coatings by Pin Corona Plasma', *Plasma Process. Polym.,* **8**(3), pp. 230–238.

Hocker H. (2002). 'Plasma Treatment of Textile Fibers', *Pure and Applied Chemistry,* **74**(3), pp. 423–427.

Holme I. (2000). 'Challenge and Change in wool dyeing and finishing', 10th International Wool Textile Research Conference, Aachen, KNL-9, 1.

Holme L. and Mc Intyre J.E. (1998). "Electrostatic charging of textile", Ed J. M. Layton, *Textile progress,* **28**(1), pp. 1–68.

Hossain M.M., Herrmann A.S., and Hegemann D. (2006). 'Plasma Hydrophilization Effect on Different Textile Structures', *Plasma Process. Polym.* **3**(3), pp. 299–307.

Jahagirdar C.J. and Tiwari L.B. (2004). "Study of plasma polymerization of dichloromethane on cotton and polyester fabrics", *J. App. Poly. Sci,* **94**(5), pp. 2014–2021.

Jahagirdar C.J. and Tiwari L.B. (2007). 'Plasma treatment of polyester fabric to impart the water repellency property', *Pramana,* **68**(4), pp. 623–630.

Jimenez A.B., Bistritz B., Schulz E., and Bismarck A. (2008). 'Atmospheric Air pressure plasma treatment of lignocellulosic fibres: Impact on mechanical properties and adhesion to cellulose acetate butyrate', *Composites Science and Technology,* **68**(1), pp. 215–227.

Joshi V.K. (1996). 'Antistatic Fibers and Fabrics', *Manmade Textile in India,* **7**, pp. 245–251.

Kale K. and Palaskar S. (2010). Studies on atmospheric pressure plasma treatment of polyester/cotton blended fabric', Paper presented in 51st Joint Technological Conference of ATIRA, BTRA, SITRA & NITRA (NITRA, Ghaziabad), 29th June 2010.

Kalia S., Kaith B.S., and Kaur I. (2009). 'Pretreatments of natural fibers and their application as reinforcing material in polymer composites—A review', *Polymer Engineering and Science,* **49**(7), pp. 1253–1272.

Kang E.T. and Neoh K.G. (2009). 'Surface Modification of Polymers', *Encyclopedia of Polymer Science and Technology,* Wiley InterScience, N.Y.

Karmakar S.R. (1999). *'Chemical Technology in the pre-treatment processes of textiles',* *Textile Science and Technology,* **12,** Elsevier.

Krump H., Simor M., Hudec I., Jasso M., and Luyt A.S. (2005). 'Adhesion strength study between plasma treated polyester fibres and a rubber matrix', *App. Surf. Sci.,* **240,** pp. 268–274.

Labay C., Canal J.M., and Canal C. (2012). 'Relevance of Surface Modification of Polyamide 6.6 Fibers by Air Plasma Treatment on the Release of Caffeine', Plasma Process. Polym., 9, pp. 165–173.

Latschar, C.E.L. (1969), U. S. Patent 3428481.

Leroux F., Campagne C., Perwuelz A., and Gengembre L. (2009). 'Atmospheric air plasma treatment of polyester textile materials, textile structure influence on surface oxidation and silicon resin adhesion, *Surf. Coat. Technol.*, **203**(20–21), pp. 3178–3183.

Malek R.M.A. and Holme I. (2004). 1st International conference of Textile Research Division, NRC, Cairo, Egypt.

Morent R. (2007). 'Adhesion enhancement by a dielectric barrier discharge of PDMS used for flexible and stretchable electronics", *J. Phys. D: Appl. Phys.*, **40**, pp. 7392–7401.

Muthukumar M., Sargunamani D., Selvakumar N., and Rao J.V. (2004). 'Optimisation of ozone treatment for colour and COD removal of acid dye effluent using central composite design experiment', *Dyes and Pigments*, **63**, pp. 127–134.

Naebe M., Cookson P.G., Rippon J., Brady R.P., Wang X., Brack N., and Van R. (2010). 'Effects of Plasma Treatment of Wool on the Uptake of Sulfonated Dyes with Different Hydrophobic Properties', *Textile Res. J.*, **80**(4), pp. 312–324.

Nasadil P. and Benešovsky P. (2008). 'Plasma in Textile Treatment', II Central European Symposium on Plasma Chemistry, *Chemicke Listy*, **102**, pp. 1486–1489.

Nobuyuki Z., Hiroto I., and Kazuya Y. (2008). 'Plasma-chemical surface functionalization of flexible substrates at atmospheric pressure', *Thin Solid Films*, **516**(19), pp. 6683–6687.

Nourbakhsh S. and Ebrahimi I. (2012). 'Different Surface Modification of Poly (Ethylene Terephthalate) and Polyamide 66 Fibers by Atmospheric Air Plasma Discharge and Laser Treatment-Surface Morphology and Soil Release Behavior', *Journal of Textile Science and Engineering*, **2**(2), p. 109.

Özdogan E., Saber R., Ayhan H., and Seventekin N. (2002). 'A new approach for dyeability of cotton fabrics by different plasma polymerization methods', *Coloration Technology*, **118**(3), pp. 100–103.

Panda P.K., Rastogi D., Jassal M., and Agrawal A.K. (2012). Effect of atmospheric pressure helium plasma on felting and Low temperature dyeing of wool, *J. App. Poly. Sci.*, **124**(5), pp. 4289–4297.

Parida D. (2010). 'Functionalization of textiles using atmospheric pressure plasma reaction technology', *M.Tech. thesis,* Indian Institute of Technology (IIT)-Delhi, India.

Park D.J., Lee M.H., Yeon I.W., Han D.W., Choi J.B., Kim J.K., Hyun S.O., Chung K., and Park J. (2008). 'Sterilization of microorganisms in silk fabrics by microwave-induced argon plasma treatment at atmospheric pressure', *Surf. Coat. Technol.*, **202**, pp. 5773–5778.

Píchal J. and Klenko Y. (2009). 'ADBD plasma surface treatment of PES fabric sheets', Plasma Physics and Technology Topical issue: 23rd Symposium on Plasma Physics and Technology, *European Phyisical Journal D*, **54**(2), pp. 271–280.

Qiuran J., Ranxing L., Jie S., Chunxia W., Shujing P., Feng J., Lan Y., and Yiping Q. (2009). 'Influence of ethanol pretreatment on effectiveness of atmospheric pressure plasma treatment of polyethylene fibers', *Surf. Coat. Technol.* **203**(12), pp. 1604–1608.

Ren Z.F., Tang X.L., Chen X.L., and Qiu G. (2007). Dyeing behavior of atmospheric dielectric barrier discharge Ar-O2 plasma treated poly (ethylene terephthalate) fabric', *Pulsed Power Conference, 16th IEEE International*, **2**, pp. 1399–1402.

Riccobono R. (1973). *Textile Chemist and Colorist.*, **5**, p. 219.

Samanta K.K., Jassal M., and Agrawal A.K. (2008). 'Formation of nano-sized channels on polymeric substrates using atmospheric pressure glow discharge cold plasma', *Nanotrends: A journal of Nanotechnology and its Application,* **4**(1), pp. 71–75.

Samanta K.K., Jassal M., and Agrawal A.K. (2009). 'Improvement in water and oil absorbency of textile substrate by atmospheric pressure cold plasma treatment', Surf. Coat. Technol., **203**, pp. 1336–1342.

Samanta K.K., Jassal M., and Agrawal A.K. (2006). 'Atmospheric pressure glow discharge plasma and its applications in textile', *Indian J Fibre Text Res.,* **31**(1), pp. 83–98.

Samanta K.K., Jassal M., and Agrawal A.K. (2010). 'Antistatic effect of atmospheric pressure glow discharge cold plasma on textile substrates', Fibers and polymers, **11**(3), pp. 431–437.

Samanta K.K. (2010). 'Surface functionalization of textile substrates using atmospheric pressure glow plasma', PhD thesis, Indian Institute of Technology (IIT)-Delhi, India.

Sarmadi A.M., Ying T.H., and Denes F. (1993). 'Surface modification of polypropylene fabrics by acrylonitrile cold plasma', *Textile Res. J.,* **63**(12), pp. 697–705.

Seki Y., Sever K., Sarikanat M., Gülec H.A., and Tavman I.H. (2009). 'The influence of Oxygen Plasma treatment of Jute fibers on mechanical properties of jute fiber reinforced thermoplastic composites', *5th International Advanced Technologies Symposium (IATS'09),* May 13–15, Karabuk, Turkey.

Shin Y., Son K., Yoo D., Hudson S., McCord M., Matthews S., and Whang Y.J. (2006). 'Functional finishing of nonwoven fabrics. I. Accessibility of surface modified PET spunbond by atmospheric pressure He/O2 plasma treatment', *J. App. Poly. Sci,* **100**(6), pp. 4306–4310.

Souto A.P., Oliveira F.R., Fernandes M., and Carneiro, N. (2012). 'Influence of DBD Plasma Modification in the Dyeing Process of Polyamide', *Journal of Textiles and Engineer,* **19**(85), pp. 20–26.

Sparavigna A. (2008). 'Plasma treatment advantages for textiles', arxiv.org/pdf/0801.3727.

Sung-Spitzl H. (2003). 'Plasma Pre-treatment of Textiles for Improvement of Dyeing Processes', *International Dyer,* **188**(5), p. 20.

Topala I., Dumitrascu N., and Popa G. (2008). 'Properties of Acrylic acid polymers by APP polymerization',. Nucl. Instr. And Meth. In Phys. Res. B., **10**.

Tsriskina A.L., Guschchina J.N., Gorberg B.L., and Ivanov A.A. (1991). *Referativnyi Zhurnal. Kotlostroenie,* July 42 B, 7.

Vaideki K., Jayakumar S., and Rajendran R. (2009). 'Investigations on the enhancement of antimicrobial activity of neem leaf extract treated cotton fabric using DC air and oxygen plasma', Plasma Chem Plasma Process. **29**(6), pp. 515–534.

Vaideki K., Jayakumar S., Rajendran R., and Thilagavathi G. (2008). 'Investigation on the effect of RF air plasma and neem leaf extract on the surface modification and antimicrobial activity of cotton fabric', *App. Surf. Sci.,* **254**(8), pp. 2472–2478.

Vesel A. and Mozetic (2009). 'Surface Functionalization of Organic Materials by Weakly Ionized Highly Dissociated Oxygen Plasma', 2nd International Workshop on

Non-equilibrium Processes in Plasmas and Environmental Science, *Journal of Physics: Conference Series* 162: 012015.

Wakida T., Tokino S., Niu S., Kawamura H., Sato Y., Lee K., Uchiyama H., and Inagaki H. (1993). 'Characterization of Wool and Poly(ethylene terephtalate) Fabrics and Film Treated with Low Temperature Plasma under Atmospheric Pressure,' *Textile Res. J.,* 63, 433-438.

Whitacre J.R. and Bulloff J.J. (1964). U. S. Patent 3129487.

Yasuda T. (1982). *Mukogawa Joshi Daigaku Kiyo Shokumotsu-Hen* 30, A9.

Zhang C. and Fang K. (2009). 'Surface modification of polyester fabrics for inkjet printing with atmospheric-pressure air/Ar plasma', *Surf. Coat. Technol.,* 203, pp. 2058–2063.

Zubaidi A. and Hirotsu T. (1996). 'Graft Polymerization of hydrophilic monomers onto Textile fibers treated by Glow Discharge plasma', *J. App. Poly. Sci,* **61**, pp. 1579–1584.

6

Plasma processing of textiles to enhance their dyeing and surface properties

N.V. Bhat and R.R. Deshmukh

Abstract: This chapter reviews the advantages of plasma technology used for processing of textile materials in details. The plasma process for textile treatment falls under the category of "green technology" and scores over the existing wet processing used at present. Plasma processing can be used to modify surface by etching, functionalization, plasma coating (grafting) and desizing. The modified surfaces can be made either hydrophilic or hydrophobic depends on the selection of appropriate gas or precursors to form plasma as well as plasma reactor conditions. It has been found that the dye uptake can be enhanced when fabrics are pre-treated with plasma on account of formation of hydroxyl, carboxyl and carbonyl groups on the surface.

Key words: Plasma processing, textile fabrics, dyeing, hydrophobicity, hydrophilicity, antibacterial

6.1   Introduction

Although 99% of the entire universe exists in plasma state, i.e. ionised form of gas; its existence on the earth is less than 1%. However, the lightening in the sky and aurora borealis, the natural plasmas on the earth, did catch the imagination of human being. Systematic efforts to understand the phenomenon and produce it in the laboratory started since 1879 by Crook, Irving Langmuir and others have expanded the knowledge and applications in several areas. Today plasma encompasses several disciplines such as astrophysics, fusion, lasers, electronics, fluorescent lamps, plasma TV, etc.

Plasma consists of electrons, ions, excited atoms/molecules, neutrals and radiation. When electric field is applied, ionisation of gaseous atoms occurs – fragmentation takes place – and the energy is lost to the surroundings due to collisions and radiation process (Hollahan and Bell, 1984). In order to sustain the plasma states, therefore, energy has to be supplied continuously. Further we have to understand the density of states, electron/ion temperature, etc. A low pressure plasma chamber used in the present work (at Department of Physics, ICT, Mumbai) having glow discharge is shown in Fig. 6.1.

Figure 6.1 A low pressure plasma reactor which was used in the present work.

6.2   Classification of plasmas

Classification of plasmas can be done on the basis of temperature, pressure and the source used. When the temperature of entire gas is high, it is called as "hot plasma". Due to incessant collisions, the temperature of electrons and ions/molecules gets raised and the whole system is in local thermal equilibrium. Such plasma is used for cutting, welding, fusion, etc. On the other hand "cold plasma" is in non-equilibrium state, and although the temperature of electrons is high, the system remains cold (Roth, 2001a). Such cold plasmas are used in neon signs, fluorescent lamp, etc. Such low temperature plasma can be used in applications of textiles and plastic industries.

Plasma can also be classified on the basis of pressure of the system. When a system is evacuated the gaseous atoms at low pressure can be ionised easily and uniform plasma can be created. However creating and maintaining vacuum conditions for large volumes possess some problems, particularly in textile applications. For continuous process and high speeds, therefore, working at atmospheric pressure is advantageous. Atmospheric pressure plasma can be classified into (a) corona discharge, (b) dielectric barrier discharge (DBD) and (c) atmospheric pressure glow discharge (APGD) depending on the method of producing it.

Further, electric field may be applied in the form of direct current (DC), pulsed DC, low frequency (<100 kHz), radio frequency (RF 2–100 MHz) and microwave frequency (2.45 GHz) to produce the discharge.

Typical electrode configurations for corona and DBD plasmas are shown in Fig. 6.2(a) and (b). In DBD sufficiently high AC voltages are applied to the electrodes (~8 kV, 500 Hz to 500 kHz). DBD cannot be generated by DC voltages due to the requirement of capacitive coupling which needs AC voltages. As electrodes are covered with dielectric layer (glass, ceramics or polymers), the etching and corrosion of electrode is avoided. In DBD a large number of current filaments (micro discharges) occur. The micro discharge filaments in DBD are of very short durations ~ in nanoseconds (Herbert, 2009). APGD is generated by applying relatively low voltages across symmetrical electrodes at high frequency (MHz). Electrodes are bare which gives higher electron densities (~10^{12} cm^{-3}). Plasma is homogenous across the electrodes and gives one current pulse for every half cycle. Helium is the preferred gas in APGD as well as DBD (Roth, 2001b; Alexandrov, 2009). A typical atmospheric pressure plasma reactor operational at Bombay Textile Research Association (BTRA) is shown in Fig. 6.3.

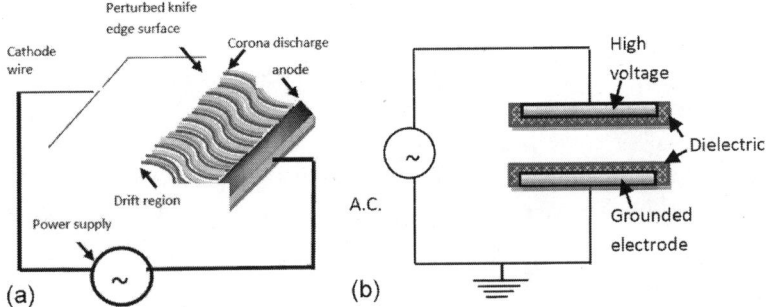

Figure 6.2 Typical electrode configuration for producing
(a) corona, and (b) DBD plasmas

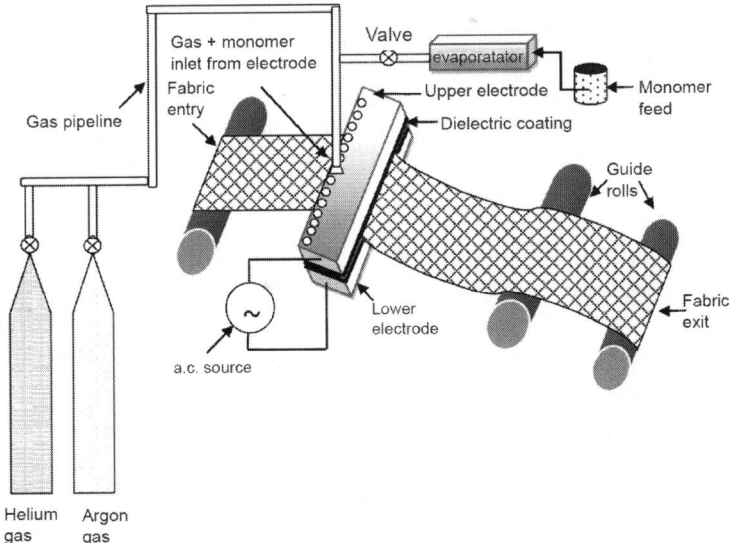

Figure 6.3 Atmospheric pressure plasma reactor operative at BTRA
(Make: Grinp, Italy)

6.3   Plasma interactions with substrate

In order to use plasma for any applications, in particular for textiles, it is essential to understand the interactions of plasma with the substrate. Bombardment of ions, electrons and high energy neutrals take place on the surface and interactions are very complex. This results in the alteration of surface chemistry and topography. In case of non-polymerizing/polymerizing gases, four different types of interactions take place (Yasuda, 1985; Clark, 1978), which are as follows:

6.3.1   Surface cleaning (removal of contaminants)

The first step of plasma processing is cleaning of surfaces. It is believed that the interaction of ions, electrons, and energetic species of neutral atoms causes rapid removal of low molecular weight contaminants such as oligomers, additives, loosely bound material, some residuals, ubiquitous contaminants, processing aids, grease, dirt, and adsorbed species, which is also called as "plasma cleaning."

6.3.2   Ablation / Etching

After the cleaning of the surface is over, ablation or etching of surface begins. Ablation is the physical removal of material from the surface by the action of plasma. Ablation of materials by plasma can occur by means of two principal processes: one is physical sputtering and the other is chemical etching. The sputtering of materials by chemically non-reactive plasma, such as argon (Ar) gas plasma, is a typical example of physical sputtering, which is essentially a momentum-exchange process. The energy of the impinging argon ion (Ar+) is transferred to the colliding atom and the atom is knocked out from the structure and goes into the vapour phase. Chemical etching occurs in chemically reactive types of gaseous plasma. It involves chemical reaction of impinging species with the substrate such as oxidation by O_2 plasma. When the polymer/textile surface is subjected to plasma treatment with reactive or non-reactive gases as mentioned above, an etching or scrapping of material is observed and this etched material is then vaporized or pumped out of the plasma chamber as effluent gas. The second possibility is that the ablated material may be re-deposited as plasma polymer at different locations. Ablation can be monitored and reported in terms of weight loss of the substrate. Chain scission, rupturing of bonds, removal of loosely bound (mostly amorphous) material take place during ablation. It has been confirmed through various studies that only the amorphous portion gets degraded and etched away in the initial stage of etching (Bhat et al., 2002; Thomas et al., 1998; Yoon et al., 1996). On the other hand, the crystalline regions are more compact and hard as compared to amorphous regions, and therefore the amorphous region gets removed easily in the plasma etching. As a result, the percentage crystallinity relatively increases to some extent. This has been concluded by previous studies using X-ray diffraction (XRD) and Fourier transform infrared (FTIR) for silk (Bhat and Nadiger, 1978). Pandiyaraj and Selvarajan (2008) have also observed that the effect of plasma on the amorphous zone is more predominant than the crystalline zone of the cotton fabrics. In plasma etching, formation of free radicals and increase in effective surface area take place. It enhances certain properties such as wettability, dyeability, etc., of the textile materials.

6.3.3   Plasma polymerization

Many reactions occur simultaneously during plasma polymerization. There are two simultaneous processes that occur, i.e. ablation, which leads to the removal of material from the surface and polymer formation, which results in the deposition. The interaction of these two opposite processes and their co-existence in plasma had been known for a long. The overall scheme of plasma polymerization encompasses the principle of competitive ablation and polymerization (CAP) mechanism described in detail by Yasuda (Yasuda and Hsu, 1978). Schematic representation of competitive ablation and polymerization, chain scission, free radical formation is shown in Fig. 6.4.

Plasma ablation competes with the polymer formation (deposition) in almost all the cases when plasma is used to treat the surfaces of solid materials (Shijian et al., 2002). The rate of deposition on substrate depends on several parameters (Hollahan and Bell, 1984). Since etching and deposition are simultaneous processes, it is very difficult to determine actual deposition rate on polymer substrate. Deshmukh et al. (2007a) studied the percentage weight change of plasma-polymerized tetraethylorthosilicate (PPTEOS) and plasma-polymerized hexamethyldisiloxane (PPHMDS) deposited on PE films for various durations of time. They have observed that the percentage weight rises with increase in processing time. Though the etching and deposition occur simultaneously, deposition rate is much faster than that of etching and results in weight gain.

Plasma chemical reactions can be categorised in three groups: (i) chemically non-reactive plasma, (ii) chemically reactive plasma and (iii) polymer forming plasma. The first two types are explained in the ablation section as they are responsible for physical sputtering and chemical etching. Polymer-forming plasma can be generated using various organic or inorganic monomer vapours capable of polymerization. Polymer-forming plasma is obviously chemically reactive and forms a polymeric solid deposit by itself in the environment of glow discharge. Deposition occurs when the impinging species fails to bounce back from the colliding surface resulting in loss of kinetic energy and formation of chemical bond with the substrate. In low pressure plasmas, there are several factors affecting the deposition rate such as type of monomer, monomer flow rate, flow rate of carrier gas if it is used, distance between electrodes, power, working pressure, position of the substrate in the plasma chamber etc. The structure of plasma-deposited films is highly complex and is influenced by some process parameters such as source frequency, power, substrate temperature, monomer structure, working pressure, monomer and carrier gas (if it is used) flow rate, etc. Two types of polymerization reactions can occur simultaneously, namely

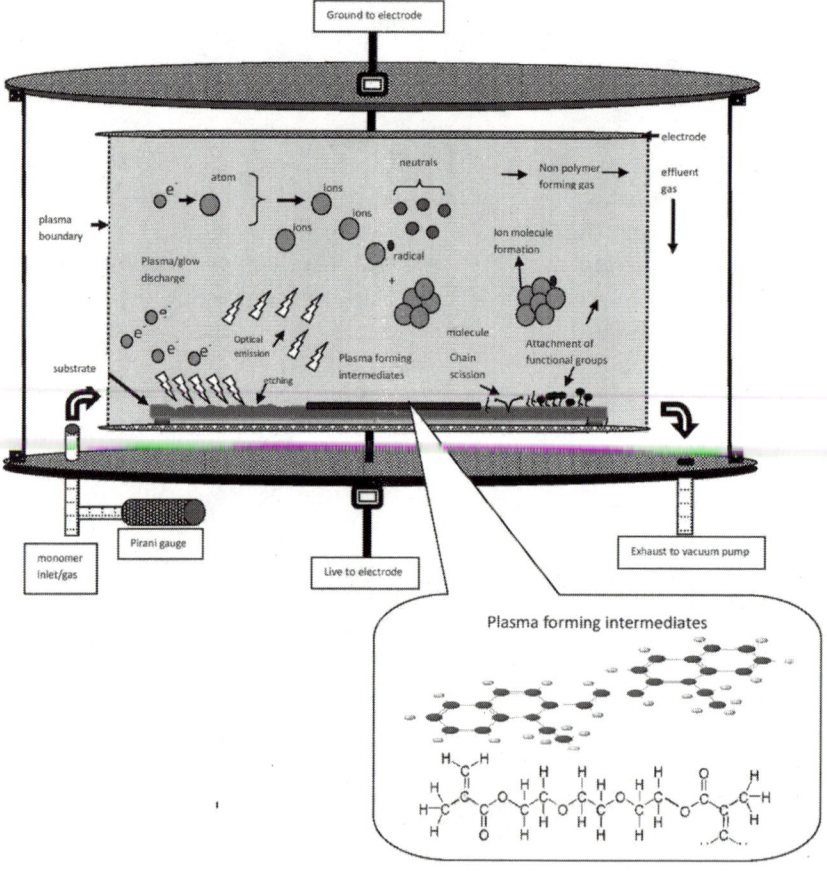

Figure 6.4 Schematic of competitive ablation and polymerization process in plasma state

plasma-induced polymerization and plasma-state polymerization. The main route of polymer formation occurs under the influence of plasma, i.e. via plasma-state polymerization. Plasma-induced polymerization represents the polymerization of condensable monomer, e.g. vinyl-type monomer, on the surface by its interaction with plasma.

6.4   Plasma processing of textiles

The textile industry is considered as one of the most polluting industries because of the various processes involved such. as singeing, desizing, scouring, bleaching, mercerisation, dyeing, etc. These processes pollute the environment and water resources. It uses variety of chemicals for different textile wet processes. In all textile wet processes, 100% utilization of chemicals is not possible, so all excess and unutilized chemicals are directly drained out causing environmental pollution. Now-a-days, due to the increase in environmental awareness, governments of many countries have restricted the production of many textile industries due to toxic effluent discharge. Hence, there is a need to replace polluting textile wet processes by eco-friendly ones. Owing to the effect of pollution caused by various chemical treatments, plasma treatments have been introduced recently as they are capable of achieving many objectives of textile wet processing. Since 1960s, scientists have successfully exploited plasma techniques in materials science and engineering. The plasma technologies have also been fully utilized to improve the surface properties of fibres in many applications. The fibres that can be modified by plasma include almost all kinds such as textile fibres, metallic fibres, glass fibres, carbon fibres, fabrics and other organic fibres (Kan and Yuen, 2007). The review article on "plasma treatment on textile fibres" by Hocker (2002) has covered usefulness of plasma processing for textile materials. Usefulness of pre-treatment of textile by plasma prior to dyeing has been covered by Deshmukh and Bhat (2012). A review article by Mehta (2010) highlights the advantages of this technique over conventional wet processing. The book edited by Shishoo (2007) highlights the huge potential of plasma treatment for textile processing. An overview of the literature on potential uses of non-thermal plasmas for the modification of textile products and benefit from the plasma treatment is given by Morent et al. (2008). Kale and Desai (2011a) have reviewed atmospheric pressure plasma technology for surface modification of textiles.

Surface modification using low temperature low pressure plasma processing has advantages such as environmentally friendly, uniformity, good control over process parameters thereby giving reproducibility. Variety of gases and monomer vapours can be used for altering surface properties without compromising the bulk properties. The quantity of gases/monomers required is very less; there is no question of disposal of polluted water. Thus plasma processing is a powerful surface-modification technique to cater environmental needs. Surface modification techniques are mainly carried out to remove loosely bound materials, foreign particles/impurities and to improve hydrophilic nature and thereby improving dyeability. Tailoring of surface characteristics of textile materials is of fundamental importance in the production of advanced functional textiles. Low temperature plasma generated using inert or reactive gases have been used as an effective and environment-friendly option for the modification of surface properties of polymers and textiles, thereby causing both physical and chemical changes. The changes could be temporary or permanent, i.e. providing more reactive surface layers, without adversely affecting the desirable mechanical properties and quality of the material.

Low pressure plasmas (LPP) are in use in textiles for a long time for surface modification and grafting. The pressure used in low pressure systems could be in the range of 0.01–1.0 mbar. The advantages of low pressure plasma include uniform glow, low voltages, high concentration of charged species, higher plasma volumes to treat three-dimensional objects, etc. However, low pressure plasma systems operate in batch mode as every time new samples need to be loaded and vacuum is created before plasma treatment (Kale and Desai, 2011a; Deshmukh and Bhat, 2012). Thus in textile industries, where high speed is necessary, the low pressure plasma may not fulfil processing requirements of industries. Therefore atmospheric pressure plasma (APP) has emerged as a method to be adopted in industry. This allows online continuous treatment of fabrics. It is possible to make this adaptive with the existing machinery. Shenton and Stevens (2001) have compared surface modifications of polymers carried out using atmospheric pressure and vacuum plasma. These surface modifications include surface cleaning and degreasing, oxidation, reduction, grafting, cross-linking (carbonization), etching and deposition.

Most of the textile processes such as sizing, desizing, scouring, bleaching, dyeing, etc., are wet processes. These processes need a lot of water and cause pollution. On the other hand plasma is a dry process and eco-friendly and is "green technology" (Kangti and Saramadi, 2004). Following processes can partially or fully be replaced by plasma treatment:

6.4.1   Plasma-assisted desizing

Removal of sizing material from the grey fabrics is necessary for the processes like bleaching, dyeing, etc. The conventional desizing requires 24 hours of time, needs lot of water at high temperature (~90°C) and repeated washing.

This water is released into drain which causes pollution. It has been observed that plasma processing helps in desizing step by,

(i)   removal of size material by etching process,

(ii) loosening of size material,

(iii) degradation of size materials (like starch, PVA) (Peng et al., 2009; Cai et al., 2003a),

(iv) removal of size from fabric by water at relatively low temperature (60°C), during short time (~6 h) (Cai et al., 2003b; Bhat et al., 2011a; Cai et al., 2006).

It has been shown that even polyacrylate size material can be removed using APP (with helium/oxygen). It has been shown by Bhat and Benjamin (1999) that the desizing becomes effective when polyester fabric is treated by employing low pressure plasma.

6.4.2   Plasma for functionization of textile surfaces

Reactive species of plasma interact with any surface placed in the plasma. These species induce charged states and give rise to a number of reactions. In particular when polymer and fibre is subjected to plasma, it can introduce charged species due to ejection of electron or ion and due to scission of molecular chains. Malkov and Fisher (2010) have used pulsed RF plasma-enhanced chemical vapour deposition to coat various natural fibres such as silk, wool and cotton with poly(allyl alcohol) film. They found that regardless of type of fibres, the coatings were conformal and uniform. Using pulsed plasma processing to deposit coatings on textiles, it is possible to tailor chemical composition of the deposited film by minimising monomer fragmentation by selecting suitable duty cycle. Timmons et al. (2004) have shown that the chemistry of plasma polymers can be systematically adjusted in terms of retention of monomer functionalities by varying duty cycle in the pulsed plasma technology (Bhattacharya et al., 2010). Further exposing the treated surface to ambient air can initiate reactions with oxygen and water vapour at room temperature conditions. Gulrajani and Gupta (2011) have described the use of plasma technology for nano-level finishing and functionalization of textile materials. Thin polymeric films can be deposited on fabric surfaces which are quite continuous, uniform and contain additional reaction sites. Hydrophobic (or hydrophilic) textile surface can be created due to such functionalization and polymerization reactions have been discussed in next section.

6.4.3   Plasma for enhancing hydrophilicity

Various research reports suggests that low pressure as well as atmospheric pressure plasma treatments (for cotton, wool, polymer) modifies the upper surface of fibres (Höcker, 2002; Hossain et al., 2009; Rakowski, 1997, Shah et al., 2013). This enhances the hydrophilic property depending on the treatment conditions. Etching causes roughness of fibre surface leading to larger surface area in addition to pores and cracks. This allows more pick up of water. In addition, creation of hydrophilic groups like –OH, C=O allows attachment of water molecules enhancing the hydrophilicity. For example grey cotton fabric takes about 1 minute for wetting, but with plasma treatment it occurs in less than 1 second. Even polyester, nylon, PP, which has a wetting time of more than 1 hour, got wetted in less than 1 min. Bhat et al. (2011a) have shown that contact angle for cotton fabrics reduces from 64° to 25° and the rate of wetting becomes almost double (Bhat et al., 2011a). The increase in hydrophilicity/ wettability after the plasma treatment is basically due to the incorporation of polar functional groups such as –OH, –CO, –COO, –COOH, etc., onto the polymer/fibre surfaces (Deshmukh and Shetty, 2007b). The change in surface free energy and its polar and dispersion components after the plasma treatment can be obtained using contact angle measurements. There are various methods to calculate surface free energy from contact angle data. Each method is having its peculiarity. Recently some review papers have compared and given various methods of estimating surface energy from contact angle data (Etzler, 2003; Etzler, 2013; Zenkiewicz, 2007; Deshmukh and Shetty, 2008).

6.4.4   Plasma for enhancing hydrophobicity: Lotus effect

Plasma treatment can also be employed to make textile surfaces more hydrophobic for certain applications, such as, water/dirt-repellent fabrics for umbrella, swim suit, roof and tents. Anti-grease surfaces are also required for workers in various companies and motor vehicle garages. Surgical gown have special requirement of antibacterial and blood stain proof fabric. For all such applications, it is necessary to modify the surfaces to become more hydrophobic. Conventionally hydrophobicity is enhanced by chemical coating, grafting with monomers, etc. These wet methods do not have control of processing and causes pollution; waste of chemicals and long durations of treatment time is necessary. On the other hand process using plasma can be carried in short time, in clean environment and can be optimised. Various conventional or non-conventional chemicals/monomers can be used. TEOS, TMOS and HMDSO monomers are used with and without oxygen as a carrier gas to deposit SiOx layer on fabric material to improve hydrophobicity. Recent work in our laboratory carried out for P/C and PES fabrics using HMDSO (Kale et al., 2011b; 2012a) has shown that excellent hydrophobic surface can be developed.

The surface of the apparel which is in contact with the body of an athlete/ sport person should always remain dry and at the same time it should also absorb excessive sweat. Such one-side hydrophobic and other-side hydrophilic surfaces can be prepared using plasma surface modification technique. It is well known that plasma processing modifies surfaces suitably without altering bulk properties for which the material is selected for certain application. Cotton is comfort fabric and has excellent water-absorbing capacity. The one of the surfaces of cotton can be made hydrophobic using plasma treatment of above-mentioned monomers. By controlling plasma parameters, if nano-structure/nano-roughness morphology is introduced onto the fabric, it will behave like a super hydrophobic surface. It is called Lotus effect. When water drop is put on the hydrophobic surfaces, it balls up and makes larger angle of contact as shown in Fig. 6.5(a). We have modified cotton fabric using dichlorodifluoromethane (DCDFM) plasma for creating such super hydrophobic surfaces (Bhat et al., 2011b). The contact angle increased dramatically for the DCDFM plasma-treated fabrics up to 140° as shown in Fig. 6.5(b).

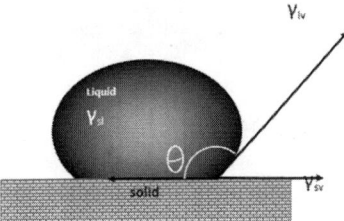

Figure 6.5 (a) Schematic representation of Angle of contact

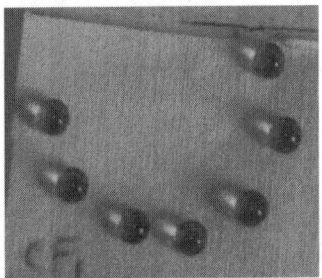

Figure 6.5 (b) Angle of contact measured for a water drop spread over a cotton fabric which was made hydrophobic by plasma coating of DCDFM

Super hydrophobic textile surfaces can be obtained using fluorine-containing plasma. Hocker (2002) showed that low temperature plasma of hexafluoroethane can be successfully used for imparting hydrophobic effect onto the cotton fibre. He further showed that the water droplets are able to effectively remove dirt particles from the surface of the cotton fabric (Hocker, 2002). Hydrophobic finishing was obtained using atmospheric glow discharge of He/Butadiene onto the viscose rayon fabric (Samantha et al., 2012). Plasma polymerization of Butadiene yielded in incorporation of species containing $-CH_x$, such as $-CH_2$ and $-CH_3$, in cellulose and removal of $-OH$ bonds from the top layers of cellulosic fibre. It was further observed that the wicking time increased to more than 60 min and contact angle to 142° compared to <1 S and 0°, respectively. The super hydrophobic surfaces with specific surface topography obtained using plasma processing are excellent water- and dirt-repellent fabrics. In addition to this, the plasma-treated surface also becomes repellent to bacteria and fungi.

6.5   Plasma treatment for enhancing dyeing

Studies on surface modification of fabrics using plasma showed that the surface roughness and introduction of functional groups facilitates dye uptake. Early work on low pressure plasma and recent work with APP has shown that the dyeing gets enhanced, after suitable plasma treatments. Wool fabrics have been mostly thoroughly explored for various modifications on subjecting it to plasma.

6.5.1   Wool processing

Textile fibres can be classified in two main groups: natural and synthetic. Natural fibres are extensively used now-a-days for their comfort properties. Natural fibres are of either animal origin or plant origin such as silk, wool, angora fibres, hairs, jute, cotton, cellulose, etc. The textile industry uses different natural as well as man-made fibres as a raw material. Wool consumption in the field of textiles is second largest after cotton in case of natural fibres.

Basically, wool is a nano-composite fibre, i.e. composite of reinforcing fibril (have a diameter of about 10 nm) – matrix material with both the fibrils and the matrix consisting of polypeptides (thus of chemically similar nature), interconnected physically and chemically (Simpson and Crashaw, 2002). Wool is a natural highly crimped protein hair fibre derived from sheep. The fineness, structure and properties of the wool depend on the breeds of sheep, variety from which it is derived such as Merino, Lincoln, Leicester, Sussex, Cheviot, and other (Needles, 1986). The great value of wool as a fibre lies in the fact that it is strong, elastic, soft, and being woven, furnishes a great number of air spaces, rendering clothing made from it very warm and light (Watson, 1907).

Wool is a protein fibre and has very complex physical structure. The detailed structure of wool is described elsewhere (Simpson and Crashaw, 2002; Mendhe, 2012). Even after the natural wool grease has been removed by scouring with a detergent, wool fibres are relatively difficult to wet compared with other textile materials. This natural water repellency makes wool fabrics 'shower-proof' and able to resist water.

The outermost cuticle layer and the scales make it hydrophobic. The scales are relatively hard and have sharp edges which are responsible for causing fibre directional movement and shrinkage during felting. Furthermore, the scales also serve as a barrier for diffusion processes which adversely affect the sorption behavior. Chemical processing of wool is much difficult due to its hydrophobic nature caused by the presence of scales on its surface. Therefore, these scales need to be removed for the improvement of absorbency leading to proper interaction with dyes, chemicals and finishes and for improvement in other surface related properties such as anti-felting, spinnability, absorbency, dyeability, etc. In recent years, there has been an increase in the modification of wool surface scales by physical means such as mechanical, thermal and ultrasonic treatments, and chemical methods such as oxidation, reduction, enzyme and ozone treatments which can solve the felting and sorption problems to a certain extent. Chlorination is extensively used for the surface modification of wool which removes the scales and hydrophobic cuticle layer from the surface of wool fibre/fabric resulting in the reduction of its felting property and improved dyeability. The process of chlorination has a problem of producing toxic AOX compounds liberating non-eco-friendly chlorine to the effluents (Nebojsa et al., 2010). Hence, this procedure has to be replaced by alternative eco-friendly processes. Among all physical processes, plasma treatment is gaining importance for the surface treatment of textile material because of number of advantages associated with it as mentioned above.

Gaseous plasma treatment leads to the etching – partial removal of hydrophobic outermost cuticle lipid layer and oxidation/incorporation of polar functional groups onto the wool fibre surface resulting in shrink resistance. Another effect of oxygen-containing plasma treatment is related to a significant reduction of the cross-link density of the epicuticle layer in wool. The epicuticle layer is highly cross-linked via disulfide bonds. The oxygen-containing plasma processing of wool leads to the oxidation and breaking of the disulfied bonds, results in improved properties (Hocker, 1995). The usefulness of plasma technology is not limited to well-known and well-understood positive effect of plasma in dyeing and printing of textiles. Plasma is also effectively used for more specific treatment of wool fibres for enhancement of anti-felting, shrink resistance. Plasma technology is found to be effective for surface modification of wool. Lot of work has been carried out by researchers on application of plasma technology in surface modification of wool. Researchers have showed that the plasma processing of wool with various gases/monomers have improved properties of wool and are listed in Table 6.1.

Table 6.1 Studies carried out on plasma processing of wool with various gases/ monomers to improve wool properties

Type of plasma	Monomers/ gases used	Working parameters	Properties investigated	References
Corona discharge	Air	Power: 800 W; Electrode gap: 4 mm	Wettability, dyeability, colour fastness	Nebojsa R et al., JAPS (2010)
DC plasma sputtering	Argon	2×10^{-2} Torr Voltage: 2000 V DC	Low temperature dyeing, natural dyeing and substitute for Mordent, colour strength, antimicrobial property	Ghoranneviss et al. (2011)
DC plasma	Argon	2×10^{-2} Torr Voltage: 2000 V DC Current: 220 mA	Disperse and metal complex dyes used, polyester/wool (55/45) fabric was used. Dye uptake, K/S	Motaghi Z. and Shahidi S. (2012)
DC glow discharge plasma	O_2, N_2 and Argon	Press: 0.02 Torr Discharge voltage: 500 V and current: 200 mA	Wicking time and dyeability	Shahidi S et al. Proceedings 3rd int. conf. (2009)
DC glow discharge plasma	O_2	Press: 0.4 Torr Power: 20 W	Hydrophilicity, wettability, dyeability (three different dyes such as acid dye, metal complex dye and reactive dye was used), washing and light fastness	Gawish et. al (2011)
Low freq. 40 kHz plasma	Acrylic acid, Oxygen	40–100 W Flow rate: 180 cm^3/min	wrinkle recovery, breaking strength, tensile properties, hydrophilicity	Kutlu et al. (2010)
Low temp RF plasma	N_2	10 pa press, 80 W power	Surface energy and surface morphology	Kan et al. Fibres and Polymers (2009)
Low temperature RF plasma	Oxygen	Flow rate: 20 SCCM, press: 0.1 mbar Power: 30–80 W	Wettability to reduce dyeing temperature without affecting dye bath exhaustion and colour fastness	Rombaldoni et al. (2010)
RF (13.56 MHz) plasma	O_2, N_2 and mixture of H_2 and N_2	Power: 80 W Flow rate 20 cc/min Press: 10 Pa	Dye absorbency	Kan et al., Fibres and Polymers (2004)
RF (13.56 MHz) plasma	Air	Power: 100 W Press: 20 mbar	Shrink resistance, dyeability, colour fastness, contact angle,	Dragan J et al., JAPS (2005)
RF (13.56 MHz) plasma	Air	Power: 150 W Press: 50 Pa	Anti-felting performance, wetting time, and dyeability at low temperature	Wang X, et al., JAPS, (2011)

Contd...

Contd...

Type of plasma	Monomers/ gases used	Working parameters	Properties investigated	References
RF (13.56 MHz) plasma	O_2	Power: 300 W O_2 flow rate: 0.3 SLM	Hydrophilicity, contact angle, wicking, scouring, dyeing	Sun D and Stylios G K, TRJ (2004)
Atmospheric plasma	Air/He, O_2/ He	Freq: 7.5 kHz Voltage: up to 7.5 kV	Shrink resistance, anti-felt, water penetration, wicking time, dyeing rate (acid dye), tensile strength	Cai Z et al., JAPS (2008)
Atmospheric plasma	He – Air	Voltage: 2 kV Freq: 20 kHz	Low temperature dyeing using reactive dye, increase in dye uptake, colour strength (K/S), fastness, anti-felting, hydrophilicity	Panda et al. (2012)
APDBD	Air, Argon	Atmospheric pressure, Power: 130 W, 2 mm distance between two electrodes	Surface morphology, reactive groups, water absorbancy, antimicrobial property	Demir et al. (2010)
APDBD	Air	Voltage: 4, 4.5, 5, 5.5 kV Electrode spacing: 2,3 and 4 mm	Tensile, bending, shear, elongation properties studied	Goud V.S., IJFTR (2012)
APDBD	He/Ar and Acetone/Ar	Power: 80 W Freq: 3 kHz Voltage: 4–4.2 kV	Acid dyes used, dyeing rate, saturation dye exhaustion	Wakida T. et al. TRJ (1993b)

Kan et al. (2009) have presented comparative study on the physico-chemical properties of wool fibre modified by low temperature plasma (LTP) as well as chemical treatment (chlorination). Kan et al. (2000, 2007) listed many applications of plasma technology for the modification of wool surface such as spinnability, shrink-proofing, improvement in air permeability, printability, dyeability, etc. Lot of work has been carried on plasma treatment of wool to make it shrink resistant and to impart anti-felting property to the wool (Shahidi et al., 2010a; Panda et al., 2012). Coulson et al. (2011) have patented the method of wool treatment to make it shrink resistant during laundering. Thomas (2007) has described applications of plasma for the modification of wool. Many authors have reported air plasma treatment of wool to improve its mechanical properties (Udakhe et al., 2011; Mori et al., 2011; Karahan et al., 2009a). Kan et al. (2010) have subjected wool fibres and fabrics to nitrogen plasma prior to polymer deposition. They revealed that plasma treatment alone could achieve the best anti-felting effect, but the hand feel was adversely affected. Further, with the deposition of the polymer on the fibre surface, the anti-felting properties were retained with an acceptable feel. The enhanced hydrophilicity due to nitrogen plasma treatment achieved a better polymer deposition on wool fibre.

Wool surface has a typical structure in terms of scales. Plasma etching causes these scales to be partially removed and also causes grooves and cracks. These facilitates penetration of water and dye molecules, and therefore the rate of dyeing is significantly enhanced (Kan et al., 1998; Wakida et al., 1996). Acid and basic dyes both show increase in rate of dyeing of wool. Argon gas alone and also together with helium and acetone have been used for plasma treatments showing increase in dyeing rate. C I acid orange 7, C I acid red 18, C I acid blue 113, C I acid blue 83 have been used successfully (Wakida et al., 1993a). Of the two mixtures of gases He/Argon was found to be more effective. Detailed studies, in some cases, however showed that although the rate of dyeing increases significantly, the exhaustion equilibrium uptake did not show much significant change (Kan et al., 2006; Cai et al., 2008). El-Zawahry et al. (2006) studied the effect of different gases on physical, chemical, morphological and dyeing properties of wool (El-Zawahry et al., 2006). The final exhaustion of dyeing depended on the type of gas and the order was found to be $N_2 > N_2 + O_2 > O_2 > Ar$.

One of the important aspects of plasma treatment is to try dyeing at lower temperature, if possible, to save energy. Results by Rombaldoni et al. (2010) showed that pre-treatment with O_2 and subsequent dyeing at different temperatures like 98°C, 85°C, 80°C showed that initial increase in wetting and dyeing takes place even at low temperature, better diffusion of dye at 85°C without affecting the exhaustion and colour fastness properties of the dyed material. Plasma processing for another important animal fibre, i.e. Angora rabbit wool, has been reported by Danish et al. (2007), wherein they observed that the dyeability improves together with spinning properties. However, the whiteness of the fibres reduces subsequent to plasma processing. Although the results for wool have been encouraging, the situation is somewhat different for cotton, polyester, PP and P/C blends. Therefore it is necessary to look at other reasons which lead to enhanced/decreased dyeing, apart from surface morphology.

6.5.2   Cotton processing

We have modified cotton fabric in low temperature plasma generated in bell jar-type reactor as shown in Fig. 6.1. The bell jar-type plasma chamber had top and bottom cover plates with ports to measure pressure, supply power, gas, etc. A mass flow controller (Unit Model URS-100) was used to control the gas flow. The pressure of 20 Pascals was maintained during the plasma treatments. Two stainless steel plates inside the chamber were capacitively coupled with the RF source to generate plasma. The frequency of the power source was 13.56 MHz and power could be varied from 20 to 100 W. In the present study, power of 40 W was used. A sample of size 20 cm × 20 cm was inserted in the chamber and treated by plasma. The samples of cotton fabrics A to D were exposed to air plasma and then dyed using three different types of dyes viz. reactive, direct and natural. The dye absorption was determined spectroscopically by measuring the absorption band maximum of the dye-bath solution before starting and after exhaustion. In addition, the amount of dye uptake was determined also from the color matching instrument Nova Scan. For each variety of fabric, the dye uptake of plasma-treated sample was compared with the untreated sample of the same variety. Our results on dyeing of cotton fabrics after subjecting to air plasma showed interesting results. Table 6.2 gives the comparative values of the colour strength for only one type of fabric D, as such and only for one time of plasma treatment of 5 minutes. The values express the colour strength, as measured by Nova Scan; this gives the strength of the shade. From the data in Table 6.2, it can be seen that the changes in dye absorption are dependent on the type of the dye used. When reactive dye was used, the colour strength increased after the plasma treatment; whereas for the direct dye there was a decrease in the strength. In the case of one natural dye (Amazon), it was found that the strength increased after the plasma treatment. Thus it is seen that the type of dye is important. Attempts were made to find out if the amount of dye absorbed depended on the time of plasma treatment, but unfortunately no direct correlation could be found except in the case of reactive dye. This is depicted in Fig. 6.6 where the percent exhaustion is expressed with the time of exhaustion and different times of plasma treatments are plotted. It was found that when the time of plasma treatment was increased up to 20 min duration, the dye uptake was enhanced.

Table 6.2 Comparative study of dye uptake of different dyes for cotton fabric (all samples desized, scoured and bleached, plasma treatment done for 5 min in air, color strength using Nova Scan)

Type of dye	Color strength control	Color strength after plasma treatment
Reactive dye	100	102.7
Direct dye	100	97.9
Natural dye (Amazon)	100	103.2

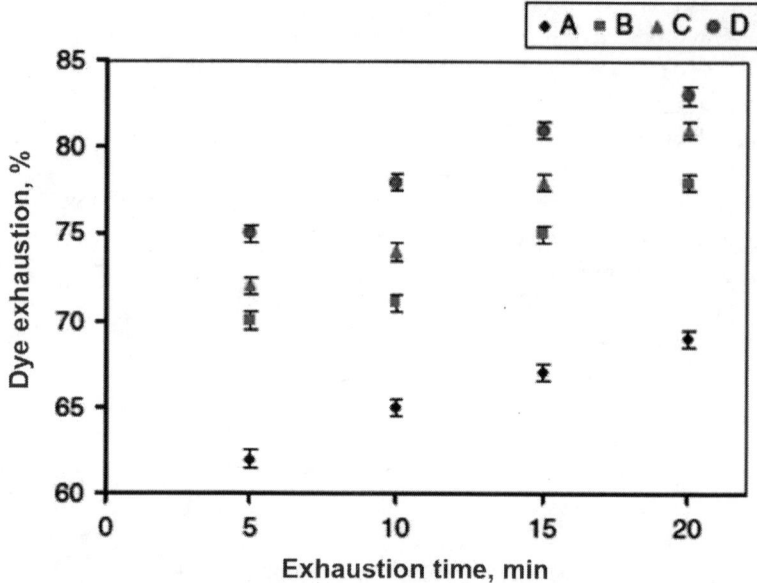

Figure 6.6 Relation between dye bath exhaustion and time of exhaustion for various durations of air plasma treatments (for Reactive Dye). A – untreated, B – 5 min plasma, C – 10 min plasma, D – 25 min plasma treatment.

The increase in the dyeing in the initial stage could be due to the fact that plasma etches out the surface of the cotton fibres removing the waxy layer and creating rougher surface with irregularities as shown in Fig. 6.7. The etching of the surface has been confirmed by other researchers (Ozdogan et al., 2009; Karahan et al., 2009b). The removal of waxy layer and creation of rougher surface makes the diffusion of the dye molecules easy. Additionally, there may be simultaneous breaking of bonds of the cellulose molecules which may make the –OH groups available for the reaction with the dye. For the reactive dye, there are two SO_3^- groups at the benzene ring and this group can bond at –OH site by abstraction of H due to plasma interaction. On prolonged exposure to plasma, additional sites of –C–O– are created where dye molecules can interact and dyeing is enhanced.

In case of direct dye the mechanism of dyeing is mainly through the diffusion phenomenon. For direct dye, there appears to be a slight decrease in the dye-uptake which can be due to cross-linking on the surface of the cellulose fibre. This will not allow the diffusion of the dye into the fibre structure and thus the decrease could occur.

 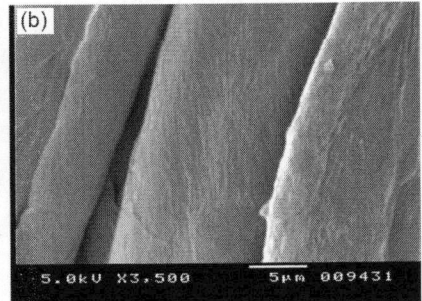

Figure 6.7 SEM photomicrographs of (a) control (desized and scoured) cotton fabric, (b) Two min air plasma-treated fabric

Thus our results showed that whereas dye up-take increased with reactive and natural dye (Amazon), it did not for direct dye. In the case of reactive dye, it was found that the amount of dye absorbed depended on the plasma treatment. Although many hydrophilic groups get created on cotton fibre, the direct dye cannot react with these groups and therefore dye up-take did not increase. Direct dye rather needs to diffuse into the fibre structure and on account of surface cross-linking does not allow diffusion into the interior (Bhat et al., 2011a). The detailed analysis of these results revealed that the factors play an important role whether dye uptake will increase or not are outlined in Section 6.5.6. Thus although application of plasma processing to dyeing looks very promising, optimization has to be carried out. Therefore, it was thought that some additional functionalization is needed to get enhanced dyeing for cotton. Ozdogan et al. (2003) carried out plasma treatment in ethylenediamine and triethylenetetradiamine before dyeing with Ramazol Blue reactive dye. The K/S values for plasma treated samples were significantly higher than that of untreated cotton (Ozdogan et al., 2003). In addition to the application of plasma processing for facilitating dyeing of cotton fabric, this technology is also used to enhance other properties and is listed in Table 6.3.

Table 6.3 Studies on plasma processing of cotton to improve various properties

Type of plasma	Monomers/ gases used	Working parameters	Properties investigated	References
Corona discharge	Air	Power: 1.02 kW Electrode gap: 3 mm	Hydrophilicity, dyeability, exhaustion, breaking load	Carneiro N. et al., Color Technology (2001)
Corona discharge	Air	Power: 1.29 kW	Reactive dye used, hydrophilicity, desizing, washing fastness, pick-up of dyeing bath, water drop absorption time, K/S	Carneiro N. et al. (2007)

Contd...

Contd...

Type of plasma	Monomers/ gases used	Working parameters	Properties investigated	References
Corona discharge	Air	Power: 660 and 880 W	Indigo dye used, colour strength, colour difference, colour fastness, K/S	Shirin N. et al. (2008)
DC glow discharge plasma	Ar	Voltage: 250–500V Current: 7–19 mA Press: 0.3–0.7 mbar Electrode separation: 3 cm	Wicking, free radical formation, etching behaviour and weight loss	Inbakumar S. et al. (2012)
Microwave plasma (2.46 GHz)	Ar, CF_4, acrylate phosphate and phosphonates derivatives	Power: 100 W and 300 W	Water repellence and flame retardancy	Tsafack M.J. and Grützmacher J.L. (2007)
RF (13.56 MHz) plasma	Ar	Power: 40 W Press: 100 mtorr	Moisture regain, wettability, wicking time, rate of dyeing	Jung H.Z. et al., TRJ, (1977)
RF (13.56 MHz) plasma	SF_6	Power: 25–75W Press: 0.005–0.5 Torr	Weight loss, hydrophobicity after washing cycles, ageing effect, mechanical properties	Kamlangkla K. et al., (2010)
RF (13.56 MHz) plasma	Air plasma followed by DCDMS grafting	Pressure: 0.1 mbar Power: 30 W	Water repellence, wetting time, mechanical properties, colour strength	Jahagirdar C.J. et al. (2004)
RF (13.56 MHz) plasma	Air, O_2	Power: 70–120 W Gas flow: 30 ml/ min	Wicking, dye uptake, dye exhaustion, direct dye used, whiteness and yellowness, free radicals, weight loss, ageing effect	Malek R.M.A. and Holme I. (2003)
RF (13.56 MHz) plasma	O_2	Power: 300 W O_2 flow rate: 0.3 SLM	Hydrophilicity, contact angle, wicking, scouring, dyeing	Sun D. and Stylios G.K., TRJ (2004)
RF (13.56 MHz) plasma	He/O_2	Atmospheric press; Optimum conditions are power: 160W O_2 conc.: 1% Jet distance: 3 mm	Effect of treatment time, power, concentration of O_2, jet distance, etc., was studied on desizing of grey cotton denim fabric followed by enzymatic color fading wettability	Kan C.W. and Yuen C.W.M., Coloration Technology (2012)
APDBD	Air, Ar	Power: 50, 100, 130 W Distance between two electrodes: 2 mm	Hydrophilicity, wickability, surface friction, air and water vapour permeability, colour fastness, etc.	Demir A. et al., JAPS, (2011)

Contd...

Contd...

Type of plasma	Monomers/ gases used	Working parameters	Properties investigated	References
APDBD	N_2, O_2	Power: 50 W F = 20 kHz	Wettability, dye exhaustion, K/S	Ibrahim N.A. et al., TJIT (2010a)
APDBD	Air	Gap between two electrodes: 3 mm Freq: 1.5 kHz Discharge voltage: 28 kV	Enhances grafting efficiency of acrylic acid monomer and subsequently dyeability was improved	Ren C.S. et al. (2008)
APDBD	Air, O_2, N_2	Power supply A: 50 Hz; 8 kV; 30 W Power supply B: 20 kHZ; 5 kV; 50 W Flow rate: 3 l/min	Hydrophilicity, bleachability, reactive and basic dyes used, dyability, colour strength (K/S), whiteness and yellowness index, dye bath exhaustion	Ibrahim N.A. et al., JIT (2010b)
APDBD	He/Air	Power: 8 kV, Freq: 50 Hz Gap between electrodes: 2 mm	Desizing, wettability, wicking action, contact angle	Bhat N.V. et al., IJFTR (2011a)

Cotton textiles with a superhydrophobic coating constitute an attractive potential in many applications as water proof and self-cleaning apparel or with better adsorption capacity and bioactive properties (Marciano et al., 2009). Various monomers containing silicone and fluorine have been used to impart hydrophobicity. However, they suffer from the lack of stability of the hydrophobic state and poor durability. Therefore, diamond-like carbon (DLC) coatings are gaining considerable attention due to their well-known physical properties such as mechanical stability, extreme hardness, low friction coefficient, high corrosion resistance, and biocompatibility (Matsumoto et al., 2008). Caschera et al. (2013) have shown that the pre-treatments had a significant impact on wettability behaviour resulting from an induced nano-scale roughness combined with an incorporation of selected functional groups. Upon subsequent deposition of diamond-like carbon (DLC) films, the cotton fibers yield to a highly controlled chemical stability and hydrophobic state and could be used for self-cleaning applications (Caschera et al., 2013). They have used H_2, O_2, or Ar as precursor gas. Carbon fibres pre-treated with O_2 and Ar plasma followed by DLC deposition showed enhanced hydrophobicity than just DLC-deposited/functionalized fibres. The super hydrophobic effect was found to be very robust and durable. Guzenda et al. (2013) have modified cotton fabric using RF-PECVD technique. They have deposited stable TiO_2 coating onto the cotton fabric to make it hydrophobic.

6.5.3   Polyester processing

Since polyester fabric has a lot of special characteristics, such as superior strength and resilience, the molecular structure of the poly(ethylene terephthalate) (PET) lacks polar groups, which causes it to have low surface-free energy and poor wettability (Chen et al., 1999). Ferrero (2004) used the acrylic-acid-based plasma treatment of polyester, PP and polyamide fabrics. After plasma treatments, fabrics were dyed with basic dyes. Significant improvement in the colour strength was observed. This improvements are probably due to coating of surface by polyacrylic acid (or other polymer gas) and the penetration of dye molecule in it. Colour fastness to washing studies showed that whereas results were good for polyamide, it was not so for polyester and PP. This shows lack of interaction of dye and acrylic acid with the parent fibre. Oktem et al. (2000) treated polyester and polyamide fabrics in acrylic acid, water, air, argon, and O_2 gas plasmas. They evaluated the treated fabrics according to wettability, dyeing, and soiling behaviour (Oktem et al., 2000). They concluded that all in situ plasma polymerization types improve wettabilities, and therefore, dyeability and soil resistance of the fabrics. Blends of polyester/cotton fabrics when treated in plasma using acrylic acid and water showed good wettability and dyeability (Oktem et al., 2002). A single bath method for dyeing of polyester/cotton blends was replaced by Nobojsa et al. (2009), where in fabrics were treated in corona discharge followed chitosan treatment. The colour strength was found to be increased with the amount of chitosan. Ammonia-Acetylene plasma treatment was used for enhancing the dyeing of PES fabrics. The additional advantage of the treatment was that dyeing could be carried out at low temperature (Hossain et al., 2009). Mixture of Ar and O_2 were used to modify surface of polyester fabrics. The dyeability improves dramatically on account of functionalization of surface (Zhongfu et al., 2007). Dyeability of PES is related to hydrophilicity and increase in micro roughness and effective surface area; the plasma treatment can improve dyeability with disperse dyes (Raslan et al., 2011a). Okuno et al. (1992) studied the correlation between the crystallinity and dyeability of PET fibers by using non-polymerizable gases with low-temperature plasma. They found that plasma-treated samples significantly reduced the dyeability because of the etching of macromolecules constituting the dyeable amorphous phase. Raffaele-Addamo et al. (2006) reported that the color depth of air radio frequency plasma-treated PET fibers is related to their topographical characteristics and to their chemical surface composition. They observed that the K/S value with a disperse dye at a dyeing temperature of 100°C can be increased by a reduction of the fraction of light reflected from treated surfaces. Samantha et al. (2008), employed atmospheric pressure glow discharge plasma technology for etching and creating nano channels on nylon and polyester fabrics using He/Air plasma.

The hydrophilic nature of nylon and polyester (obtained after plasma treatment) facilitates water absorbency at the same time nano channels helps in spreading oil (hydrophobic liquid); thus they have shown that a single processing is helpful for wicking and spreading of hydrophilic as well as hydrophobic liquids. The degree of plasma-induced hydrophilization and the stability of the treatment are closely linked to the textile structure and the weave construction. Hossain et al. (2006a) have investigated the plasma activation of polyester (PES) fabric structures to improve its wettability, as well as the aging effects. The hydrophilic modification was carried out by low pressure plasma treatments using oxygen-containing gaseous mixtures (Ar/O$_2$ and He/O$_2$). Fabric was characterized by contact angles measurements and capillary rise tests with water. In all cases, the wettability of plasma-treated PES fabrics was improved significantly due to the formation of polar groups on the surface. In particular, the hydrophilicity of looser structured fabrics is improved remarkably as compared to tightly woven fabrics. Furthermore, the capillary phenomenon in fibrous assemblies is also described in this study.

Effect of plasma treatment on polyester fibres using R.F. discharge was investigated by Bhat and Benjamin (1999) which showed that the surface resistivity decreased dramatically; further such plasma-treated polyester fabrics could be grafted with acrylonitrile and acrylamide. Table 6.4 summarises various plasma technologies and plasma parameters used to modify polyester fabrics.

Table 6.4 Various plasma technologies and plasma parameters used to modify polyester fabrics world over

Type of plasma	Monomers/ gases used	Working parameters	Properties investigated	Refs.
Corona discharge	Air	Power: 0–10 kV	Wicking, hydrophilicity, dyeing speed, dye uptake	Xu W. and Liu X., (2003)
DC pseudo-glow discharge plasma	N$_2$	Power: 1–20 W Press: 0.06, 0.1 and 0.2 Torr	Hydrophilicity, disperse dye was used, dyeability, mechanical properties	Nagar Kh El et al. (2006)
Low frequency, 40 kHz	Acrylic acid	Power: 10, 30, 60 W Monomer flow rate: 200 cm³/ min	Hydrophilicity, wetting time, wrinkle recovery angle, and breaking strength	Cireli A et al., JAPS (2007)
Low temp RF (13.56 MHz) plasma	Air	Power: 5–10 W Press: 0.1–1 Torr Electrode gap: 5 cm	Wettability, contact angle, dyeability	Lehocky M. and Mracek A. (2006)

Contd...

Contd...

Type of plasma	Monomers/ gases used	Working parameters	Properties investigated	Refs.
RF (13.56 MHz) plasma	Monomer DCDMS was used for post plasma grafting	Power: 30 W Electrode gap: 3 cm	Water repellancy, mechanical properties	Jahagirdar C.J. and Tiwari L.B. (2007) Pramana
RF (13.56 MHz) plasma	Air, CO_2, water vapour, Ar/O_2, He/O_2	Power: 10–16 W Press: 10–16 Pa	Wettability, contact angle ageing effect	Hossain et al., JAPS (2006b)
RF (13.56 MHz) plasma	Ar/O_2 Ammonia/ Acetylene	Power: 400 W Press: 10 Pa	Dyeability, colour fastness	Hossain et al., JAPS (2009)
RF (13.56 MHz) plasma	O_2	Power: 40–120 W O_2 flow rate: 0.1 l/min Press: 20–40 Pa	Weight loss, inkjet printing, colour yield, K/S, anti-bleeding	Wang Chunying and Wang Chaoxia (2010)
RF-PECVD	O_2 followed by HMDSO	Bias voltage: 400 V Press: 10 mtorr and 4 mtorr, respectively Flow rate: 20 cm^3/min	Nanostructured super hydrophobicity, contact angle	Shin B et al., Soft Matters (2012)
APDBD	Air	Power: 350 mW Freq: 50 Hz Electrode gap: 10 mm	Hydrophilicity, ageing effect	Pichal J. and Klenko Y., (2009)
APDBD	Air	Voltage: 7 kV, Freq: 50 Hz Electrode gap: 3 mm,	Natural dye used, dye uptake, K/S, wash and light fastness, wettability, contact angle, ageing	Davo H. et al. (2012)
APDBD	HMDSO	Power: 2000 to 5000 W Electrode gap: 0.5 mm	Water repellancy, mechanical properties	Kale K.H. et al., IJFTR (2012b)
APDBD	Air	Power: 500 W, Freq: 1 kHz Electrode gap: 3 mm	Wetting time, anti-bleeding property in inkjet printing, penetration depth of ink, colour strength, K/S, rubbing fastness	Zhang C.M. and Fang K.J. (2011)
APDBD	He, Air, O_2, Ar	Power: <50 W Voltage: ~17 kV Freq: 10 kHz Electrode gap: 1–2 mm Flow rate: 2–3 l/min	Wicking, capillary action, surface energy, improvement in absorbency and spreading of both hydrophilic (water) and hydrophobic fluids (mustard oil) after the plasma treatment	Samanta K.K. et al., Sur & Coat Tech (2009)

Contd...

Contd...

Type of plasma	Monomers/ gases used	Working parameters	Properties investigated	Refs.
APDBD	He, Air	Power: <50 W Voltage: ~17 kV Freq: 21 KHz Electrode gap: 1–2 mm	Antistatic property, surface energy	Samanta K.K. et al. Fibres and Polymers (2010)
Microwave plasma	O_2, N_3	Power: 100 W and 600 W Press: 3.6×10^{-3} mbar	Wetting time, intra-yarn porosity	Calvimontes A. et al. AUTEX, (2011)

6.5.4   Silk processing

Silk fibres are protein fibres but their structure is different than wool. Therefore the observation of large increase in dyeability of silk after plasma treatment may not be observed. Iriyama et al. (2002) showed treatment of silk in O_2, N_2 and H_2 plasma did give some improvement in dyeing using reactive black S dye; however, the fastness to rubbing was found to be poor. Fang et al. (2008) used oxygen plasma for silk fabrics prior to inkjet printing using pigments; although dye uptake did not increase significantly, the anti-bleeding performance showed appreciable improvement (Fang et al., 2008). Plasma surface treatment of silk carried out in air for different powers and times showed that the wetting time decreased and printability with reactive dye improved (Gorafa, 1980).

Detailed work was also carried out on silk fibers using N_2 plasma at low pressures (frequency 13.56 MHz, power 40 W and pressure 20 Pascal). Time of treatment was increased from 30 seconds to 4 minutes. Dyeing was carried out using acidic dyes (CI Red 361) in a beaker dyeing machine with 1% dye and at 85°C. Dyeing was carried out for different durations (from 15 to 60 minutes) and exhaustion time of dye was measured using UV/Vis spectrophotometer. K/S value was also measured using Macbeth colour dye 7000A. K/S values for 15 min dyed sample for various plasma treatments are shown in Fig. 6.8. It may be seen that it increases significantly with the time of treatment. Higher dyeing can be attributed to removal of waxy (Serine) layer, creation of rough surfaces and better diffusion of dye in the fibre.

6.5.5   Nylon fabric processing

In another study we have modified nylon fabric with the aim to improve wettability and dyeability. A tubular-type plasma reactor having electrodes

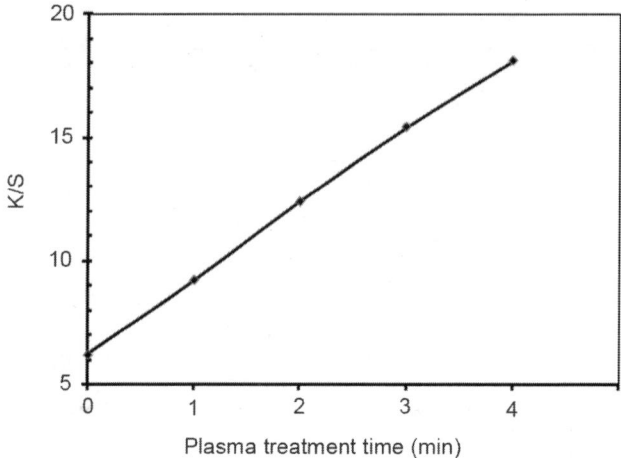

Figure 6.8 Variation of K/S values for silk fabric pre-treated with N$_2$ plasma

outside the glass tube was connected to RF (13.56 MHz) power supply capable of delivering 100 W power. The system was evacuated to 0.05 mtorr with rotary pump before inserting the gas/monomer vapours. The system was purged three times with the relevant gas/monomer vapours and the desired working pressure 0.2 mtorr was obtained with the help of fine control needle valve. The sample was mounted on a glass sample holder inside the chamber. The details of the plasma system is described elsewhere (Deshmukh and Bhat, 2012). Nylon fabric was treated in O$_2$ and N$_2$ plasma for different durations of time. Similarly plasma polymerization of acrylic acid was carried out onto the nylon fabric for different durations of time. These samples were tested for their hydrophilicity, dyeabilty and surface morphology. Nylon is quite hydrophobic in nature. The contact angle of untreated nylon fabric was observed to be 83° (±2) corresponding to the surface energy 33.6 mJ/m^2. However, we could not measure contact angle for any plasma-processed samples. It shows that the surface energy of all the samples increases rapidly after the plasma processing. The surface morphology was studied using SEM. It can be seen from Fig. 6.9 that the surface morphology changed after the plasma treatment. The untreated nylon fibres are smooth as shown in Fig. 6.9(a). The etching, roughening effect of oxygen plasma on the surface (Fig. 6.9c) is more as compared with the nitrogen plasma as shown in Fig. 6.9(b). It is probably due to oxygen, which is more reactive than nitrogen. The treated surfaces look damaged or abraded. This is due to the removal of some material by etching as explained in Section 6.3.2.

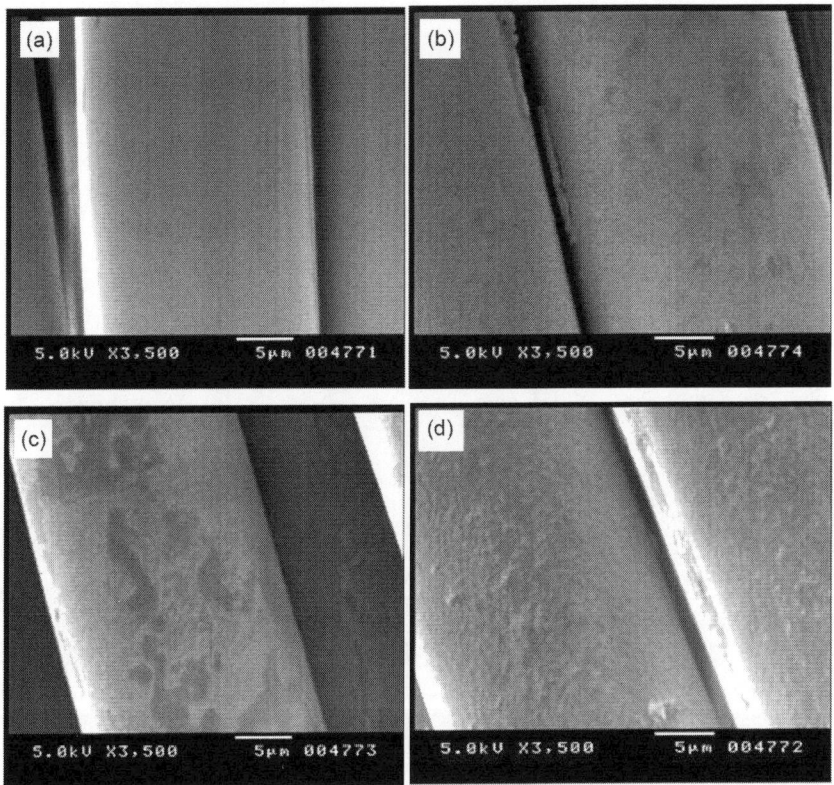

Figure 6.9 SEM micrograph of (a) untreated nylon fabric, (b) 4 min. N$_2$ plasma-treated nylon fabric, (c) 4 min. O$_2$ plasma-treated nylon fabric, (d) 4 min. PPAA deposited on nylon fabric.

Similarly when nylon fabric was subjected to the plasma polymerization of acrylic acid (PPAA), we observed deposition as shown in Fig. 6.9(d). This deposition leads to decreased capillaries present in the texture. The choice of acrylic acid monomer was based on its hydrophilic properties. In order to study the characteristics of plasma-polymerized acrylic acid film, KBr disc was kept in the plasma chamber while carrying out deposition onto the fabrics. Such deposited film was used for FTIR study. The IR spectra of plasma-polymerized acrylic acid (PPAA) is given in Fig. 6.10.

The FTIR spectrum of a PPAA film prepared using the technique of plasma polymerization was very similar to the spectrum of poly(acrylic acid) prepared by conventional polymerization techniques and shows all the characteristic bands (Deshmukh and Bhat, 2012). In particular, the FTIR spectrum shows that the film contains a high density of C(O)OH groups.

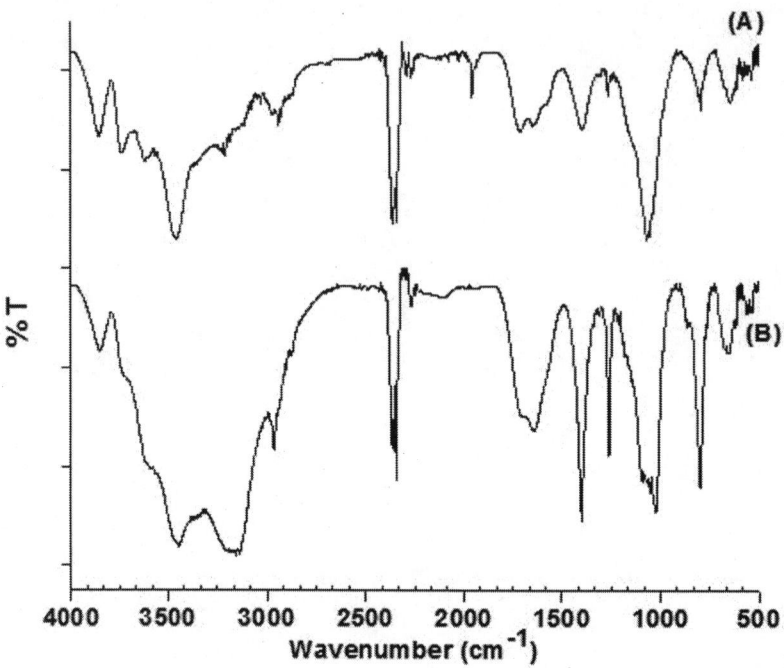

%T

Figure 6.10 FTIR spectra of PPAA (A) 2 min deposition, (B) 4 min deposition.

Dyeing studies of plasma-processed nylon fabrics: 1% shade using acid dye (blue coloured) was prepared and its pH was adjusted to 4.8 by adding formic acid. The liquor to fabric ratio was maintained at 50:1. The temperature of the dyeing bath was adjusted to 85°C. The nylon fabric (untreated and plasma processed) were immersed in dye-bath for 15 min. Then the fabric was washed in the soda soap solution with 5 gpl soap and 2 gpl soda for 30 min followed by distilled cold water wash (three times) and then the samples were dried in air. Such samples were used for colour measurement. Colour measurements were performed on Spectra-flash SF 3000 (Datacolour International). Dye exhaustion was measured using UV-visible spectrophotometer. The amount of dye absorbed by the sample was determined by measuring the optical densities of the initial solution and also the exhausted one by using an UV-visible spectrophotometer. It was noted that the dye bath exhaustion was observed in a sequence of PPAA > O_2 > N_2 plasma-treated samples for all the treatment time. The dye bath exhaustion was just 15% for control samples. The maximum dye bath exhaustion was more than 65% for PPAA deposited for 4 min. It may be also be noted from the SEM micrographs that the formation of etch-pits and voids is very predominant for O_2 plasma-treated nylon fabrics, whereas PPAA-treated samples reveal some deposition on the surface due to formation of PPAA. The increase in the dyeing could be due to the fact that gaseous plasma etches out the surface of the nylon fibres, creating a rougher surface with irregularities as discussed earlier. The effective surface area increases after the plasma treatment. Thus the interaction and diffusion of the dye molecules is facilitated. The chemical changes in the nylon fibre surface can lead to the possibility of the formation of free radicals on the nylon chains and the subsequent formation of hydroxyl, carbonyl and carboxyl groups. It is also important to mention that the dyeing process mainly occurs through the amorphous regions. The etching away of the amorphous regions during the plasma treatment (particularly for longer treatment time) can lead to lowering of the dye uptake. Therefore the optimum treatment time needs to be found. Yoon et al. (1996) and Thomas et al. (1998) have found that longer plasma treatment decreases dye uptake. It may be noted that the oxygen is more reactive than nitrogen, and hence the dye uptake is slightly more for oxygen-treated samples. The etching caused due to oxygen plasma is also more as shown in Fig. 6.9(c). Plasma polymerization of acrylic acid (PPAA) onto the nylon fabric incorporates good amount of hydroxyl, carbonyl and carboxylic acid groups onto the surface as evident from FTIR studies. These functional groups are responsible for wettability and hence dye uptake. The dye bath exhaustion is slightly more in case of PPAA-deposited nylon fabric as compared to that of gaseous plasma-treated samples, because of the functional groups which are incorporated during the deposition of PPAA. The increase in the dye uptake due to plasma treatment is also evident from the measurement of K/S values shown in Fig. 6.11. The bar-graph shows K/S values for control and plasma-treated samples for 4 min. in different gases. The highest gain is seen to be for PPAA-deposited nylon fabrics followed by oxygen-treated sample. This trend is similar to that observed for the data of dye bath exhaustion.

6.5.6   Polypropylene processing

Polypropylene (PP) fibres have high crystalline index, excellent properties, high strength, thermal resistance, etc., but does not contain polar groups; and hence it is hydrophobic in nature. Polypropylene fibres have poor wettability and hence dyeability is low. Therefore dyeing of PP fibre has been challenging by conventional methods. Air and argon plasma treatment of PP fibres followed by dyeing with vat dyes showed significant increase in colour strength (Yaman et al., 2011). Plasma produces free radicals and unsaturated bonds on the surface. Our work using oxygen and nitrogen plasma at low pressures has shown that dyeing increases with disperse and direct dyes. Nitrogen gas plasma introduces N-H groups which are useful for direct dyes where as O-H

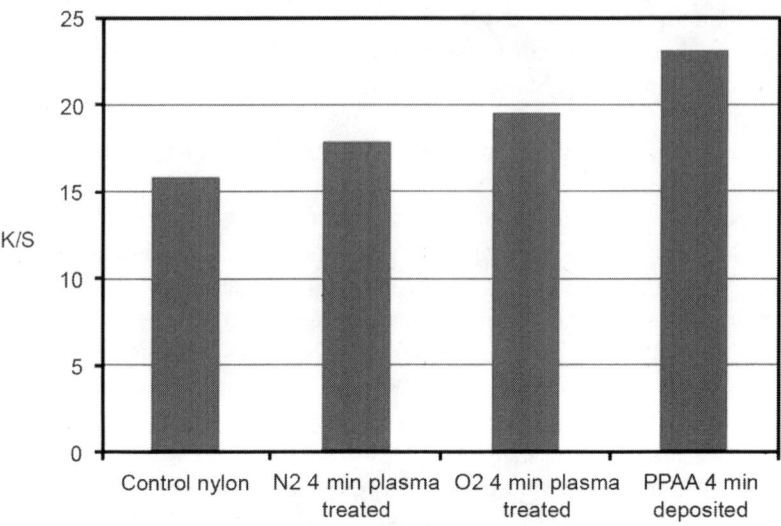

Figure 6.11 K/S values of nylon fabric subjected to RF plasma treatment.

groups are produced during O_2 plasma which helps in basic and disperse dyes. FTIR spectra show the presence of OH, C=O and N-H groups. Further SEM studies revealed the roughness created on the surface which effectively increases the area and hence the dye uptake. Apart from the measurement of exhaustion of dye, K/S values on the fabric which indicates more dye uptake, the actual observation and comparison of shades also indicates achievements of the deeper shades. Figure 6.12 shows PP dyeing Absorbance Vs Treatment time. It is clear from Fig. 6.12 that dye both absorbance decreases with increase in plasma treatment time, which confirms improved dye uptake by the PP fabric with increase in plasma treatment time. Similarly, Shahidi et al. (2007) have reported a significant increase in color depth upon dyeing after treating PP fabric with low temperature plasma of O_2 and N_2. It may be noted from the results described and Tables 6.1, 6.3, and 6.4, that the dyeing behaviour depends on various factors such as,

(i)   plasma treatment conditions (pressure, time of treatment)
(ii) plasma gas (air, N_2, O_2, NH_3, organic vapour)
(iii) type of dye (reactive, acidic, vat, disperse, etc.)
(iv) dyeing procedure (temp., solvent, etc.)

Therefore, the general anticipation of textile technologists that the plasma processing enhances the dye up take is not generally true as several factors influence the results. The earlier results using low pressure plasma, particularly for wool and nylon, indicated the saving of the dye due to plasma processing was not observed for cotton, polyester, etc.

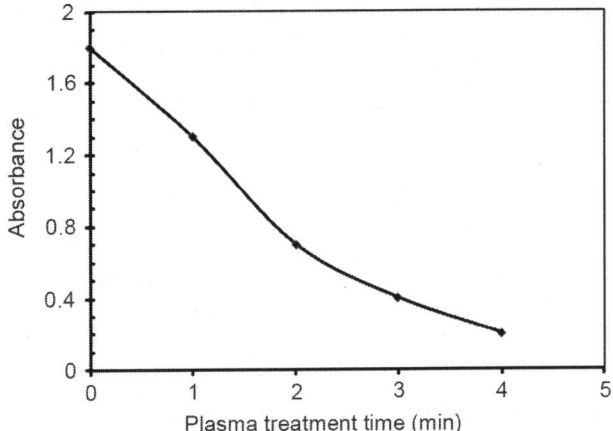

Figure 6.12 Dyeing enhancement of PP fibres subjected to N$_2$ plasma measured by decreasing absorbance in the dye bath.

6.6   Plasma for creating anti-bacterial and anti-flammable fabrics

Recently plasma treatments have shown that some exotic properties can also be achieved. Since plasma consists of charged particles and its bombardment on surfaces leads to "cleaning", the existing bacteria may be killed or etched away leading to anti-bacterial properties. However, this alone has a limitation in the sense that it acts rather as "biostat". Biocidal property can be obtained if we can graft an agent on the functionalized fibre surface.

Reports using chitosan, a natural polymer, in combination with plasma can make polyester fabrics highly anti-bacterial (AB). When chitosan oligomers were used, the results were more encouraging (Chang et al., 2008). Use of chitosan gives surface grafts and its eco-friendly approach for creating anti-bacterial materials. Excellent AB activity and surface properties of polyamide 6 films/ fibres modified with Ar-plasma and methyl diallyl ammonium grafts were reported (Yao et al., 2008). Innovative dual anti-microbial (AM) and anti-creasing finishing of cotton fabrics has been reported by Aly et al. (2007). Atmospheric plasma-aided biocidal finishing for PP fabrics has been reported by Wafa et al. (2007).

Use of nanoparticles (NP) deposition on the fabrics by aid of plasma has been reported (Shahidi et al., 2010b; Gorensek et al., 2010; Maja et al., 2008). Silver nanoparticles give excellent AB properties. In some cases, fabrics were first treated by plasma followed by treatment with silver nitrate or Ag NP solution. Morphological changes caused due to etching of polyester fabrics lead to better adhesion at Ag NP. AB property was found to have been improved with respect to *Staphylococcus aureus* and *Escherichia coli*. The rate of bacterial inactivation was found to be over a considerable period and also up to 5 washes.

Wang (2007) reported plasma irradiation followed by Ag-coated SiO2 NP. Increasing AB property was found with increasing locking of NP. Imparting of anti-bacterial and flame-retardant (Raslan et al., 2011b) properties to polyester by using Ag NP and TiO$_2$ NP has been reported.

Ibrahim et al. (2012) have studied the anti-bacterial activity of plasma-treated (argon or N$_2$) and/or nanosilver-treated PET against gram positive bacteria (*Staphylococcus aurous*) and gram negative bacteria (*Escherichia coli*). Their results showed that both the hydrophilicity and antibacterial behaviours against gram positive bacteria (*Staphylococcus aurous*) and gram negative bacteria (*Escherichia coli*) were highly improved by the treatment of fabric by either individual or combined DC cold plasma or nanosilver treatments.

Plasma-induced graft polymerization either in-situ or after pre-treatment can be used for grafting anti-flammable monomers. Recently (Tsafack et al., 2006) acrylic monomers containing phosphorus such as DEAEP diethyl(acryloxy ethyl) phosphate, DEMEP diethyl2-(methyl acryloyloxy ethyl) phosphate and DMAMP dimethyl(acryloyloxy methyl) phosphate were used to calculate limiting oxygen index (LOI) values which were reported ~28.5–29%. Washing durability though reduced was found to be satisfactory. Attempts to improve the flame resistance of underlying fabrics by plasma treatments have recently been reviewed by Horrocks (2008).

6.7   Industry relevance of our work

The plasma technology, cold plasma that we have been using, has immense potential in several industries such as textiles, plastic, paper and medicine. Our main emphasis at BTRA as well as ICT was to evolve applications in the area of textile and polymers. Our research publications have established beyond doubt that RF-generated plasma using nitrogen, oxygen, helium, etc., gases and/or monomer (acrylates, fluorine containing, silane containing) vapours can be effectively used for making the fabrics and polymer surfaces either hydrophilic or hydrophobic and in the process there is etching/ ablation or deposition. Similarly by use of organic vapours as medium, it is possible to coat the surfaces to make them suitable for various purposes. These principles were used while demonstrating the suitability of plasma technology to the industry. In fact the textile industry is looking out for new technology and in keeping with the modern trend it has to be eco-friendly. The plasma technology is "Green technology" with added advantage that it is non-polluting, water saving and easy to operate with the existing type of man power.

At BTRA we have set up a 50 cm width plasma-processing machine to demonstrate the easy working conditions of the technology. Since it was possible to use different gases or vapours along with He gas and apply different powers using BTRA machine, it was easy to use it for a variety of applications. Researchers have demonstrated that the BTRA system can be used to improve water absorbancy of Turkish towels simultaneously making it very soft and extra white. The same system was used to increase water repellency for cloth to be used for umbrellas and raincoats. In addition, the plasma treatment has increased dye uptake giving deeper shades with saving in dye and water. The plasma-based technology is emerging rapidly and in near future it would be replacing many conventional processes.

6.8   Acknowledgement

Thanks are due to Ms. Gauree Arolkar, Ms. Shital Palaskar and Ajinkya Trimukhe for their support in this work.

6.9   References

Alexandrov S.E. (2009). 'Chemical vapour deposition: precursors, processes and applications' in Anthony C. Jones and Michel L. Hitchman, Royal Society of Chemistry.

Aly A.S., Mostafa A.B.E., Ramadan M.A., and Hebeish A. (2007). 'Innovative dual antimicrobial & anticrease finishing of cotton fabric', Polymer plastics technology and engg., **46**(7), pp. 703–707.

Bhat N.V. and Nadiger G.S. (1978). 'Effect of nitrogen plasma on the morphology and allied textile properties of tassar silk fibres and fabrics', Text Res J, **48**, pp. 685–691.

Bhat N.V. and Benjamin Y.N. (1999). 'Surface resistivity behavior of plasma treated and plasma grafted cotton and polyester fabrics', Textile Research Journal, **69**, pp. 38–42.

Bhat N.V. and Deshmukh R.R. (2002). 'X-ray crystallographic studies of polymeric materials', Indian Journal of Pure & Applied Physics, **40**(5), pp. 361–366.

Bhat N.V., Bharati R.N., Gore A.V., and Patil A.J. (2011a). 'Effect of plasma treatment on desizing & wettability of cotton fabrics', Indian Journal of Fibres and Textile Research, **36**(1), pp. 42–46.

Bhat N.V., Gore A.V., Sathyanarayan M.P., Netravali A.N., Arolkar G.A., and Deshmukh R. R. (2011b). 'Surface modification of cotton fabrics using plasma technology', *Textile Research Journal*, **81**, pp. 1014–1026.

Bhattacharyya D., Xu H., Deshmukh R.R., Timmons R.B., and Kytai T.N. (2010). 'Surface chemistry and polymer film thickness effects on endothelial cell adhesion and proliferation', Journal of Biomedical Materials Research Part A, **94A**(2), pp. 640–648.

Cai Z., Qiu Y., Zhang C., Hwang Y-J, and Marian M. (2003a). 'Effect of atmospheric plasma treatment on desizing of PVA on cotton', *Textile Research Journal*, **73**(8), pp. 670–674.

Cai Z., Qiu Y., Hwang Y-J., Chuyang Z., and Marian M. (2003b). 'The use of atmospheric pressure plasma treatment in desizing PVA on viscose fabrics', Journal of Industrial Textiles, 32(3), pp. 223–232.

Cai Z. and Qiu Y. (2006). 'The mechanism of air/oxygen/helium atmospheric plasma action on PVA', Journal of Applied Polymer Science, 99(5), pp. 2233–2237.

Cai Z. and Qiu Y. (2008). 'Dyeing properties of wool fabrics treated with atmospheric pressure plasmas', Journal of Applied Polymer Science, 109(2), pp. 1257–1261.

Calvimontes A., Saha R., and Dutschk V. (2011). 'Topographical effects of O2 and NH3 plasma treatment on woven plain polyester fabric in adjusting hydrophilicity', AUTEX Research Journal, 11(1), pp. 24–30.

Carneiro N., Souto A.P., Silva E., Marimba A., Tena B., Ferreira H., and Magalhães V. (2001). 'Dyeability of corona-treated fabrics', Color. Technol. 117, pp. 298–302.

Carneiro N., Souto A.P., and Nogueira C. (2007). 'Reactive Pad-Batch dyeing in corona discharged fabrics', Journal of Natural Fibers, 4(2), pp. 51–65.

Caschera D., Cortese B., Mezz A., Brucale M., Ingo G.M., Gigli G., and Padeletti G. (2013). 'Ultra hydrophobic/superhydrophilic modified cotton textiles through functionalized diamond-like carbon coatings for self-cleaning applications', Langmuir, 29, pp. 2775–2783.

Chang Y.B., Tu P.C., Wu M.W., Hsueh T.H., and Hsu S.H. (2008). 'A study on chitosan modification of polyester fabrics by atmospheric pressure plasma and its antibacterial effects', Fibers and Polymers, 9(3), pp. 307–311.

Chen J.R., Wang X.Y., and Tomiji W. (1999). 'Wettability of poly (ethylene terephthalate) film treated with low-temperature plasma and their surface analysis by ESCA', J. Appl. Polym. Sci, 72(10), pp. 1327–1333.

Cireli A., Kutlu B., and Mutlu M. (2007). 'Surface modification of polyester and polyamide fabrics by low frequency plasma polymerization of acrylic acid', Journal of Applied Polymer Science, 104, pp. 2318–2322.

Clark D.T., Dilks A., and Shuttleworth D. (1978). 'Polymer surfaces' in Clark D.T. and Feast W.J. John Wiley, New York, pp. 185–210.

Coulson S., Khondddama A., and Carr C. (2011). 'Plasma polymerization for coating wool with a fluoropolmer to prevent shrinkage due to felting during laundering step', U.K. Patent GB 2475685.

Danish N., Garg M.K., Rane R.S., Jhala P.B., and Nema S.K. (2007). 'Surface modification of Angora rabbit fibers using dielectric barrier discharge', Applied Surface Science' 253(16), pp. 6915–6921.

Dave H., Ledwani L., Chandwani N., Kikani P., Desai B., Chowdhuri M.B., and Nema S.K. (2012). 'Use of dielectric barrier discharge in air for surface modification of polyester substrate to confer durable wettability and enhance dye uptake with natural dye eco-alizarin', Composite Interfaces, 19(3–4), pp. 219–229.

Demir A., Ark B., Ozdogan E., and Seventekin N. (2010). 'The comparison of the effect of enzyme, peroxide, plasma and chitosan processes on wool fabrics and evaluation for antimicrobial activity', Fibers and Polymers, 11(7), pp. 989–995.

Demir A., Ozdogan E., Ozdil N., and Gurel A. (2011). 'Ecological material and methods in the textile industry: Atmospheric-plasma treatment of naturally colored cotton', Journal of Applied Polymer Science, **119**, pp. 1410–1416.

Deshmukh R.R. and Shetty A.R. (2007a). 'Modification of polyethylene surface using plasma polymerization of silane', Journal of Applied Polymer Science, **106**, pp. 4075–4082.

Deshmukh R.R. and Shetty A.R. (2007b). 'Surface characterization of polyethylene films modified by gaseous plasma', Journal of Applied Polymer Science, **104**, pp. 449–457.

Deshmukh R.R. and Shetty A.R. (2008). 'Comparison of surface energies using various approaches and their suitability', Journal of Applied Polymer Science, **107**, pp. 3707–3717.

Deshmukh R.R. and Bhat N.V. (2012). 'Pre-treatments of textiles prior to dyeing: Plasma processing' in 'Textile dyeing', pages 33–56, Ed Hauser P J., InTech Publisher, Croatia.

Dragan J., Susana V., Tatjana T., Ricardo M., Antonio N., Petar J., Maria R.J., and Pilar E. (2005). 'Effect of low-temperature plasma and chitosan treatment on wool dyeing with Acid Red 27', Journal of Applied Polymer Science, **97**, pp. 2204–2214.

El-Zawahry M.M., Ibrahim N.A., and Eid M.A. (2006). 'The impact of nitrogen plasma treatment upon the physical-chemical and dyeing properties of wool fabric', Polymer and Plastics Technology and Engineering, **45**(10), pp. 1123–1132.

Etzler F.M. (2003). 'Characterization of surface free energies and surface chemistry of solids, in: Contact Angle, Wettability and Adhesion, Vol. 3, K.L. Mittal, (Ed.), pp. 219–264, VSP, Utrecht.

Etzler F.M. (2013). 'Determination of the surface free energy of solids: A Critical Review', Rev Adhesion Adhesives, Scrivener Publishing, 1(1), pages 3–45.

Fang K., Wang S., Wang C., and Tian A. (2008). 'Inkjet printing effects of pigment inks on silk fabrics surface-modified with O_2 plasma', J. Applied Polymer Science, **107**, pp. 2949–2955.

Ferrero F., Tonin C., Peila R., and Pollone R. (2004). 'Improving the dyeability of synthetic fabrics with basic dyes using in situ plasma polymerization of acrylic acid', Coloration Technology, **120**(1), pp. 30–34.

Gawish S.M., Saudy M.A., El-Ola A.S.M., and El-Kheir A. (2011). 'The effect of low temperature plasma for improving wool and chitosan-treated wool fabric properties', The Journal of Textile Institute, **102**(2), pp. 180–188.

Ghoranneviss M., Shahidi S., Anvari A., Motaghi Z., Wiener J., and Slamborova I. (2011). 'Influence of plasma sputtering treatment on natural dyeing and antibacterial activity of wool fabrics', Progress in Organic Coatings, **70**(4), pp. 388–393.

Gorafa A.M. (1980). Textile Chemistry and Colaration **12**(4).

Gorensek M., Gorjanc M., Bukosek V., Kovac J., Petrovic Z., and Pauc N. (2010). 'Functionalization of polyester fabric by Ar/N_2 plasma and silver', Textile Res .J., **80**(16), pp. 1633–1642.

Goud V.S. (2012). 'Influence of plasma processing parameters on mechanical properties of wool fabrics', IJFTR **37**(3), pp. 292–298.

Gulrajani M.L. and Gupta D. (2011). 'Emerging techniques for functional finishing of textiles', Indian J Fibre and Textile Res, **36**, pp. 388–397.

Guzenda A.S., Szymanowski H., Jakubowski W., Błasińska A., Kowalski J., and Lipman M.G. (2013). 'Morphology, photocleaning and water wetting properties of cotton fabrics, modified with titanium dioxide coatings synthesized with plasma enhanced chemical vapor deposition technique', **217**, pp. 51–57.

Herbert T. (2009). 'Plasma technologies for textiles' in Shishu R., Cambridge, Woodhead Publishing Ltd.

Höcker H. (1995). Int. Text. Bull. Veredlung **4**, p. 18.

Höcker H. (2002). 'Plasma treatment of textile fibres', Pure Appl. Chem., **74**(3), pp. 423–427.

Hollahan J.R. and Bell A.T. (1984). Techniques and applications of plasma chemistry, John Wiley & Sons, New York.

Horrocks A.R. (2008). 'Advances in fire retardant materials', (Eds. Horrocks A R,

Price D), Woodhead Publishing Ltd., Cambridge, UK, pp. 181–184.

Hossain M.M., Hegemann D., and Herrmann A.S. (2006a). 'Plasma hydrophilization effect on different textile structures', Plasma Processes and Polymers. **3**, pp. 299–307.

Hossain M.M., Hegemann D., Herrmann A.S., and Chabrecek P. (2006b). 'Contact angle determination on plasma-treated Poly(ethylene terephtalate) fabrics and foils, Journal of Applied Polymer Science **102**, pp. 1452–1458.

Hossain M., Mussig J., Heremann A., and Hegemann D. (2009). 'Ammonia / acetylene plasma deposition: An alternative approach to the dyeing of poly(ethylene terephthalate) fabrics at low temperatures', J Applied Polymer Sci, **111**(5), pp. 2545–2552.

Ibrahim N.A., Eid B.M., Hashem M.M., Refai R., and EL-Hossamy M. (2010b). 'Smart options for functional finishing of linen-containing fabrics', Journal of Industrial Textiles **39**, pp. 233–265.

Ibrahim N.A., Hashem M.M., Eid M.A., Refai R., Hossamy M. El and Eid B.M. (2010a). 'Eco-friendly plasma treatment of linen-containing fabrics', The Journal of The Textile Institute, **101**(12), pp. 1035–1049.

Ibrahim S.F., Essa D.M., Abdel-Razik A.M., Nagar Khaled El, Saudy M.A., and Abdel-Rahman A.A.H. (2012). 'Application of DC plasma discharge and /or nanosilver treatments to Poly(ethylene terephthalate) fabrics to induce hydrophilicity and antibacterial activity', Elixir Appl. Chem. **50**, pp. 10370–10377.

Inbakumar S. and Anukaliani A. (2012). 'Chemical and physical changes in surface of argon plasma treated cotton fabrics', Composite Interfaces, **19**(3–4), pp. 209–218.

Iriyama Y., Mochizuki T., Watanabe M., and Utada M. (2002). 'Plasma treatment of silk fabrics for better dyeability', Journal of Photopolymer Science and Technology, **15**(2), pp. 299–306.

Jahagirdar C.J. and Tiwari L.B. (2004). 'Effect of Dichlorodimethylsilane on plasma treated cotton fabtric', Pramana- J Physics, **62**(5), pp. 1099–1109.

Jahagirdar C.J. and Tiwari L.B. (2007). 'Plasma treatment of polyester fabric to impart the water repellency property' Pramana-J Physics, **68**(4), pp. 623–630.

Jung H.Z., Ward T.L., and Benerito R.R. (1977). 'The effect of argon cold plasma on water absorption of cotton', Textile Research Journal, **47**, pp. 217–223.

Kale K.H. and Desai A.N. (2011a). 'Atmospheric pressure plasma treatment of textiles using non-polymerizing gases', Indian J of fibre and textile research, **36**, pp. 289–299.

Kale K H and Palaskar S, (2011b), 'Atmospheric pressure plasma polymerization of hexamethyldisiloxane for imparting water repellency to cotton fabric', Textile Research Journal, **81**(6), pp. 608–620.

Kale K.H. and Palaskar S. (2012a). 'Plasma enhanced chemical vapor deposition of tetraethylorthosilicate and hexamethyldisiloxane on polyester fabrics under pulsed and continuous wave discharge', Journal of Applied Polymer Science, **125**(5), pp. 3996–4006.

Kale K.H., Palaskar S.H. and Kasliwal P.M. (2012b). 'A novel approach for functionalization polyester and cotton textiles with continuous online deposition of plasma polymers', IJFTR **37**, pp. 238–244.

Kamlangkla K., Paosawatyanyong B., Pavarajarn V., Hodak J.H., and Hodak S.K. (2010). 'Mechanical strength and hydrophobicity of cotton fabric after SF6 plasma treatment', Applied Surface Science, **256**, pp. 5888–5897.

Kan C.W., Chan K., Yuen C.W.M., Miao M.H. (1998). 'Surface properties of low-temperature plasma treated wool fabrics', Journal of Material Processing Technology, **83**(1–3), pp. 180–184.

Kan C.W., Chan K., and Yuen C.W.M. (2000). 'Application of low temperature plasma on wool – part I: Review', The Nucleus, **37**(1/2), pp. 9–21.

Kan C.W., Chan K., and Yuen C.W.M. (2004). 'Surface characterisation of low temperature plasma treated wool fibre – The effect of the nature of gas', Fibers and Polymers, **5**(1), pp. 52–58.

Kan C.W. and Yuen C.M (2006). 'Dyeing behaviour of low temperature plasma treated wool', Plasma Processing and Polymers, **3**(8), pp. 627–635.

Kan C.W. and Yuen, C.W.M. (2007). 'Plasma technology in wool', Textile Progress, **39**(3), pp. 121–187.

Kan C.W. and Yuen C.W.M. (2009). 'A Comparative study of wool fibre surface modified by physical and chemical methods, Fibers and Polymers', **10**(5), pp. 681–686.

Kan C.W., Yuen C.W.M., Tsoi W.Y.I., and Tang T.B. (2010). 'Plasma pre-treatment for polymer deposition - improving antifelting properties of wool', Plasma Science, IEEE Trans. **38**(6), pp. 1505–1511.

Kan C.W. and Yuen C.W.M. (2012). 'Effect of atmospheric pressure plasma treatment on the desizing and subsequent color fading process by cotton denim fabric', Coloration Technology, **128**(5), pp. 356–363.

Kangti Yun and Saramadi M. (2004). AATCC reviews, **4**, p. 28.

Karahan H.A., Ozdogan E., Demir A., Kocum I.C., Oktem T., and Ayhan H. (2009a). 'Effects of atmospheric pressure plasma treatments on some physical properties of wool fibres', Textile Research Journal, **79**(14), pp. 1260–1265.

Karahan H.A., Ozdogan E., Demir A., Ayhan H., and Seventekin N. (2009b). 'Effects of atmospheric pressure plasma treatments on certain properties of cotton fabrics', Fibers Text East Eur, **73**, pp. 19–22.

Kutlu B., Aksit A., and Mutlu M. (2010). 'Surface modification of textiles by glow discharge technique: part ii: low frequency plasma treatment of wool fabrics with acrylic acid' Journal of Applied Polymer Science, **116**, pp. 1545–1551.

Lehocky M. and Mracek A. (2006). 'Improvement of dye adsorption on synthetic polyester fibers by low temperature plasma pre-treatment' Czechoslovak Journal of Physics, **56**, pp. B1277–B1282.

Liu Y.C., Xiong Y., and Lu D.N. (2006). Surface characteristics and antistatic mechanism of plasma-treated acrylic fibers', Applied Surface Science, **252**(8), pp. 2960–2966.

Maja R., Vesna I., Vesna V., Suzana D., Petar J., Zoran S., and Jovan M.N. (2008). 'Antibacterial effect of silver nanoparticles deposited on corona-treated polyester and polyamide fabrics', Polymers Advanced Technologies, **19**(10), pp. 1816–1821.

Malek R.M.A. and Holme I. (2003). 'The effect of plasma treatment on some

properties of cotton' Iranian Polymer Journal, **12**(4), pp. 271–280.

Malkov G.S. and Fisher E.R. (2010). Pulsed plasma enhanced chemical vapor

deposition of poly(allyl alcohol) onto natural fibers', Plasma Process Polymer, **7**, pp. 695–707.

Marciano F.R., Bonetti L.F., Da-Silva N.S., Corat E.J., and Trava Airoldi V.J. (2009). 'Wettability and antibacterial activity of modified diamond-like carbon films', Appl. Surf. Sci, **255**, pp. 8377–8382.

Matsumoto R., Sato K., Ozeki K., Hirakuri K., and Fukui Y. (2008). 'Cytotoxicity and tribological property of DLC films deposited on polymeric materials' Diamond Related. Materials, **17**, pp. 1680–1684.

Mendhe Pankaj K, M Tech Thesis, (2012), 'Surface modification of textile materials by using plasma technology', Institute of Chemical Technology.

Mehta R. (2010). "Plasma Treatment" in the Textile Industry: An eco-friendly approach to wet processing, Colourage, **57**(7), pp. 45–48.

Morent R., Geyter N.D., Verschuren J., Clercke K.D., Kiekens P., and Leys C. (2008). 'Non-thermal plasma treatment of textiles', Surf. Coat. Technol. **202**(14), pp. 3427–3449.

Mori M., Von Arnim V., Dinkelmann A., Matsudaira M., and Wakida T. (2011). 'Modification of wool fibres by atmospheric pressure plasma treatment', Journal of the Textile Institute, **102**(6), pp. 534–539.

Motaghi Z. and Shahidi S. (2012). 'Development of polyester-wool fabrics dye ability using plasma supttering', RMUTP International Conference: Textiles & Fashion, July 3-4, Bangkok Thailand.

Nagar Kh El, Saudy M.A., Eatah A.I., and Masoud M.M. (2006). 'DC pseudo plasma discharge treatment of polyester textile surface for disperse dyeing', The Textile Institute JOTI, **97**(2), pp. 111–117.

Needle Howard L. (1986). 'Textile fibers, dyes, finishes, and processes', Noyes publication, New Jersey USA.

Nebojsa R., Petar J., Cristina C., and Dragan J. (2009). 'One-bath one-dye class dyeing of PES/cotton blends after corona and chitosan treatment', Fibres and Polymers, 10(4), pp. 466–475.

Nebojsa R., Petar J., Cristina C., and Dragan J. (2010). 'Influence of corona discharge and chitosan surface treatment on dyeing properties of wool', Journal of Applied Polymer Science, 117, pp. 2487–2496.

Ozdogan E., Saber R., Aghan H., and Seventekin N. (2003). 'A new approach for dyeability of cotton fabrics by different plasma polymerization Methods', Coloration Technology, 118(3), pp. 100–103.

Ozdogan E., Demir A., Karahan A.A., Ayhan H., and Seventekin N. (2009). 'Effects of atmospheric plasma on the printability of wool fabrics', Tekstil ve Konfeksiyon, 19, pp. 123–127.

Oktem T., Seventekin N., Ayhan H., and Piskin E. (2000). 'Modification of polyester and polyamide fabrics by different in situ plasma polymerization methods', Turk J Chem, 24, pp. 275–285.

Oktem T., Seventekin N., Aylian H., and Piskin E. (2002). 'Improvement in surface-related properties of poly(ethylene terephthalate)/cotton fabrics by glow-discharge treatment', Indian Journal of Fiber and Textile Research, 27, pp. 161–165.

Okuno T., Yasuda T., and Yasuda, H. (1992). 'Effect of crystallinity of pet and nylon 66 fibers on plasma etching and dyeability characteristics', Text Res J, 62(8), pp. 474–480.

Panda P.K., Rastogi D., Jassal M., and Agrawal A.K. (2012). 'Effect of atmospheric pressure helium plasma on felting and low temperature dyeing of wool', Journal of Applied Polymer Science, 124(5), pp. 4289–4297.

Pandiyaraj N. and Selvarajan V. (2008). 'Non-thermal plasma treatment for hydrophilicity improvement of grey cotton fabrics', Journal of Materials Processing Technology, 199(1–3), pp. 130–139.

Peng S., Gao Z., Sun J., Yao L., and Qiu Y. (2009). 'Influence of argon/oxygen atmospheric dielectric barrier discharge treatment on desizing and scouring of poly (vinyl alcohol) on cotton fabrics', Applied Surface Science, 255(23), pp. 9458–9462.

Pichal J. and Klenko Y. (2009). 'ADBD plasma surface treatment of PES fabric sheets', The European Physical Journal D 54, pp. 271–279.

Raffaele-Addamo A., Selli E., Barni R., Riccardi C., Orsini F., Poletti G., Meda L., Massafra M.R., and Marcandalli B. (2006). 'Cold plasma-induced modification of the dyeing properties of poly(ethylene terephthalate) fibers', Applied Surface Science, 252(6), pp. 2265–2275.

Rakowski W. (1997). 'Plasma treatment of wool today. Part 1 – Fibre properties, spinning and shrinkproofing', JSDC, 113, pp. 250–255.

Raslan W.M., Rashed U.S., El. Sayed H., and El. Halwagy A.A. (2011a). ECAPC11, Sharjah AUE.

Raslan W.M., Rashed U.S., El. Sayed H., and El. Halwagy A.A. (2011b). 'Ultraviolet protection, flame retardancy and antibacterial properties of treated polyester fabric using plasma-nano technology', Materials Science and Applications, 2, pp. 1432–1442.

Ren C.S., Wang D.Z., and Wang Y.N. (2008). 'Improvement of the graft and dyeability of linen by DBD treatment in ambient air', Journal of materials processing technology, **206**, pp. 216–220.

Rombaldoni F., Montarsolo A., Mossotti R., Innocenti R., and MazzuChetti G. (2010). 'Oxygen plasma treatment to reduce the dyeing temperature of wool fabrics', Journal of Applied Polymer Science, **118**(2), pp. 1173–1183.

Roth J.R. (2001a). 'Industrial Plasma Engineering' **1**, IOP, Bristol.

Roth J.R. (2001b). 'Industrial Plasma Engineering' **2**, IOP, Bristol.

Samanta K.K., Jassal M., and Agrawal A.K. (2008). 'Formation of nano-sized channels on polymeric substrates using atmospheric pressure glow discharge cold plasma ', Nanotrends: A Journal of Nanotechnology and its Applications', **4**(1), pp. 71–75.

Samanta K.K., Jassal M., and Agrawal A.K. (2009). 'Improvement in water and oil absorbency of textile substrate by atmospheric pressure cold plasma treatment', Surface & Coatings Technology **203**, pp. 1336–1342.

Samanta K.K., Jassal M., and Agrawal A.K. (2010). 'Antistatic effect of atmospheric pressure glow discharge cold plasma treatment on textile substrates', Fibers and Polymers, **11**(3), pp. 431–437.

Samanta K.K., Joshi A.G., Jassal M., and Agrawal A.K. (2012). 'Study of hydrophobic finishing of cellulosic substrate using He/1,3-butadiene plasma at atmospheric pressure', Surface & Coatings Technology **213**, pp. 65–76.

Shah J.N. and Shah S.R. (2013). 'Innovative plasma technology in textile processing: a step towards green environment', Research Journal of Engineering Sciences ISSN 2278–9472, **2**(4), pp. 34.–39.

Shahidi S., Ghoranneviss M., Moazzenchi B., Rashidi A., and Dorranian D. (2007). 'Effect of using cold plasma on dyeing properties of polypropylene fabrics', Fibers and Polymers, **8**(1), pp. 123–129.

Shahidi S., Ghoranneviss M., Bahareh M., Rashidi A., and Dorranian D. (2009). Study of surface modification of wool fabrics using low temperature plasma, Proceedings of the 3rd International conference on the frontiers of Plasma Physics and Technology (PC/5099) ISBN:978-92-0-159608-6.

Shahidi S., Rashidi A., Ghoranneviss M., Anvari A., and Wiener J. (2010a). Plasma effects on anti-felting properties of wool fabrics', Surface and Coatings Technology, **205**(1), pp. 349–354.

Shahidi S., Rashidi A., Ghoranneviss M., Anvari A., Rahimi M.K., Bameni M.M., and Wiener J. (2010b). 'Investigation of metal absorption and antibacterial activity on Cotton fabric modified by low temperature plasma', Cellulose, **7**(3), pp. 627–634.

Shenton M.J. and Stevens G.C. (2001). 'Surface modification of polymer surfaces: atmospheric plasma versus vacuum plasma treatments', IOP- Journal of Physics D: Applied Physics, **34**, pp. 2761–2768.

Shijian L., Van O. and Wim J. (2002). 'Surface modification of textile fibres for improvement of adhesion to polymeric matrices: A review', Journal of Adhesion Science and Technology, **16**(13), pp. 1715–1735.

Shin B., Lee K-R., Moon M-W., and Kim H-Y. (2012). 'Extreme water repellency of nanostructured low-surface-energy non-woven Fabrics', Soft Matter, **8**, pp. 1817–1823.

Shirin N. and Mohammad E.Y. (2008). 'Effect of corona discharge treatment on indigo dyed cotton fabric', Society of Dyers and Colourists, Coloration Technology, **124**, pp. 43–47.

Shishoo R. (2007). 'Plasma Technologies for Textiles, Woodhead Publishing Ltd, UK.

Simpson W.S. and Crawshaw G.H. (2002). Wool: Science and Technology (2002). 'Fibre morphology', Woodhead Publishing Limited.

Sun D. and Stylios G.K. (2004). 'Effect of low temperature plasma treatment on the scouring and dyeing of natural fabrics', Textile Res J., **74**(9), pp. 751–756.

Thomas H., Denda B., Hedler M., Kasermann M., Klein C., and Merten T. (1998). 'Textile finishing with low temperature plasma', Melliand Textiliber, **79**, pp. 350–352.

Thomas H. (2007). Chap: 'Plasma modification of wool' in 'Plasma Technologies for Textiles', in Shishoo R. pp. 228–246.

Timmons R.B. and Griggs A.J. (2004). 'Pulsed plasma polymerizations' in polymer plasma films', Biederman H., Ed., Imperial College Press, pp. 217–245.

Tsafack M.J. and Grutzmacher J.L. (2006). 'Flame retardancy of cotton textiles by plasma-induced graft-polymerization (PIGP)', Surface and coating technology, **201**(6), pp. 2599–2610.

Tsafack M.J. and Grützmacher J.L. (2007). 'Towards multifunctional surfaces using the plasma-induced graft-polymerization (PIGP) process: Flame and waterproof cotton textiles' Surface and Coatings Technology, **201**(12), pp. 5789–5795.

Udakhe J. and Tyagi S. (2011). 'Effect of plasma density on surface morphology and mechanical properties on wool fibres' Man-Made Textiles in India, **54**(4), pp. 137–140.

Wafa D.M., Breidt F., Gawish S.M., Mathews S.R., Donohue K.V., Roe R.M., and Bourham M.A. (2007). 'Atmospheric plasma-aided biocidal finishes for nonwoven polypropylene fabrics. II. Functionality of synthesized fabrics', J Applied Polymer Science, **103**(3), pp. 1911–1917.

Wakida T., Tokino S., Niu S., Lee M., Uchiyama H., and Kaneko M. (1993a). 'Surface characteristics of wool and poly (ethylene terephthalate) fabrics and film treated with low-temperature plasma under atmospheric pressure' Textile Research Journal, **63**(8), pp. 433–438.

Wakida T., Tokino S., Niu S., Lee M., Uchiyama H., and Kaneko M. (1993b). 'Dyeing Properties of Wool Treated with Low-Temperature Plasma Under Atmospheric Pressure', Textile Res J., **63**(8), pp. 438–442.

Wakida T., Lee M., Sato Y., Ogasawara S., Ge Y., and Niu S. (1996). 'Dyeing properties of oxygen low-temperature plasma-treated wool and nylon 6 fibres with acid and basic dyes', Colouration Technology, **112**(9), p. 233.

Wang S., Hou W., Wei L., Jia H., Liu X., and Xu B. (2007). 'Antibacterial activity of nano-SiO2 antibacterial agent grafted on wool surface', Surface Coating Technology, **202**(3), pp. 460–465.

Wang C. and Wang C. (2010). 'Surface pretreatment of polyester fabric for ink jet printing with radio frequency O2 plasma', Fibers and Polymers, 11(2), pp. 223–228.

Wang X. and Peng Y. (2011). 'Comparative study of the structure and properties of

wool treated by a chicken-feather keratin agent, plasma, and their combination', Journal of Applied Polymer Science, 119, pp. 1627–1634.

Watson Kate Heintz (1907). 'The Project Gutenberg eBook of Textiles and Clothing', again released in 2007.

Xu W. and Liu X. (2003). 'Surface modification of polyester fabric by corona discharge irradiation', European Polymer Journal 39, pp. 199–202.

Yaman N., Özdo□an E., and Seventekin N. (2011). 'Atmospheric plasma treatment of polypropylene fabric for improved dyeability with insoluble textile dyestuff', Fibers and Polymers', 12(1), pp. 35–41.

Yao W.H., Chen J.C., and Chen C.C. (2008). 'Excellent anti-bacterial activity and surface properties of polyamide-6 films modified with argon-plasma and methyl diallyl ammonium salt-graft', Polymers Advanced Technologies 19(12), pp. 1513–1521.

Yasuda H. and Hsu T. (1978). 'Plasma polymerization investigated by the comparison of hydrocarbons and perfluorocarbons' Surface Science, 76, pp. 232–241.

Yasuda H. (1985). 'Plasma Polymerization', Academic Press Inc, Orlando, New York, London.

Yoon N.S., Lim Y.J., Tahara M., and Takagishi T. (1996). 'Mechanical and dyeing properties of wool and cotton fabrics treated with low temperature plasma and enzymes', Text Res J; 66, pp. 329–336.

Zenkiewicz M. (2007). 'Methods for the calculation of surface free energy', J. Achievements Mater. Manuf. Eng. 24, pp. 137–145.

Zhang C.M. and Fang K.J. (2011). 'Influence of penetration depth of atmospheric pressure plasma processing into multiple layers of polyester fabrics on inkjet printing', Surface Engineering, 27(2), pp. 139–144.

Zhongfu R., Xiaolieng T., Houglen W., and Gao Q. (2007). 'Continuous modification treatment of polyester fabric by Ar-O$_2$(10:1) discharge at atmospheric pressure', J Industrial Text 37, pp. 43–53.

7

Next to skin easy care woolens by plasma and other eco-friendly technologies

M. K. Bardhan and J. S. Udakhe

Abstract: Finer wool clothing is used for warmth, which can be attributed to the fibre structure. Apart from this, woollen garments cannot be worn next to the skin unlike other garments. This is due to the pricking and itching sensation caused by the wool fibres. Woollens also suffer from the disadvantage that these garments cannot be washed in domestic washing machines. There are many commercial treatments available to reduce the prickling or itching and to make woolens washable. Most of these commercial treatments use chemicals and are hazardous to the environment. Researchers worldwide are working on the eco-friendly methods of making wool soft and supple as well as easy to care. These methods include use of plasma, enzyme, UV/ozone and biopolymers alone or in combinations. This chapter discusses about these newer methods in detail.

Key words: Shrink proofing of wool, plasma, enzyme, biopolymers, UV/ozone treatments

7.1 Introduction

Apparel grade wool like Merino is used for warm clothing throughout the world. Internationally, all the producers of such woollen textiles are finally beginning to differentiate wool fibre by what the consumers need in their woollen clothing. Most relevant is the consumer's need for woollen garments which are worn next to skin to be soft, sensuous, luxurious, absolutely itch-free and above all should be machine washable and hence easy to care. One of the major problems of woollen garment lies in the fact that they require an intermediate garment made of cotton or other fibres next to the skin to avoid itching when worn. Even the finest wool like merino, pashmina, cashmere, etc., can suffer from inadequacies of creating uncomfortable itching to the wearer when worn next to skin. As India do not produce apparel grade wool and import about 75 million kg of merino wool, the value addition due to non-itchy and easy-to-care washable woolen textiles will boost the Indian woolen brands internationally.

As we know that next to skin garments are washed frequently and these garments must be machine washable. Commercial shrink proofing techniques of wool are known to make woollens soft and sensuous, but these treatments releases environmentally hazardous organic halogens (AOX) and most of the time the level exceeds the permitted limit of 40 mg/l. Hence there is an immediate need to replace these commercial shrink-proofing treatments for wool by eco-friendly processes like plasma, enzymes and biopolymer application.

This chapter details the problem of itching caused by woolen garments, and the ways to solve this problem using eco-friendly techniques. This chapter also briefs other dry technology available for the same, e.g. UV ozone method. We shall discuss about the worldwide research on the topic of itch proofing and shrink proofing of wool, with reference to plasma, enzyme and biopolymers.

7.2   Fabric-evoked itch and prickle

The skin is the largest organ of the human body. Some portion of an individual's skin is in constant contact with fabric, be it in the form of clothing, sheets, or towels. Fabric and skin work as a system to establish a thermal and sensorial state of comfort and to maintain a person's physically healthy state. Comfort has been defined as a 'state of satisfaction indicating physiological, psychological, and physical balance among the person, his/her clothing, and his/her environment'. Health has been defined by the World Health Organization as a 'state of complete physical, mental, and social well being'. Clothing comfort and clothing health are viewed as forming a continuum with a state of comfort and health lying at the extreme left, degrees of discomfort including being somewhat cool or warm or sensing prickliness or roughness or being psychologically uneasy forming the center of the continuum, and an unhealthy state at the extreme right (Hatch et al., 1992).

Fabric-evoked itch and prickliness sensation are one of the main factors for clothing discomfort. Initially many people were thinking that itching due to woollen garments is allergy to wool. Latter on researchers established a relationship of fabric-evoked prickle with neurophysiology and psychophysiology. Itch is defined as a sensation that evokes a desire to scratch, and prickling or stinging is described as sharp sensations that evoke a reflex reaction to withdraw rapidly. Fabric-evoked prickle is the result of low-grade activity in nociceptors and the stimuli are protruding fiber ends exerting loads of approximately 75 mgf or more against the skin (Garnsworthy et al., 1988). Prickle is the sensation of many gentle pinpricks and is solely fibre diameter dependent. Fine (19 μm) wool fibre fabrics are non-prickly, and prickle increases significantly as the mean fiber diameter increases (Naylor et al., 1992). Industrially, the comfort factor of the wool is calculated on the basis of percentage of fibres above 30 μm diameter, as it is proved that such fibres prickles much more and garments made of such fibres are not wearable (Li, 2001; Weiyu et al., 2002; Das et al., 2010). Smooth and less prickly perceptions are associated with fibers having diameter of less than 23 μm and fiber bulk greater than 31cm³/g; rougher and more prickly perceptions occurs when fiber diameter is greater than 34 μm and fiber bulk less than 21cm³/g (Wilson et al., 1995). As the fabric presses against the skin during wear, protruding fiber ends appear to obey Euler's simple buckling theory (Fig. 7.1), i.e. under compression between the fabric surface and the skin, a fiber end remains rigid up to a threshold load, the buckling load. Above this threshold, buckling occurs. Critical buckling load (P) indicates the amount of force a single fibre end can apply against the skin and can be determined mathematically as per Eq. [7.1].

$$P = \frac{n^2\pi^2 EI}{Le^2} \hspace{2cm} [7.1]$$

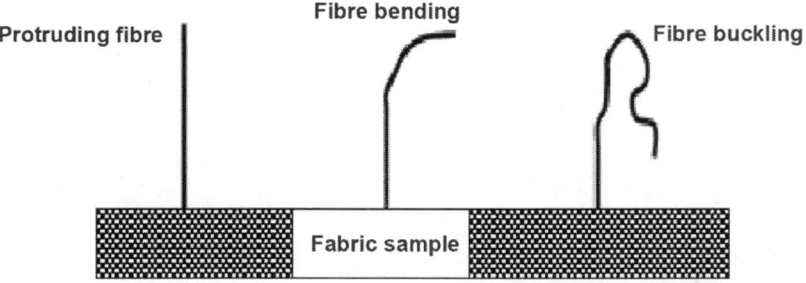

Figure 7.1 Euler buckling model for textile fibres protruding from the fabric surface

Where, n – mode of buckling, for this case the lowest mode is significant, i.e. n = 1, E – Young's modulus, I – Moment of Inertia (I= $\pi d^4/64$, d – fiber diameter), and L – effective length of fiber specimen. While Le = 0.699L and L can be assumed to be 1 mm for convenience (He et al., 2002). The critical buckling load (Fig. 7.2) of wool fibre increases exponentially with increase in average fibre diameter with R^2 of 0.99 (Udakhe et al., 2011b; Bardhan et al., 2012). Fabric-evoked prickle also depends on single fiber diameter irregularities as the critical buckling load changes and thus the fiber buckling behavior (Naylor et al., 1997b). Calculation of critical buckling load using finite element method (FEM) has shown that critical buckling load reduces with increase in the fibre irregularities. Comparison of classical critical buckling and FEM of buckling has shown that agreement between these two methods is quite good, with an approximate error of only 2% (He et al., 2002).

To model the mechanics of buckling of a fiber in relation to fabric-evoked prickliness, the fiber–skin friction, the elastic stiffness, and the initial inclined angle of the fiber are introduced into the simple Euler model of buckling of a slender rod. The results demonstrated that the fiber–skin friction and the elastic stiffness have a significant effect on the buckling behavior of fiber end prickling skin and the stimulus intensity to skin (Hu et al., 2011). Reducing the buckling load of protruding fiber ends contributes to a reduction in fabric prickle, and that the fiber diameter and protruding length of fiber ends are the major factors affecting the buckling load (Mayfield, 1987). Prickle sensation is due to the level of coarse fibres present in the fabric rather than the average fibre diameter (Dollong et al., 1990). Fiber diameter distribution (coefficient of variation of diameter CVd) is a more important factor than mean fiber diameter. Knitted fabric with a mean diameter of 23.2 µm and CVd of 16.4% was less prickly than a similar fabric with a mean fiber diameter of 21.5 µm and CVd of 21.7% (Dolling et al., 1992). Fabric comfort also depends on the time of shearing as autumn shearing leads to better skin comfort due to finer ends while wide ends from spring shearing make the garments uncomfortable (Naylor et al., 1997a; Naylor et al., 2004). Prickling property of fabric varies from loom state to finished fabric as each step brings in certain variations in the fabric (Matsudaira et al., 1990). The prickle threshold mean fibre diameters for woven fabrics was considered to be about 3 µm finer than that for knitted fabrics and reducing fabric cover factor was found to be helpful for prickle reduction (Naylor et al., 1997d). Acceptability of single jersey fabric for next-to-skin wear changes rapidly from approximately 90% for 20.5 µm wool to 50% for 23.5 µm wool (Naylor et al., 1997c).

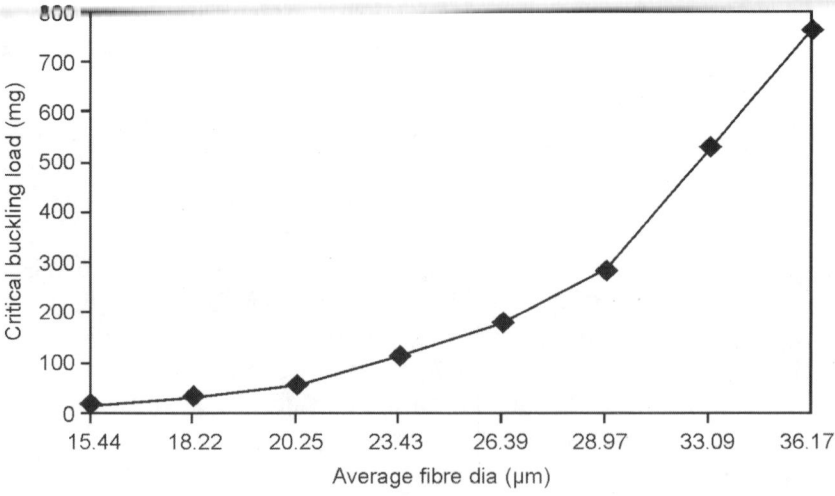

Figure 7.2 Relationship between average fibre diameter and critical buckling load

Study of the skin mechano-sensitive nociceptors A fiber and C fiber has shown that for a given set of stimuli, A fiber nociceptors exhibits a greater response rate than the C fiber and the A fiber also exhibits a greater differential response related to probe size (stimulant size) than the C fiber nociceptors (Garell et al., 1996). The neural responses of triggered A δ (A-fibre) nociceptors deceleratingly increases with the spatial density of fiber ends in a power law, and sigmoidally increases with the mean diameter of fiber ends and their standard deviation whereas the increasing standard deviation of fiber ends length acceleratingly decreases triggered neural responses by an increasing number of fiber ends in contacts with skin (Jiyong et al., 2011).

Fabric-evoked itching is dependent on the sex (male or female); it is well known that female skin is more sensitive and hence more prone to itching due to woollen garments. Itching also depends on the position of nociceptors, which are very close for hairy skin but not in gabrous skin; hence, woollen garments will itch more on the gabrous skin than non-hairy skin. Itching is age dependent, as aged person have rough skin hence itching to aged person will be less than that of young child. Environment condition decides the amount of itching, if relative humidity increases the skin becomes soft, friction of fabric with human body will be more and thus the itching (Wang et al., 2003; Das et al., 2010). Parameters such as fibre diameter and fabric stiffness are important considerations in the comfort of active sportswear. As the moisture content of skin increases, it becomes softer making it easier for textiles on the skin surface to activate the underlying nerve receptors. As a consequence, the skin is more sensitive during active sports or in warm climates than when dry and cold. For wool products to be worn next to skin during active sports, the issue of fibre diameter is quite important. Wool is commercially available in a wide range of diameters and diameter distributions and the fine wools required to ensure universal acceptance in active sportswear (typically 18.5 μm and finer) come at a significant cost disadvantage. This is one of the factors that have discouraged the use of wool fabrics in active sportswear, particularly in the high-volume look-alike sector where base fabric cost is a critical issue (Holcombe, 2009).

Prickle in wool fabrics can be reduced or eliminated by blending with fine, non-prickly fibres. This is achieved by diluting the concentration of prickly fibres in fabrics to a level below the prickle threshold. Natural fibres such as cotton and most synthetic apparel fibres are finer than their respective prickle threshold diameters, with lower variability of diameters. Most wool blends achieve some degree of prickle reduction. Another strategy to eliminate or reduce fabric-evoked prickle is physically keeping prickly wool fibres out of direct contact with the skin. Bilayer fabrics can be produced, in which the wool layer is shielded from direct skin contact by a non-prickly layer. The main disadvantages of such bilayer fabrics are their heavy weight and limitations to fabric design (Miao, 2009). Scientists in Western Australian Department of Food & Agriculture have been working on genetic improvement of sheep to produce softer wool. Studies on genetic modifications have shown that it is possible to reduce prickle property by increasing the flexibility of the fibres (Rogers et al., 2009). CSIRO has developed a process to slender the wool fibres (Optim™ fine fiber process). This process reduces the average fibre diameter compared to parent fibres and hence gives more comfort (Naylor, 2010). Carefully controlled treatments with proteolytic enzymes can reduce the buckling load and collapse energy of wool yarns. These treatments were shown to improve the softness and reduce the subjectively perceived prickle of wool fabric knitted from the treated yarns (Bishop et al., 1998). In the past decade, scientists have made a great effort to use different wool pre-treatments prior to enzyme treatment in order to limit enzymatic degradation to the cuticle scales and thus to achieve machine-washable wool without causing significant fibre damage. Recently a two-step process which combines bleaching, shrinkage prevention and biopolishing was suggested as a way to make wool feel silky smooth. This involved a pre-treatment using hydrogen peroxide enhanced by dicyandiamide and stabilised by gluconic acid for powerful oxidation, and followed by enzyme treatment with proteases in the presence of sodium sulphite in triethanolamine buffer (Shen, 2009).

Recently Wool Research Association, Thane, completed R&D project entitled 'Development of itch-free woolens using plasma and enzyme technology' sponsored by Ministry of Textiles, Govt. of India. In this project a dielectric barrier discharge (DBD) plasma chamber (Fig. 7.3) developed by FCIPT was employed for plasma treatment of knitwear, and the treatment was carried out at 2 mm electrode spacing, 5.5 KV and 0.5 A for 5 min on each side of the knitwear panel. Air was used as the non-polymerizing gas for plasma treatment. Enzyme treatment was carried out on untreated and plasma-treated knitwear with 1%, 2%, 3%, 4% and 5% (owf) with alkaline protease enzyme for 1 hour. Effect of plasma (P) and enzyme (E) treatment on the wool fibre properties is shown in Table 7.1. It was found that average fibre diameter of wool fibre reduces with increase in enzyme concentration and if the knitwear is plasma pretreated then the average fibre diameters reduces more compared to knitwear treated with enzyme only. Comfort factor of wool also increases with enzyme concentration. Mechanical properties, for e.g. breaking strength, breaking elongation and Young's modulus, also found to reduce with increase in enzyme concentration. Critical buckling load also found to reduce drastically with increase in enzyme concentration and for the samples pretreated with plasma the critical buckling load reduces more as compared to enzyme only.

Figure 7.3 Lab scale plasma reactor at Wool Research Association, Thane

Table 7.1 Effect of plasma and enzyme treatment on wool fibres properties

Sample	Avg. fibre diameter μm (SD)	Comfort factor	Brk. load g (SD)	Elg. % (SD)	Young's Modulus (MPa)	Critical Buckling Load (gf)
UT	23.46 (5.08)	91.42	7.02 (2.87)	49.67 (4.69)	3075	86.27
1%E	23.39 (5.02)	91.53	6.85 (3.12)	47. 72 (5.71)	2854	79.11
2%E	23.33 (5.15)	91.75	6.18 (4.10)	45.19 (3.61)	2703	74.16
3%E	23.21 (5.05)	92.14	5.89 (4.15)	44.23 (5.87)	2539	68.24
4%E	23.10 (5.23)	92.49	5.52 (3.76)	43.28 (5.91)	2287	60.31
5%E	22.93 (5.03)	93.45	5.12 (3.11)	42.73 (5.61)	2152	55.10
P	23.44 (4.89)	91.39	6.95 (3.84)	50.13 (4.12)	3070	85.83
P1%E	23.32 (5.20)	91.58	6.51 (3.92)	44.29 (4.23)	2756	75.49
P2%E	23.19 (5.40)	92.35	6.13 (4.26)	44.58 (4.57)	2573	68.92
P3%E	23.09 (4.94)	92.73	5.68 (3.45)	43.95 (4.02)	2346	61.76
P4%E	22.95 (5.05)	93.17	5.27 (4.08)	43.49 (3.75)	2234	57.40
P5%E	22.86 (5.01)	93.82	4.89 (3.86)	41.56 (4.23)	1975	49.95

Itching propensity of the treated and untreated knitwear was studied by user trials. The knitwear was given for wearing to the selected population of 50 users along with questionnaire. The observations were noted in the questionnaire. All the wearers were from textile field and well aware of the tactile characteristics of fabrics. The chosen population was 22–56 years old and this has taken care of skin types with varying age limits. It was found that itching to skin increases with increase in yarn hairiness index. For untreated woolen knitwear, red skin rashes were observed on the stomach and these fabrics were unbearable as next to skin garments. Enzyme treatment has a major impact on reducing the itching propensity of the fabric. Critical buckling load reduces with increase in the enzyme concentration and the reduction is much more if the fibres were plasma pretreated. SEM images (Figs. 7.4 and 7.5) shows that level of surface modification with enzyme is higher if the fabrics were plasma pretreated; this indicates that plasma and enzyme treatment work synergistically. All the users observed that knitwear both untreated and plasma-treated knitwear with 5% enzyme are very soft and can be worn next to the skin and are comparable to regular cotton innerwears. Classical critical buckling load for the plasma- and enzyme-treated samples was found to be less than 60 mgf which is much less than the threshold buckling load of 75 mgf responsible for prickly sensation (Udakhe et al., 2012; Bardhan et al., 2012).

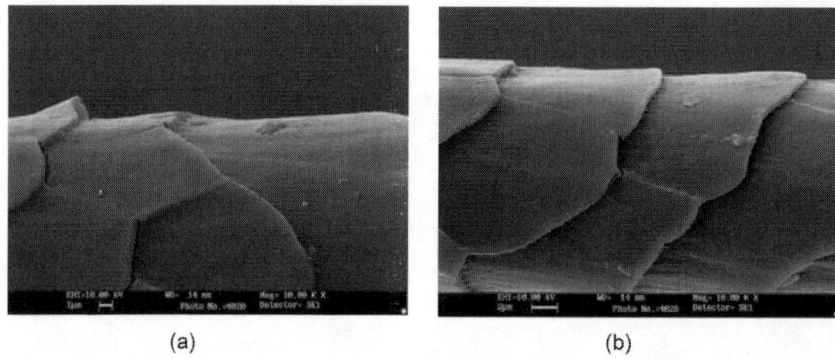

(a)   (b)

Figure 7.4 SEM images of (a) untreated (b) plasma-treated wool fibres

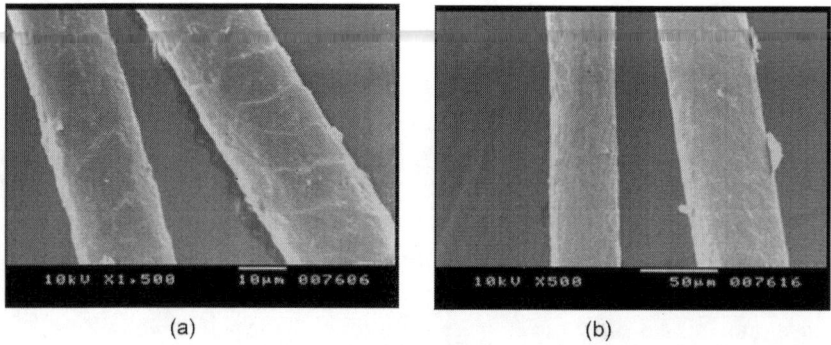

(a)   (b)

Figure 7.5 SEM images of (a) enzyme treated (b) plasma followed by enzyme-treated wool fibres

Australian wool innovation recently launched the MerinoPerform™ range of next to skin products for sports and outdoor activities. The range consist of three different garments namely MerinoPerform™ *Next* to *Skin,*

MerinoPerform™ *Advantage* and MerinoPerform™ *Pro*. These products are made from chemically processed fine (19.5 µm) merino wool and are machine washable/tumble dryable without shrinkage (wool.com).

7.3   Shrink proofing processes

Apart from dyeing, shrink proofing processes are the most common chemical treatments applied to wool. In the absence of any preventative treatment, almost all types of woven and knitted wool products will shrink, although the propensity to do so varies widely (Simpson, 2002). For most of the past century, the need to reduce or completely eliminate shrinkage of fabrics as a result of felting has become a necessity for the woollen fabrics to make them easy care and machine washable. For truly machine washable wool fabrics both relaxation shrinkage and felting shrinkage should be eliminated. Relaxation shrinkage is due to the release of mechanical stresses introduced into the fabric during manufacture, and is reversible and governed by tensions imposed during spinning, weaving, knitting and making-up. Felting shrinkage is due to the fibre movement and interlocking during mechanical agitation, due to the directional frictional effect. Hence felting shrinkage is irreversible and is the important factor in producing machine washable wool (Holt, 1975). State-of-the-art process so far available is the Chlorine/Hercosett treatment, continuously operating combined process uses chlorination as a preliminary step followed by polymer coating. This process uses large amounts of water as well as polluting substances, which leads to significant wastewater pollution with adsorbable halogen compounds (AOX) (superwool.com). Since the AOX generation during conventional shrink proofing of wool exceeds the permitted levels by up to 40 mg/l, an environmentally acceptable, chlorine-free process for imparting full machine washability is required (Thomas, 2007). Commercially successful shrink-resist processes used by the textile industry in the past, and technologies currently being developed, can be divided into five groups:

1.   Combination of oxidation and additive polymer processes
2.   Additive polymer only processes
3.   Plasma treatment followed by resin polymer or softener finishing
4.   Enzymatic processes followed by resin polymer or softener finishing (Shen, 2009)
5.   UV ozone followed by resin polymer or softener finishing (Gupta et al., 2013; Udakhe et al., 2011).

Extensive efforts have been made to find an environmentally acceptable alternative to the chlorination treatment. We discuss in details about these techniques.

7.3.1   Plasma treatment

Shrink proofing using corona discharge plasma treatment has shown that shrink resistance increases with increase in voltages and frequencies. To achieve the shrink-resistant properties, top stage treatment was suggested instead of fabric stage, because in fabric the gaseous reactants did not penetrate adequately. Incorporation of Chlorine gas in the air-corona field markedly increases shrink-resistance properties of wool (Thorsen et al., 1966; Thorsen, 1968; Thorsen et al., 1970). Plasma treatment leads to formation of cystine oxides, i.e., S-(O)-S, S-(O)$_2$-S, carboxyl and sulphonic acid groups on the wool fiber surface. These groups have a strong affinity with water and may increase surface hydration. Cohesive force is exerted by hydrogen bonding between these groups and water molecules on the fiber surfaces. This interaction results in a decrease in the flexibility of the individual fibers in the assembly and thus improves the anti-felting behavior of the wool fibers (Mori et al., 2006). Plasma treatment is surface specific and does not affect the bulk and mechanical properties of wool fibres. The extent of surface etching increases with increase in treatment time (Udakhe et al., 2011c). Dielectric barrier discharge (DBD) plasma treatment improves dyeability, mechanical properties and shrink proofing of wool fabric; this has been attributed to the increased hydrophilic sites and surface roughness (Fig. 7.6) created on the fibre surface compared to untreated reference (Udakhe et al., 2013; Udakhe et al., in press). The DBD plasma treatment of wool fabric improves the shrink proofing of wool fabrics. Shrinkage of wool fabric (Fig. 7.7) after 1X7A followed by 3X5A washing cycle is lesser than the untreated reference.

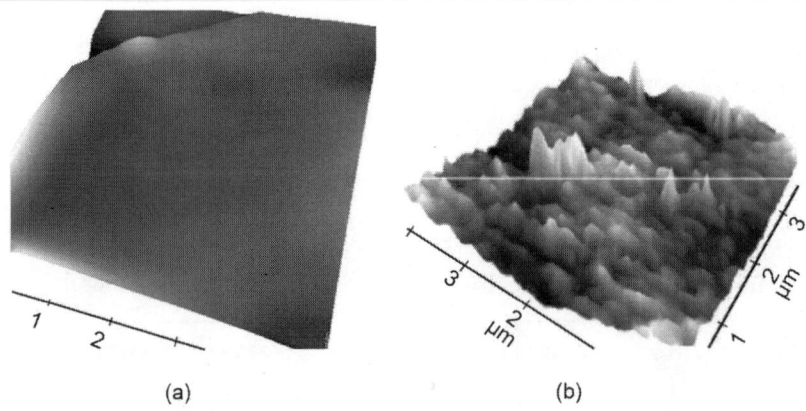

(a)   (b)

Figure 7.6 AFM images of (a) untreated and (b) 15 min plasma treated wool fibre sample (scanned area is 5 × 5 μm)

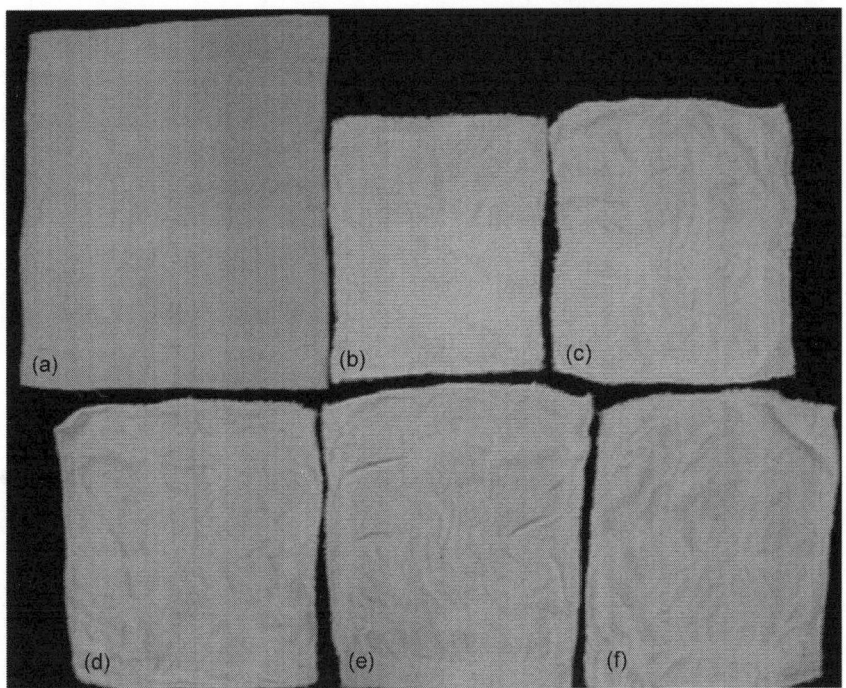

Figure 7.7 Shrinkage in wool fabrics (a) reference sample, (b) untreated, (c) 2.5 min plasma, (d) 5 min plasma, (e) 10 min plasma, (f) 15 min plasma-treated sample

For shrink proofing of wool, a plasma-induced surface oxidation as well as a direct coating of the fibres by plasma polymerization reduces, but not eliminate, felting (Thomas, 2007). The glow discharge (GD) treatment of wool in non-polymerizing gases like air, oxygen, and nitrogen is presented as a new zero-AOX pretreatment and almost complete shrink resistance of wool fabric and top is achieved using the collagen / Araldit resin (Pavlath et al., 1975; Hesse et al., 1995; Hesse et al., 1995b). Plasma treatment increases the surface area of wool fibres from 0.1 m²/g to 0.35 m²/g. These physiochemical changes decrease the felting/shrinkage behavior of wool from more than 0.2 g/cm³ to less than 0.1 g/cm3 (Hocker, 2002). Glow discharge plasma treatment did not impair the physicomechanical, heat-insulation, and thermal properties of wool (Sadova, 2006) rather plasma treatment improves thermal resistance and anti-pilling performance as compared to reference untreated (Goud et al., 2011). Glow discharge makes wool yarn shrink resistant, stronger and induces grafting of polymer (Lee et al., 1975).

Plasma modification of the fibre surface is known to cause a significant increase in the coefficients of friction between the wool fibre, in both with-

and against-scale directions, resulting in a rather harsh handle (Jhala et al., 2009; Shen, 2009). Application of amino-functional polydimethylsiloxane to plasma pre-treated wool fabric enhances the shrink resistance with improved handle (Naebe et al., 2011).

Complete shrink proofing of woollen garments cannot be achieved using plasma treatment alone, and hence this requires an additional coverage of the fibre surface to further decrease the directional friction effect. Since commercially available resins do not show the expected positive influence on the felt-free performance of plasma-treated wool, new resins have been tailored for the plasma-treated fibre surface. Two different wool-compatible resin types have been developed by BAYER, allowing the generation of machine-washable wool after GD or BD treatment. One type consists of a water-dispersible isocyanate-bearing resin which is able to permanently coat each individual fibre. The differences in the degree of felting after different plasma-only treatments can be equalized by the resin application, to an extent which meets the value for Chlorine/ Hercosett-treated samples. The second resin is a polyurethane type which does not coat the fibre completely, but it deposits at the scale edges instead. This also results in significantly decreased shrinkage behavior. DBD treatment of tops followed by application of one or both resins guarantees the specifications set by the Woolmark Company in their Technical Method 31 (Thomas, 2007). Many researchers have been trying to find alternative to the commercial polymers. Natural biopolymers like chitosan, sericine and ceisin are good alternatives to the commercial polymers. Plasma treatment of wool and wool/nylon blend-knitted fabrics followed by softener and chitosan biopolymer treatment was found to satisfy the industrial standard of shrink proofing (Canal et al., 2008). Plasma treatment increases the surface adsorption of chitosan polymer and machine washable wool can be achieved by combination of plasma treatment and chitosan application (Erra et al., 1999; Molina et al., 2002). Plasma and chitosan can reduce the dyeing time with acid dyes and reduces the felting shrinkage in wool (Jocic et al., 2005). Recently, Richter F&A a German company completed a R&D project supported by European Commissions Life Programme entitled *'SuperWool – Sustainable, AOX-free Superwash Finishing of Wool Tops for the Yarn Production with the innovative AOX-free plasma technology'*. Based on the project findings, the company has launched the commercial-level Corona finish process for wool (richter-fa.de). Richter International, a Canadian company, also launched washable woollen products with plasmawool® brand (richterinternational.ca). The result has led to the development of a completely AOX-free shrink proofing process for wool tops and thus opened up a new possibility for the wool industry to have an environmentally acceptable, plasma-based shrink proofing process in the near future.

The treatment of wool tops with low-temperature plasma and alternative resin systems has turned out to be considerably better for the environment compared to the conventional Chlorine-Hercosett process; it is also economically very promising. This new approach not only leads to an AOX and wastewater-free process but also gives advantages regarding the physical health of the operators. Since no hazardous chemicals e.g. chlorine, sulphuric acid or volatile organic compounds are needed, air contamination is significantly reduced and the risk of industrial accidents is lowered. The contamination of water bodies will also be reduced in developing countries that do not possess appropriate water-treatment plants. The environment-friendly plasma process has the potential to be defined as a new Best Available Technique (BAT) and could therefore replace the chlorine-Hercosett process worldwide. This would affect some 10 installed Hercosett plants in the EU (Germany, UK, Italy, France and the Czech Republic) as well as about the same number of plants in Asia (superwool.de).

Worldwide many companies make the lab scale as well as commercial plasma reactors for plastics and textiles. Some of them are PlasmaEtch (USA), Tri-Star Technologies (USA), Lectro Engineering Company (USA), AcXys Technologies (France), Plasma Technology (Germany), Enercon Asia Pacific Systems (India), Ferrarini & Benelli Srl (Italy), Diener electronic (Germany), Europlasma (Belgium), P2i (UK), Dow Corning Plasma Systems (Ireland), Mascioni (Italy), Ahlbrandt System (Germany), Institute for Plasma Research (India).

7.3.2   Enzyme and polymer in combination

Enzymes are biological catalysts in the form of proteins that catalyze chemical reactions. An enzyme usually catalyzes only one specific chemical reaction or a number of closely related reactions. Enzymes have found wide application in the textile industry for improving production methods and fabric finishing (novozymes.com). Protiolytic enzymes can be used to replace the oxidative processes like pretreatment using chlorination and will be an eco-friendly alternative. Enzyme activity is dependent on surface energy of the fibre and the surface energy can be increased by LTP treatment (Vilchez et al., 2008; Demir et al., 2008). A new ecological method has been proposed for the shrink proof finishing of wool, based on an enzymatic pretreatment and chitosan deposition, ensuring dimensional stability after multiple washing processes (Walawska et al., 2006). Enzyme application in alkaline peroxide treatment bath enhances wool wetability and effectiveness of subsequently applied chitosan biopolymer due to formation of ionic bonds between the new sulphonic groups generated on the wool fiber surface and chitosan which contribute to the shrink resistance (Jovancic et al., 2001). Enzymatic pretreatment itself greatly improves the shrink proofing property by partial destruction of the fiber scales, and the post-application of chitosan enhances the shrink proofing and ensures an increased dyeability of wool fabrics by reactive dyes (Rybicki et al., 2000; Onar et al., 2004).

Enzymatic methods for treating wool, used alone or in conjunction with an oxidative chemical step, have had little commercial value. A fact that is attributable to their relatively high costs and their tendency to damage wool by causing weight and strength losses. To overcome these disadvantages, the pretreatment of wool with hydrogen peroxide at alkaline pH in the presence of high concentrations of salt is suggested. Use of salt suppresses the swelling of wool fibres, thus avoids the penetration of enzyme in fibre core and restricts the action on the surface of wool fibres (Lenting et al., 2006). Other possible solutions for this problem is either the enzyme has to be controlled (for example, diffusion control by enzyme immobilization) or the enzyme has to be specially designed (for example, by genetic engineering) in such a way that only a distinct part of the substrate is altered like scales on the wool fibre surface (Xie et al., 2009). Molecular weight of the enzyme can be increased to restrict the enzyme on the surface of wool fibres and to avoid penetration into the core (Silva et al., 2005; Paulo et al., 2006).

Another way of recovering the strength loss of wool is to crosslink glutamine and lysine protein residues in the wool fibre using transglutaminases (TG) enzymes to form covalent bonds. TG belong to a class of enzymes known as aminoacyl-transferases that catalyze calcium-dependent acyl transfer reactions between peptide-bound glutamine residues as acyl donors and peptide-bound lysine residues as acyl acceptors, resulting in the formation of intermolecular e-(g-glutamyl)lysine crosslinks (Griffin et al., 2002; Vlist et al., 2010). Wool fibres contain 450 (μmol/g) of glutamine -$(CH_2)_2$-$CONH_2$ and 250 (μmol/g) of lysine -$(CH_2)_4$-NH_2 (Hocker, 2002). Hence there are fare chances of crosslinking between glutamine and lysine initiated by TG. This covalent isopeptide crosslink is stable and resistant to proteolysis, thereby increasing the resistance to chemical, enzymatic, and mechanical disruption (Greenberg et al., 1991). Cytotoxicity, genotoxicity and mutagenic studies of TG have shown that this enzyme are safe for food application and has a potential for its industrial usages (de Souza et al., 2011).

Combine process of treating wool with a proteolytic enzyme, either preceding or simultaneously with a TG improves shrink resistance and is able to retain the mechanical properties of wool fibres (McDevitt et al., 2000). Plasma and electron-beam pretreatments for application of TG lead to increased cross-links or incorporation of primary amine compounds in wool fabrics, due to physical etching of epicuticle layer (Fatarella et al., 2010). TG could remediate wool damage following hydrogen peroxide and protease anti-felting finishing, resulting in an increase in wool fabric strength and a decrease in alkali solubility (Zhang et al., 2010; Ge et al., 2009; Cardamone et al., 2007; Hossain et al., 2008; Cortez et al., 2004). Routinely used biological detergents contain proteases and damages wool fibres during laundering. Wool garments previously treated with TG are likely to have increased resistance to domestic washing and thus provide increased longevity (Cortez et al., 2005).

Transglutaminase also have the ability to crosslink protein biopolymers, and many researchers used this to graft natural proteins on to wool fibres. TG initiates the cross-linking of the wool protein through covalent binding with reduced carboxymethylated κ-casein (Gembeh et al., 2005). TG-mediated incorporation of pancreatic digest of milk casein (Tryptone) and putrescine into wool fibres found to reduce felting shrinkage, retaining the fibre mechanical properties (Griffin et al., 2003). Casein, a protein found in milk, is used for surface-coating and crosslinked using TG on $KMnO_4$ pretreated wool fabric. The area shrinkage was reduced to industrial acceptable level (Cui et al., 2011). TG-mediated grafting of silk proteins leads to significant changes in wool yarn and fabric, resulting in increased bursting strength, as well as reduced levels of felting shrinkage along with improved fabric softness (Cortez et al., 2007). TG-mediated application of Keratin hydrolysates and their lyophilized powders to peroxycarboximidic acid pretreated fine jersey wool fabric minimized felting shrinkage and dry burst strength (Cardamone et al., 2007b). Gelatin has been crosslinked on $KMnO_4$ pretreated wool fabric using TG, and it was found that area shrinkage improves more effectively than that of fabric treated with gelatin alone (Cui et al., 2009).

An enzyme in wool anti-felting is a most promising area in order to achieve high-grade, comfortable and washable effects. With the environmental consideration, enzymes consumption in wool industries will have a rapid increase. This will call for a simple, low cost technique in enzyme production. Enzyme is a special protein; its activity can easily be affected by temperature and other chemical reagents; hence stability and the specificity of enzymes need to be enhanced. The wool mills should find a proper process to keep enzymatic fabrics quality stability (Dong et al., 2008).

7.3.3   Ultra violet/ozone treatment

Apart from plasma-processing technique, researchers are also working on other alternative surface modifications methods like high energy UV rays or ozone. These are dry processes for surface modification. Ultra violet ozone treatment oxidizes the surface proteinaceous di-sulphide sulphur to sulphonic acid groups and also oxidizes the surface carbon species leading to increase in surface polarity promotes aqueous wetting, dyability, chitosan adsorption and shrink resistance. The data presented indicate that the UV ozone treatment used is capable of producing surface sulphur and carbon chemistry of the type usually obtained industrially by wet chemical methods which have the disadvantage of producing chlorinated effluent (Bradley et al., 1993; Bradley et al., 1994; Bradley et al., 1997; Xin et al., 2002; Gupta et al., 2010; Osman et al., 2010). CSIRO has developed a Siroflash anti-pilling treatment for knitted fabrics. Siroflash is a process involving exposure of the fabric or garment surface to short wavelength ultraviolet radiation (UVC), followed by a mild wet oxidation treatment using, for example, hydrogen peroxide or salts of permonosulphuric acid (Millington, 1998). Modification of wool fabric with UV-assisted treatments is effective in reducing pilling and shrinkage with an acceptable loss in weight and strength of the fabric. After treatment of wool fabric with safe oxidizing agents (hydrogen peroxide and SMPP) or proteolytic enzymes (papain or savinase 16L type EX) makes these treatments environment-friendly alternatives to chlorination of wool (El-Sayed et al., 2005). UV treatment followed by polymer treatment using synthappret BAP was found to reduce the shrinkage level to machine washable range (Gupta et al., 2013).

7.3.4   Application of biopolymers

Polymer deposition on wool fibers to coat the scales is one method to prevent laundering-shrinkage of wool fabrics by felting. Chitosan biopolymer can be applied to wool fabric either by single-step or two-step processes to confer the machine washability. In a two-step application, chitosan solutions in dilute acids can be padded on wool fabrics followed by drying. The chitosan-deposited fabrics can then be further treated with crosslinkers, such as glyoxal or glutaraldehyde, which react with chitosan as well as with functional groups on the wool, e.g., lysine, arginine, histidine, or serine residues. In a single-step application, mixtures of chitosan solution and crosslinkers, such as glyoxal, glutaraldehyde, or DMDHEU can be padded on the fabrics, dried, and then cured. Both application methods improve resistance to laundering-shrinkage. Pretreatment with hydrogen peroxide (H_2O_2) either under alkaline or acidic conditions increases the number of cysteic acid groups ($-SO_3H$) by the oxidation of the disulfide bonds on wool fibers and, consequently, increases the anionic charges on the fiber surface, which could enhance the sorption of chitosan with cationic charge. Shrink resistance and chitosan sorption are related to the hydrophilicity rather than the cysteic acid content of the wool. Alkaline H_2O_2 pretreatment generates more hydrophilic fibre surface than acidic H_2O_2 pretreatment and hence shows better shrink-resistance properties. Higher molecular weight chitosan gives better shrink resistance. To promote the sorption of chitosan on wool fabric, fabric can be pretreated with an anionic surfactant sodium lauryl sulfate (SLS). SLS-pretreated fabric surface acquires a substantial amount of negative charge. SLS pretreatment assist in chitosan adsorption and improves the laundering-shrinkage (Lim et al., 2003). Citric acid was used as a crosslinking agent for chitosan on potassium permanganate pretreated wool fabric. It was found that the surface crosslinks of the oxidized woollen fibers were relatively coarse, which was undesirable for shrink proofing and yet beneficial for the antimicrobial and antiseptic effects of the woolen fabrics (Hsieh et al., 2004). A new process for shrink proofing of wool fibers based on the sericin biopolymer is reported in which the wool fabric were pretreated with hydrogen peroxide and sodium sulphite to enhance sericin adsorption. The sericin combining capacity of wool has been even enhanced by carrying out the reaction in the presence of different crosslinkers namely dimethylol dihydroxy ethylene urea, dimethyl dihyroxy ethylene urea, and epichlorohydrin, which are able to link sericin with the available sites of wool. The felting resistance of wool tops treated with this system is comparable with the machine washable wool obtained commercially (Allam et al., 2009).

7.4   Future trends

Plasma treatment is a dry and completely environment-friendly technique of surface modification. The modification of the surface is in the range of 100 Å and does not affect the bulk properties of the fibre as against the other wet processes for surface modification, which penetrate deep into the core and lead to fibre damage. The surface modification using plasma generates hydrophilic sites along with small size pores on the epicuticle layer on wool fibre. These physicochemical changes are well reported to improve dyeing, printing, shrink proofing and subsequent chemical finishing processes. Plasma treatment of wool is a promising solution to replace chlorination stage of wet processing which is responsible for generation of AOX. Plasma alone is not able to reduce the shrinkage of wool fabric to industrially acceptable level, and many researchers have used synthetic polymers or biopolymers to coat the surface of wool fibres after plasma treatment and obtained commercially acceptable shrinkage results for woolens. For the plasma technology to be commercially viable in the industry, it should be easy to operate, high throughput, should have ability to be a part of existing process line or be able to do so with slight changes in the process line, low cost of machine and production. Many commercial reactor manufacturers are trying to achieve the same and will succeed in doing so in near future. The commercial machine with ability of doing plasma, enzyme and biopolymer treatments in a continuous manner will be welcomed, once the final product meet the industrial standards of shrink proofing. The plasma treatment technology is environment-friendly alternative to the commercial chemical processes that are being used now-a-days.

7.5   Sources of further information

The book entitled, *Wool: Science and Technology,* edited by W S Simpson and G H Crawshaw, describes in details right from the production of wool fibres, physical and chemical structure of wool and processing. Recent developments and future prospects of plasma technology in the textile arena is given in details in the book entitled *Plasma Technologies for Textiles,* edited by R Shishoo. For the readers who want to know more about the recent advances in spinning, weaving, chemical processing and nanotechnologies, book entitled Advances in Wool Technology, edited by N A G Johnson and I M Russell, would be helpful. For those who want to study comfort aspect of textiles, the book entitled *Science in Clothing Comfort,* authored by A Das and R Alagirusamy would be very helpful. For the commercial and lab scale plasma reactors, the readers can refer to the web sites given below; these websites provide latest updates about plasma technology and new machines.

 www.plasmaetch.com (accessed on 21 March 2011)
 www.tri-star-technologies.com (accessed on 21 March 2011)
 www.lectroengineering.com (accessed on 21 March 2011)
 www.acxys.com (accessed on 21 March 2011)
 www.plasmatechnology.de (accessed on 21 March 2011)
 www.enerconasiapacific.in (accessed on 21 March 2011)
 www.ferben.com (accessed on 21 March 2011)
 www.plasma.de (accessed on 21 March 2011)
 www.europlasma.be (accessed on 21 March 2011)
 www.p2i.com (accessed on 21 March 2011)
 www.dowcorning.com (accessed on 21 March 2011)
 www.mascioni.it (accessed on 21 March 2011)
 www.ahlbrandt.de (accessed on 21 March 2011)
 www.ipr.res.in (accessed on 21 March 2011)

7.6   References

Allam O.G., El-Sayed H., Kantouch A. and Haggag K. (2009). 'Use of Sericin in Feltproofing of Wool' *J of Nat Fib* **6**(1), pp. 14–26.

Bardhan M.K. and Udakhe J. (2012). 'Itch free woollen garments using eco friendly methods' in 81st IWTO Congress, New York (USA), 7–9th May 2012.

Bishop D.P., Shen J., Heine E. and Hollfelder B. (1998). 'The use of proteolytic enzymes to reduce wool-fibre stiffness and prickle' *J of The Text Inst* **89**(3), pp. 546–553.

Bradley R.H., Clackson I.L. and Sykes D.E. (1993). 'UV ozone modification of wool fibre surfaces' *Appl Surf Sci* **72**(2), pp. 143–147.

Bradley R.H., Clackson I.L. and Sykes D.E. (1994). 'XPS of oxidized wool fibre surfaces' *Surf & Interface Analysis* **22**(1–12), pp. 497–501.

Bradley R.H., Mathieson I. and Byrne K.M. (1997). 'Spectroscopic studies of modified wool fibre surfaces' *J of Mat Chem* 7, pp. 2477–2482.

Canal C., Molina R., Bertrane E., Navarro A. and Erra P. (2008). 'Effects of low temperature plasma on wool and wool/nylon blend dyed fabrics' *Fib & Poly* **9**(3), pp. 293–300.

Cardamone J. M. (2007). 'Enzyme-mediated Crosslinking of Wool Part I:Transglutaminase' *Text Res J* **77**(4), pp. 214–221.

Cardamone J. M. and Phillips J.G. (2007b). 'Enzyme-mediated Crosslinking of Wool Part II: Keratin and Transglutaminase' *Text Res J* **77**(5), pp. 277–283.

Cortez J., Bonner P.L.R. and Griffin M. (2005). 'Transglutaminase treatment of wool fabrics leads to resistance to detergent damage' *J of Biotechnology* **116**(4), pp. 379–386.

Cortez J., Bonner P.L.R. and Griffin M. (2004). 'Application of transglutaminases in the modification of wool textiles' *Enzy and Microb Tech* **34**(1), pp. 64–72.

Cortez J., Anghieri A., Bonner P.L.R., Griffin M. and Fredii G. (2007). 'Transglutaminase mediated grafting of silk proteins onto wool fabrics leading to improved physical and mechanical properties' *Enzy and Microb Tech* **40**(7), pp. 1698–1704.

Cui L., Fan X., Wang P., Wang Q. and Fu G. (2011). 'Casein and transglutaminase-mediated modification of wool surface' Eng in Life Sci **11**(2), pp. 201–206.

Cui L., Fan X., Wang P., Wang Q., Huan Q. and Fu G. (2009). 'Transglutaminase-mediated crosslinking of gelatin onto wool surfaces to improve the fabric properties' *J of Appl Poly Sci* **113**(4), pp. 2598–2604.

Das A. and Alagirusamy R. (2010). 'Science in clothing comfort' *Woodhead Publishing India Pvt. Ltd, New Delhi, India* pp. 1–172.

Demir A., Karahan H.A., Ozdogan E., Oktem T. and Seventekin N. (2008). 'The Synergetic Effects of Alternative Methods in Wool Finishing' *Fib & Text in East Euro,* **16**(2), pp. 89–94.

de Souza C.F.V., Venzke J.G., Rosa R.M., Henriques J.A.P., Dallegra V.E., Flores S.H. and Ayub M.A.Z. (2011). 'Toxicological Evaluation of Transglutaminase from a Newly Isolated Bacillus Circulans BL32'*American J of food tech* **6**(6), pp. 460–471.

Dolling M., Marland D., Naylor G.R.S. and Phillips D.G. (1992). 'Knitted Fabric Made from 23.2 µm Wool Can be Less Prickly than Fabric Made from Finer 21.5 µm Wool' *Wool Tech and Sheep Breed* **46**(7), pp. 69–71.

Dolling M., Marland D., Naylor G.R.S. and Phillips D.G. and Singleton D.J. (1990). 'Raw wool with low levels of coarse fibre processes into more comfortable knitted fabric' *Proceedings of the Australian Society of Animal Production* **18**, pp. 471.

Dong L. and Xu L. (2008). 'Enzymatic Process for the Wool Fabric Anti-felting Finishing' *Mod Appl Sci* **2**(3), pp. 91–93.

El-Sayed H. and El-Khatib E. (2005). 'Modification of wool fabric using ecologically acceptable UV-assisted treatments' *J of Chem Tech and Biotech* **80**(10), pp. 1111–1117.

Erra P., Molina R., Jocic D., Julia M.R., Cuesta A. and Tascon J.M.D. (1999). 'Shrinkage Properties of Wool Treated with Low Temperature Plasma and Chitosan Biopolymer' *Text Res J* **69**(11), pp. 811–815.

Fatarella E., Ciabatti I. and Cortez J. (2010). 'Plasma and electron-beam processes as pretreatments for enzymatic processes' Enzy & Microb Tech **46**(2), pp. 100–106.

Garell P.C., Mcgillis S.L. and Greenspan J.D. (1996). 'Mechanical response properties of nociceptors innervating feline hairy skin' *J of Neurophysiol* **75**(3), pp. 1177–1189.

Garnsworthy R.K., Gully R.L., Kenins P., Mayfield R.J. and Westerman R.A. (1988). 'Identification of the physical stimulus and the neural basis of fabric-evoked prickle' *J of Neurophysiol* **59**(4), pp. 1083–1097.

Ge F., Cai Z., Zhang H. and Zhang R. (2009). 'Transglutaminase treatment for improving wool fabric properties' *Fib & Poly* **10**(6), pp. 789–790.

Gembeh S.V., Farrell Jr H.M., Taylor M.M., Brown E.M. and Marmer W.N. (2005). 'Application of transglutaminase to derivatize proteins: 1. Studies on soluble proteins and preliminary results on wool' *J of the Sci of Food & Agri* **85**, pp. 418–424.

Goud V. and Udakhe J. (2011). 'Effect of Low Temperature Plasma Treatment on Tailorability and Thermal Properties of Wool Fabrics' *Pramana-J of Physics* **76**(6), pp. 669–677.

Greenberg C.S., Birckbichler P.J. and Rice R.H. (1991). 'Transglutaminases: multifunctional cross-linking enzymes that stabilize tissues' *The FASEB J* **5**(15), pp. 3071–3077.

Griffin M., Casadio R. and Bergamini C.M. (2002). 'Transglutaminases: Nature's biological glues' *Biochem J* **368**, pp. 377–396.

Griffin M., Cortez J.M. and Bonnes P. (2003). 'Method for enzymatic treatment of textiles such as wool' *U S Pat 2003/0154555A1,* (21 August 2003).

Gupta D. and Basak S. (2010). 'Surface functionalization of wool using 172 nm UV excimer lamp' *J of Appl Poly Sci* **117**(6), pp. 3448–3453.

Gupta D. and Mondal N. (2013). 'Shrink resist treatment of wool using VUV excimer lamp' *Colourage* **LX**(1), pp. 33–39.

Hatch K.L., Markee N.L. and Maibach H.I. (1992). 'Skin Response To Fabric. A Review of Studies and Assessment Methods' *Clothing and Text Res J* **10**, pp. 54–63.

He W. and Wang X. (2002). 'Mechanical Behavior of Irregular Fibers Part III: Flexural Buckling Behavior' *Text Res J* **72**(7), pp. 573–578.

Hesse A., Thomas H. and Hocker H. (1995a). 'Zero-AOX Shrinkproofing Treatment for Wool Top and Fabric Part I: Glow Discharge Treatment' *Text Res J* **65**(6), pp. 355–361.

Hesse A., Thomas H. and Hocker H. (1995b). 'Zero-AOX Shrinkproofing Treatment for Wool Top and Fabric Part II: Collagen Resin Application' *Text Res J* **65**(7), pp. 371–378.

Hocker H. (2002). 'Fibre morphology' in *Wool: Science and Technology*, edited by W S Simpson & G H Crawshaw, Woodhead Publishing Ltd, Cambridge, England, pp. 60–79.

Hocker H. (2002). 'Plasma treatment of textile fibers' *Pure and Appl Chem* **74**(3), pp. 423–427.

Holcombe B. (2009). 'Wool performance apparel for sport' in *Advances in Wool Technology*, Edited by N A G Johnson and I M Russell, Woodhead Publishing Ltd., Cambridge, England, pp. 265–283.

Holt R.R.D. (1975). 'Introduction to Superwash Wool' *J of the Soc of Dyers and Col* **91**(2), pp. 38–44.

Hossain K.M.G., Juan A.R. and Tzanov T. (2008). 'Simultaneous protease and transglutaminase treatment for shrink resistance of wool' *Bio and Biotra* **26**(5), pp. 405–411.

Hseih S.H., Huang Z.K., Huang Z.Z. and Tseng Z.S. (2004). 'Antimicrobial and physical properties of woolen fabrics cured with citric acid and chitosan' *J of Appl Poly Sci* **94**(5), pp. 1999–2007.

Hu J., Li Y., Ding X. and Hu J. (2011). 'The mechanics of buckling fiber in relation to fabric-evoked prickliness: a theory model of single fiber prickling human skin' *J of The Text Insti* **102**(12), pp. 1003–1018.

Jhala P.B. and Gandhi G.A. (2009) 'Single Fibre Friction Cum Strength Tester' *J of the Text Asso* **70**(3), pp. 99–102.

Jiyong H., Yi L., Xin D. and Junyan H. (2011). 'Neuromechanical representation of fabric-evoked prickliness: a fiber-skin-neuron model' *Cogn Neurod* **5**(2), pp. 61–170.

Jocic D., Vilchez S., Topalovic T., Molina R., NAVARRO A, JOVANCIC P, JULIA M R & ERRA P (2005) 'Effect of low-temperature plasma and chitosan treatment on wool dyeing with Acid Red 27' *J of Appl Poly Sci* **97**(6), pp. 2204–2214.

Jovancic P., Jocic D., Molina R., Julia M.R. and Erra P. (2001). 'Shrinkage Properties of Peroxide-Enzyme-Biopolymer Treated Wool' *Text Res J* **71**(11), pp. 948–953.

Lee K.S. and Pavlath A.E. (1975). 'The Effects of Afterglow, Ultraviolet Radiation, and Heat on Wool in an Electric Glow Discharge' *Text Res J* **45**(8), pp. 625–629.

Lenting H.B.M., Schroeder M., Guebitz G.M., Cavaco-Paulo A. and Shen J. (2006). 'New enzyme-based process direction to prevent wool shrinking without substantial tensile strength loss' *Biotech Let* **28**(10), pp. 711–717.

Lim S.H. and Hudson S.M. (2003). 'Review of Chitosan and Its Derivatives as Antimicrobial Agents and Their Uses as Textile Chemicals' *J of Macro Sci Part C—Poly Rev* **C43**(2), pp. 223–269.

Li Y. (2001). 'The science of clothing comfort' *Text Prog* **31**(1/2), pp. 1–135.

Matsudaira M., Watt J.D. and Carnaby G.A. (1990). 'Measurement of the Surface Prickle of Fabrics Part II: Some Effects of Finishing on Fabric Prickle' *J of the Text Inst* **81**(3), pp. 3000–3009.

Mayfield R.J. (1987). 'Preventing Prickle' *Textile Horizons* **7**(11), pp. 35–36.

Miao M. (2009). 'High-performance wool blends' in *Advances in wool technology, Edited by N A G Johnson and I M Russell Woodhead Publishing Ltd., Cambridge, England,* pp. 284–307.

Millington K. (1998). 'Using Ultraviolet Radiation to Reduce Pilling of Knitted Wool and Cotton' *Text Res J* **68**(6), pp. 413–421.

McDevitt J.P. and Winkler J. (2000). 'Method for Enzymatic Treatment of Wool' *U S Pat 6051003* (18 April 2000).

Molina R., Jovan P., Commelles F., Bertran E. and Erra P. (2002). 'Shrink-resistance and wetting properties of keratin fibres treated by glow discharge' *J of Adhes Sci & Tech* **16**(11), pp. 1469–1485.

Mori M. and Inagaki N. (2006). 'Relationship between Anti-felting Properties and Physicochemical Properties of Wool Fibers Treated with Ar-plasma' *Text Res J* **76**(9), pp. 687–694.

Naebe M., Cookson P.G., Denning R. and Wang X. (2011). 'Use of low-level plasma for enhancing the shrink resistance of wool fabric treated with a silicone polymer' *J of The Text Inst* **102**(11), pp. 948–956.

Naylor G.R.S. (2010). 'Fabric-evoked Prickle in Worsted Spun Single Jersey Fabrics Part 4: Extension from Wool to OptimTM fine Fiber' *Text Res J* **80**(6), pp. 537–547.

Naylor G.R.S., Oldham C.M. and Stanton J. (2004). 'Shearing Time of Mediterranean Wools and Fabric Skin Comfort' *Text Res J* **74**(4), pp. 322–328.

Naylor G.R.S. and Phillips D.G. (1997d). 'Fabric-evoked prickle in worsted spun single jersey fabrics' part II: the role of fibre length, yarn count and fabric cover factor' *Text Res J* **67**(5), pp. 354–358.

Naylor G.R.S. and Phillips D.G. (1997c). 'Fabric-Evoked Prickle in Worsted Spun Single Jersey Fabrics Part III: Wear Trial Studies of Absolute Fabric Acceptability' *Text Res J* **67**(6), pp. 413–416.

Naylor G.R.S., Phillips D.G., Veltech C.J., Dolling M. and Marland D.J. (1997b). 'Fabric-Evoked Prickle in Worsted Spun Single Jersey Fabrics, Part I: The Role of Fiber End Diameter Characteristics' *Text Res J* **67**(4), pp. 288–295.

Naylor G.R.S. and Stanton J. (1997a). 'Time of Shearing and the Diameter Characteristic of Fiber Ends in the Processed Top: An Opportunity for Improved Skin Comfort in Garments' *Wool Tech and Sheep Breed* **44**(5), pp. 243–255.

Naylor G.R.S., Veitech C.J., Mayfield R.J. and Kettlewell K. (1992). 'Fabric-Evoked Prickle' *Text Res J* **68**(2), pp. 487–493.

Onar N. and Sariisik M. (2004). 'Application of enzymes and chitosan biopolymer to the antifelting finishing process' *J of Appl Poly Sci* **93**(6), pp. 2903–2908.

Osman E.M., Michael M.N. and Gohar H. (2010). 'The Effect of Both UV\Ozone and Chitosan on Natural Fabrics' *Inter J of Chem* **2**(2), pp. 28–39.

Paulo A.C., Silva C.J.S.M. (2006). 'Treatment of animal hair fibers with modified proteases' *U S Pat 2006/0121595A1* (8 June 2006).

Pavlath A.E. and Lee K.S. (1975). 'The effect of the afterglow on felting shrinkage of wool', *Text Res J* **45**(10), pp. 742–745.

Rogers G.E. and Bawden C.S. (2009). 'Improvement of wool production through genetic manipulation' in *Advances in wool technology, Edited by N A G Johnson and I M Russell, Woodhead Publishing Ltd., Cambridge, England* pp. 3–21.

Rybicki E., Filipowska B. and Walawska A. (2000). 'Application of Natural Biopolymers in Shrink-proofing of Wool' *Fib and Text in East Euro* **8**(1), pp. 62–65.

Sadova S.F. (2006). 'The use of low-temperature plasmas in wool finishing' *High Ener Chem* **40**(2), pp. 57–69.

Shen J. (2009). 'Wool finishing and the development of novel finishes' in *Advances in wool technology, Edited by N A G Johnson and I M Russell Woodhead Publishing Ltd., Cambridge, England*, pp. 147–182.

Silva C.J.S.M., Prabaharan M., Gubitz G. and Cavaco-Paulo A. (2005). 'Treatment of wool fibres with subtilisin and subtilisin-PEG' *Enzy & Microb Tech* **36**(7), pp. 917–922.

Simpson W.S. (2002). 'Chemical processes for enhanced appearance and performance' in *Wool: Science and technology, Edited by W S Simpson and G H Crawshaw, Woodhead Publishing Ltd., Cambridge England*, pp. 215–236.

Thomas H. (2007). 'Plasma modification of wool' in *Plasma technologies for textiles, Edited by R Shishoo, Woodhead Publishing Ltd., Cambridge, England*, pp. 228–246.

Thorsen W.J. (1968). 'A Corona Discharge Method of Producing Shrink-Resistant Wool and Mohair: Part II: Effects of Temperature, Chlorine Gas, and Moisture' *Text Res J* **38**(6), pp. 644–650.

Thorsen W.J. and Kodani R.Y. (1966). 'A Corona Discharge Method of Producing Shrink-Resistant Wool and Mohair' *Text Res J* **36**(7), pp. 651–661.

Thorsen W.J. and Landwehr R.C.A. (1970). 'Corona-Discharge Method of Producing Shrink-Resistant Wool and Mohair: Part III: A Pilot-Scale Top Treatment Reactor' *Text Res J* **40**(8), pp. 688–695.

Udhakhe J. and Honade S. (2013). 'Plasma induced physico-chemical changes & dyeing behavior of wool fabrics' *Colourage* **LX**(1), pp. 41–46.

Udhakhe J. and Honade S. (in press) 'Surface modification using plasma and chemical processes - correlation to shrinkproofing & mechanical properties of wool fabric' *Man Text in Ind.*

Udhakhe J., Honade S. and Shrivastava N. (2011). 'Recent advances in shrinkproofing of wool' *J of the Text Asso* **72**(3), pp. 169–177.

Udhakhe J., Honade S., Tyagi S., Shrivastava N. and Bhute A. (2012). 'Effect of yarn hairiness, DBD plasma and enzyme treatment on itching propensity of woollen knitwear' *Colourage* **LXI**(5), pp. 46–51.

Udakhe J. and Tyagi S. (2011c). 'Effect of Plasma Density on Surface & Mechanical Properties of Wool Fibres' *Man Text in Ind* **XXXIX**(4), pp. 137–140.

Udakhe J., Tyagi S., Shrivastava N., Kularni P., Acharya C. and Bardhan M.K. (2011b). 'Development of Itch-free woollens to be worn next to the skin' in *National seminar cum workshop on recent R&D initiatives and development schemes of wool and woollens Mumbai, India* 28th May 2011.

Veitch C.J. and Naylor G.R.S. (1992). 'The Mechanisms of Fiber Buckling in Relation to Fabric-Evoked Prickle' *Wool Tech and Sheep Breed* **40**(1), pp. 31–34.

Vilchez S., Manich A.M., Jovancic P. and Erra P. (2008). 'Chitosan contribution on wool treatments with enzyme' *Carbo Poly* **71**(4), pp. 515–523.

Vlist J. and Loos K. (2010). 'Transferases in Polymer Chemistry' *Adv in Poly Sci* **237**, pp. 21–54.

Walawska A., Rybicki E. and Filipowska B. (2006). 'Physicochemical Changes on Wool Surface after an Enzymatic Treatment' *Prog in Col and Poly Sci* **132**, pp. 131–137.

Wang G., Zhang W., Postle R. and Phillips D. (2003). 'Evaluating Wool Shirt Comfort with Wear Trials and the Forearm Test' *Text Res J* **73**(2), pp. 113–119.

Wilson C.A. and Laing R.M. (1995). 'The Effect of Wool Fiber Variables on Tactile Characteristics of Homogeneous Woven Fabrics' *Clothing and Text Res J* **13**(3), pp. 208–212.

www.novozymes.com (accessed on 21 March 2011)

www.richter-fa.de (accessed on 21 March 2011)

www.richterinternational.ca (accessed on 21 March 2011)

www.superwool.de/index.htm (accessed on 21 March 2011)

www.wool.com (accessed on 15 May 2013)

Xie T., Wang A., Huang L., Li H., Chen Z., Wang Q. and Yin X. (2009). 'Recent advance in the support and technology used in enzyme immobilization' *Afri J of Biotech* **8**(10), pp. 4724–4733.

Xin J.H., Zhu R., Hua J. and Shen J. (2002). 'Surface modification and low temperature dyeing properties of wool treated by UV radiation' *Col Tech* **118**(4), pp. 169–173.

Zhang R., Cai Z. and Zhang H. (2010). 'Studies on the remedial effect of transglutaminase on protease anti-felting treated wool' *J of the Text Inst* **101**(11), pp. 1015–1021.

8

Low pressure air plasma for processing of silk textiles

Shillin Sangappa and Arindam Basu

Abstract: Raw silk fiber is composed of a fibrous protein called fibroin, which is surrounded by gum-like protein Sericin. In textile applications, the Sericin (gum) needs to be removed. Cold plasma treatment is a promising technology for partial degumming of silk without damaging the delicate fibroin fiber. In this chapter, low pressure (RF) plasma treatment of mulberry and non-mulberry (Vanya) degummed silk yarn has been presented. Effect of plasma treatment on silk surface morphology, physical properties, wicking properties, and shade depth has been discussed. It shows that there is a significant improvement in wicking and shade depth. However there is a need to fine tune the plasma process parameters for different varieties of silk for industrial applications.

Key words: RF plasma, mulberry silk, vanya silk, sericin, degumming, wicking, shade depth

8.1   Introduction

Silk, the most beautiful of all natural fibres, is acclaimed as the queen of textiles. The very word 'silk' has entered in everyday language in such phrases 'smooth as silk', 'silken hair', etc. It even goes beyond the realm of textiles which indicates that silk is not perceived as an ordinary fibre, but one which has come to represent something magical. Silk being one of the most ancient fibre among all the textile fibre occupies top place owing to its special character like softness, luster, durability, elegance, smooth texture and mechanical strength, which has made it the queen of textiles for many centuries. The beauty and elegance of this fibre will enable it to live evergreen in the textile field. Silk enjoys a growing domestic market since it is a part of Indian culture and tradition.

8.2   Classification of silk

Natural silk is broadly classified into domesticated silk and wild silk. Under 'domesticated silk' we have mulberry silk. Tropical Tasar, Oak Tasar, Muga and Eri silk fall under the category of non-mulberry *(wild or Vanya)* silk. India is the only country producing all the above-said five varieties of silk.

8.2.1   Mulberry

Mulberry foliage is the only food for the silkworm *(Bombyx mori)* and is grown under varied climatic conditions ranging from temperate to tropical. A single silk filament is the product of a series of stages derived from the cultivation of mulberry trees for feed to the propagation of the domesticated silkworm, *Bombyx mori.*

Mulberry silk *(Bombyx mori)* is considered to be superior in quality as compared to others. About 92% of the total country's production is comprised of mulberry silk and is considered to be qualitatively superior. Mulberry is classified in to Bivoltine and Multivoltine. Bivoltine silk is qualitatively superior to multivoltine silk. India produces 90% of its silk production in multivoltine form. The mulberry silk accounts for bulk production in the world, and India is the second largest silk producer after China.

8.2.2   Non-mulberry *(Vanya)*

Non-mulberry sericulture, popularly known as *wild* or *Vanya* sericulture is practiced by the tribal and other backward communities of India as their livelihood option. The varieties of Vanya silks in India are Tropical Tasar *(Antheraea mylitta D.)*, Oak Tasar *(Antheraea proylei* J., *A.roylei* Mr.), Muga *(Antheraea assamensis* Ww.), and Eri *(Cynthia samia ricini* Bsd). Mulberry, Tasar and Muga cocoons are reeled, and Eri cocoons are open-mouthed and are spun instead of reeling.

8.3   Silk formation and properties

The silkworm has a pair of silk glands (Fig. 8.1) or sericteries, one on each side of the alimentary canal. The silkworm produces the silk fibre with these two silk glands and these glands are made up of about 1000 cells formed at the embryonic stage itself; the cells only enlarge with the growth of the larvae. The silk gland has three divisions, viz. anterior, middle and the posterior. The anterior divisions of both the glands unite into one forming the spinneret near the head opening. The posterior part secretes the silk substrate known as fibroin, while the middle part serves as a reservoir for receiving the fibroin from the posterior silk gland. It also secretes round the fibroin a natural gummy substance called sericin. The anterior part which is actually a silk duct or excretory canal moulds the sericin-coated fibroin and conveys it to the spinneret. Here, pair of Fillippi's gland which is situated on either side of the common duct elaborates certain secretions which helps unite the moulded silk material from the two excretory canals and prevents the freshly moulded silk from hardening. The silk substarte hardens into silk filament or bave only on coming into contact with air. It is made up of two independent fibroins called

Brins from the two glands and is covered completely with sericin. It is the continuous long silk bave that makes up the cocoon shell.

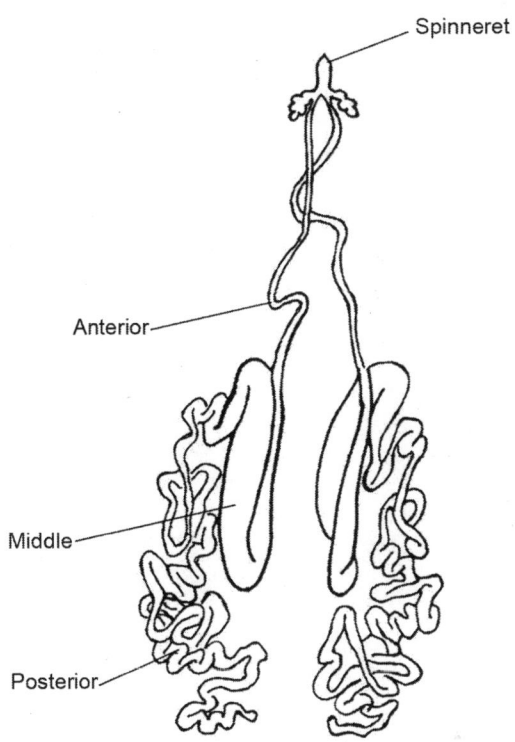

Figure 8.1 Silkworm gland

8.3.1   Silk filament

Single filament denier of mulberry, tasar, muga and eri are in the range of 2–3, 5–13, 4–5 and 2.2–2.5, respectively. The single cocoon filament cannot be used for making of the fabrics as these filaments are fine and the denier of the filament varies along the filament length. Hence, based on the denier, the silk yarn is required to be produced for any particular end-use fabrics, a known number of filaments are combined and wound together to form a single compact raw silk yarn.

8.3.2   Composition of silk

During the course of silk fibre formation, the silk is synthesized in the silk gland and converted into fibres with which the cocoons are formed. The silkworms spun cocoons act as a 'protective shell' to perpetuate the life. Man interferes this life cycle at the cocoon stage to obtain the silk, a continuous filament of commercial importance used in weaving of the dream silk fabric. The silk fibre is composed of two types of proteins: fibroin and sericin. It also contains small quantities of carbohydrate, wax and inorganic components, which also play a significant role as structural elements during the formation of silk fibres. The mulberry, tasar, muga and eri silk produced in India differ significantly in colour, lusture and other physical properties. The amino acid composition of the fibroin and the sericin of each are given in Table 8.1.

Table 8.1 Comparative amino acid composition in Indian silk

S. no.	Amino acids	Mulberry		Tasar		Muga		Eri	
		Fibroin	Sericin	Fibroin	Sericin	Fibroin	Sericin	Fibroin	Sericin
1	Glycine	43.75	9.85	24.24	11.61	25.55	13.27	26.42	18.62
2	Alanine	29.05	3.21	39.00	4.67	34.34	1.57	35.35	2.09
3	Valine	1.85	2.20	0.67	4.18	0.53	2.46	0.54	0.82
4	Leucine	0.42	1.12	0.35	2.00	0.40	0.60	0.38	1.24
5	Isoleucine	0.53	0.67	0.36	1.07	0.39	0.55	0.52	0.60
6	Serine	10.00	19.31	8.62	9.86	7.88	15.48	4.96	7.15
7	Thereonine	0.90	5.90	0.32	8.84	0.80	9.48	0.36	2.60
8	Aspartic acid	1.51	10.30	5.53	8.11	4.82	9.78	3.25	6.77
9	Glutamic acid	1.07	2.34	0.78	5.50	1.17	4.56	0.66	3.88
10	Phenylalanine	0.50	0.27	0.42	1.56	0.56	0.00	0.64	0.49
11	Tyrosine	0.42	1.51	4.71	5.40	4.61	4.75	5.37	4.34
12	Lysine	0.60	26.83	0.05	26.38	0.29	20.51	0.30	26.27
13	Histidine	0.36	5.97	1.68	5.05	1.06	2.81	1.53	3.86
14	Arginine	1.90	1.21	11.84	5.47	12.25	3.38	6.95	14.19
15	Praline	0.50	0.02	0.70	0.60	0.47	0.30	0.52	0.20
16	Tryptophan	0.39	0.24	2.04	0.43	2.8	0.19	0.39	0.16
17	Cystine	0.08	0.01	0.20	0.03	0.00	0.08	0.00	0.00

8.3.3   Molecular structure

Fibroin is having 18 amino acids, of these glycine, alanine, tyrosine and serine being the chief constituents cover about 80 percent. Eighteen amino acids molecular chain are in the form of polypeptide chain with peptide bonds and it appears to have –gly–x–gly–x as the bond order forming the main structure. Sericin is the covering layer for the fibroin fibres which forms the central core. In the fibroin fibres, the completely extended β-type molecular chains of fibroin are made of parallelly arranged crystalline parts, non-crystalline part and shifting part. It has characteristically strong tendency towards axial direction. The relative molecular agglutinating energy of silk fibroin is 6.2 kcal/mol, which attributes excellent strength to the fibroin fibres as well as enhances the moisture-absorbing capacity of the fibres.

Due to the presence of acid base and basic base in the molecular chain, it has a wide range of affinity for dyes. The high content of tyrosine along with the presence of tryptophan of fibroin allows the absorption of ultraviolet part and results in photochemical changes of the silk. Silk has a specific gravity between 1.36 and 1.39 g cm^{-2} with a specific volume of 0.75 cm^3 g^{-1} and is lighter than other natural fibres.

The higher hydrophilicity was found in *vanya* silk as compared to that of the mulberry variety silk, i.e. higher moisture content in non-mulberry silks than that of mulberry silk. The muga silk fibre has greater mechanical strength and higher stiffness as compared to eri and tasar silk fibres, as the strength of silk fibres is normally proportional to crystalline area, which is equal to ellipse, and is determined by micro-structural parameters. Through amino acid analysis, it is shown that there is no difference in chemical architecture between the outer and the inner layers of all the varieties of Indian silk cocoons.

8.3.4   Wax content

The wax content in tasar, muga and eri silk is higher in comparison to mulberry silk fibre. The wax content and its properties are given in Table 8.2. It also varies according to the family and genus. The wax is accumulated only in the sericin of said variety of silk fibre. Therefore considering the low quantity of sericin percentage in *vanya* silk fibres and the wax in unit sericin quantity appear to be high. The wax present in all silk fibres has the characteristics of fatty compounds.

Table 8.2 Wax content and its properties in silk fiber

S. no.	Silk varieties	Wax content (%)	Melting Point (°C)	Colour	Appearance
1	Mulberry	0.37	59–60	Dark brown	Opaque, waxy luster
2	Tasar	0.48	45–46	Dark brown	Transparent, grease-like
3	Muga	0.42	53–55	Dark brown	Opaque, waxy luster
4	Eri	1.7	74–75	Pale brown	Opaque, crystalline

8.3.5   Metal ion absorption

As the silk has active amino group and active carboxyl group at the two terminals, in aqueous solution it shows the behavior of both the types of ions. In the vicinity of isoelectric point it separates as follows:

$$H_2N^+ - Silk - COO^-$$

The absorption of metal cations helps in excessive diffusion of the metal cation in the silk fibres.

8.4   Silk wet processing technology

Sericin is responsible for imparting a harsh and stiff feel to the individual filaments of silk. To enable the raw silk to be ready for the subsequent operations viz., to bring out its characteristic luster, smoothness, softness and other valuable properties, it is essential to remove the sericin. Removal of sericin is termed as degumming. The degumming as shown in Fig. 8.2 is generally carried out by treating the raw silk with alkalis, enzymes, and acids or even with water at high temperature and pressure.

Figure 8.2 Degumming of Mulberry silk yarn

Silk has affinity for various classes of dyestuffs. The dyes recommended for dyeing silk include acid, acid milling, metal complex, reactive, direct and basic dyes. Majority of the silk is dyed with acid, metal complex and direct dyes. Fastness of metal complex dyed silk goods to wet treatment is relatively better than acid dyed silk goods.

8.5   Plasma technology

In the year 1879, an English Physicist, Sir William Crookes, identified ionized gas state as fourth state of matter. In the last two decades, surface modification of polymers and textile using plasma is rapidly emerging as environment-friendly solution for different industrial applications such as technical textiles, adhesion, packaging, thin films, biomaterials, and functional coatings. Among the available surface modification techniques, plasma technology provides an advantage that it does not damage or alter the bulk properties of the materials and also does not create pollution.

The most effective etching gases are air and oxygen for all the varieties of the Indian silk, while other gases such as helium, argon, nitrogen and sulphur hexafluoride (SF_6) are almost ineffective for etching of silk, which is in conformity with Riccardi et al. (2005). The etching rate can be calculated from the total weight loss in a given time. In case of raw silk, air and oxygen, plasma etching selectively removes the sericin to the extent of 10–15%, which is more than half of the sericin content in silk fibre. However, it has been observed that further etching of silk fibres starts removing fibroin as well. This may be attributed to difference in the sericin distribution along the fibres and uneven etching by plasma. So it was found difficult to remove the complete sericin from silk fibroin with air/oxygen plasma treatment in comparison to conventional chemical degumming process. However, the air or oxygen plasma treatment has assisted in saving the chemicals and treatment time. In order to achieve results equivalent to conventional degumming process, further optimization of plasma process is essential for different silk fibres. However, the chapter focuses on authors' investigation of effect of low pressure air plasma treatment on silk fibres and fabrics which is mentioned below.

8.6   Materials and methods

Mulberry (*Bombyx mori* L.), Tasar (*Antheraea mylitta* D.), Muga (*Antheraea assama* Ww.) and Eri (*Samia (Philosamia) ricini* B.) degummed silks yarns were used for the study.

8.6.1   Low pressure air plasma treatment

A glow discharge generator (Show Co. Ltd., Japan) was used for the treatment of silk fibres and fabric at M/s Bangalore Plasma Tech Pvt. Ltd., Bengaluru, India. The glow discharge apparatus was a radio-frequency etching system operating at 13.56 MHZ and used an aluminium square chamber with an internal size of 200 mm × 200 mm. The silk yarn and fabrics selected for the study are dried in an oven at 40°C for 12 hours to minimize the moisture content in fibres and fabrics before plasma treatment. Later, they were treated with plasma at low pressure of 0.5 mbar and airflow rate of 20 cc/min at room temperature. The surface morphology as well as dynamometric, wicking and dye uptake properties of degummed silk yarn were studied.

8.6.2   Tenacity and elongation

Tenacity and elongation tests were carried out on Instron Tester (Model No. 5500R6021). Tenacity is expressed as breaking load in grams per denier of yarn. Elongation is the amount of the stretch when pulled to the breaking point and expressed as percentage. The breaking loads of Mulberry, Tasar, Muga and Eri yarn were determined at an effective gauge length of 250 mm and extension rate of 250 mm/min on an Instron Universal Tester. In order to study the effect of plasma treatment on mechanical properties of silk, mainly breaking strength and elongation were investigated.

Comparative test results of tenacity and elongation are given in Tables 8.3 and 8.4. It shows that the effect of plasma treatment on tenacity and elongation of mulberry and non-mulberry degummed silk yarn is statistically not significant. This is in conformity with the findings of Chen Yu-yue et al. (2004) for *Bombyx mori* that the air plasma treatment carried out at different RF power and time duration has no significant effect on the mechanical properties of silk fibres. However, some damages to tenacity and elongation have been reported for heavily etched silk fibres; this is because of the damage to the fibroin.

Table 8.3 Average test results of tenacity and elongation properties of different silk yarn

S. no.	Yarn	Treatments	Tenacity (g/den.)	Elongation (%)
1	Mulberry	Plasma treated	3.5	14.9
2	Mulberry	Control	3.2	12.7
3	Tasar	Plasma treated	2.3	30.4
4	Tasar	Control	2.3	31.0
5	Muga	Plasma treated	4.0	42.1
6	Muga	Control	3.9	41.7
7	Eri	Plasma treated	1.5	24.3
8	Eri	Control	1.5	24.2

Table 8.4 ANOVA for tenacity (g/d) and elongation (%) of mulberry, tasar, muga and eri silk yarn studied

Mulberry (*Bombyx mori* L.)		Sum of squares	df	Mean square	F	Sig.
Tenacity	Between groups	0.167	1	0.167	2.500	0.189
	Within groups	0.267	8	0.067		
Elongation	Between groups	7.707	1	7.707	2.769	0.171
	Within groups	11.133	8	2.783		

Tasar (*Antheraea mylitta* D.)						
Tenacity	Between groups	0.001	1	0.001	0.012	0.916
	Within groups	0.551	8	0.069		
Elongation	Between groups	5.161	1	5.161	0.252	0.630
	Within groups	164.155	8	20.519		
Muga (*Antheraea assama* Ww.)						
Tenacity	Between groups	0.005	1	0.005	0.025	0.878
	Within groups	1.583	8	0.198		
Elongation	Between groups	0.359	1	0.359	0.018	0.895
	Within groups	155.547	8	19.443		
Eri (*Samia (Philosamia)* ricini B.)						
Tenacity	Between groups	0.016	1	0.016	1.467	0.260
	Within groups	0.085	8	0.011		
Elongation	Between groups	44.029	1	44.029	0.062	0.062
	Within groups	74.981	8	9.373		

Hence for raw silk a partial plasma etching is advisable and for degummed silk using RF plasma having power of 200 W and duration of 8–10 minutes were optimum parameters for retaining the silk fibres mechanical properties.

8.6.3   Aggregation structure of silk fibre

Chen Yu-yue et al. (2004) found the crystallinity of 23.56% in oxygen plasma-treated degummed mulberry silk after 10 minutes treatment at a reactor power of 50 W and pressure of 50 Pa as compared to crystallinity of 27.14% in the case of control sample. However, when the mulberry silk was exposed to plasma for 30 minutes, the crystallinity of 21.02% was observed. The results clearly indicate that the oxygen or air plasma etches the surface and breaks the polypeptide chain. The breaking of polypeptide macromolecules results in lowering of the crystallinity percent of fibroin. The oxygen and air plasma treatment oxidized macromolecules are present in the fine structure, yield the etched surface and reduce the crystallinity percentage which results in lowering the tenacity of the silk fibre. So, the FT-IR spectra of untreated and atmospheric air/oxygen plasma-treated mulberry were taken. There was a change in amide II band from 1520 cm^{-1} to 1516 cm^{-1}. Also, the peak height reduced with increase in treatment time which may be due to the damage of the random coils of silk fibres. Therefore, the plasma process optimization is essential for each variety of silk fibres.

Further, the plasma treatment study on other silk like tasar, muga and eri to determine crystallinity percentage and their effect on micro and macrostructure, etc., has not been carried out earlier in detail. In the present study, it has been concluded that the plasma treatment shows no significant change in tenacity of these silk fibres, which is shown in Tables 8.3 and 8.4.

8.6.4   Acid dyeing

A C.I. Acid Red 215 (Molecular Formula: $C_{16}H_{14}N_6O_6S$) of 3% on the weight of the mulberry silk yarn was taken and standard solution was prepared and added in the water bath of 1:30 (material to liquor ratio). Sodium sulphate and acitic acid of 40% concentration was subsequently added to water bath at 10% and 5% on the weight of the material. The silk yarns were dipped in the bath at 40°C. Then, bath temperature was raised slowly to 90°C. At this temperature, the material was kept for 40 minutes for uniform dye uptake. The bath pH was maintained at 4 to 6. The silk material was taken out, washed in cold water thoroughly, squeezed and dried under shade. The structural formula of C.I. Acid Red 215 is given below.

8.6.5   Chromaflash color matching

The Chromaflash Color Matching System (ASHCO Industries Ltd. India) was used to evaluate K/S value and wave length for comparisons of color strength between plasma-treated and untreated (control) sample. Reflectance of visible light from 400 to 700 nm was measured at every 20 nm and K/S values were determined using Kubelka-Munk equation. K and S are absorption and reflection coefficient respectively of the mulberry fabrics. Total K/S values which is the summation of K/S at every 20 nm in the range of 400–700 nm is an indication of shade depth. The high strength of the color (K/S) has been achieved at 60 Pa with maximum duration of 30 min and minimum K/S value is observed at 200 and 400 Pa. It can be concluded that the higher etching and maximum K/S can be achieved when samples are treated in plasma having higher power density and at 60 Pa process pressure. The K/S graphs (Fig. 8.3) show two different peaks at a wavelength of 440 and 600 nm, which indicate that a selected acid dye i.e. C.I. Acid Red 215 was a mixture of two different colors. At 440 nm wavelength (first peak), plasma-treated mulberry sample shows K/S value of 0.635 against untreated mulberry of 0.577; whereas, at 600 nm wavelength (second peak), plasma-treated sample was having higher K/S value of 0.754 as compared to K/S value of 0.542 in control sample. However, detailed study is required to understand the mechanism of dyeing at different dye concentrations and when different plasmas are used. Power density, treatment duration, pressure and functional groups formed on the surface play important role to enhance dye uptake and require precise analysis and study for different varieties of Indian silk.

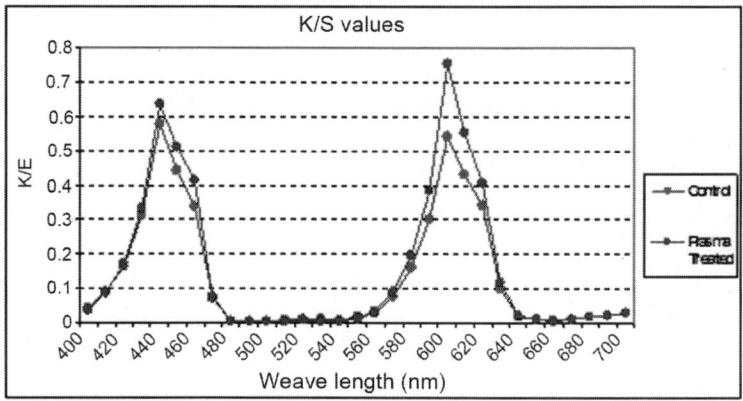

Figure 8.3 Colour (K/S) value of untreated and plasma-treated silk

8.6.6   Horizontal wicking

Horizontal wicking is the transmission of water along a horizontal threadline. It is perhaps of more importance than vertical wicking, because the mechanism of removal of liquid perspiration from the skin is involved in horizontal wicking. Therefore, the wicking behavior for hydrophilic property of the yarns of mulberry and vanya silks was studied using horizontal wicking test.

Wicking properties of plasma treated and untreated degummed silk yarns of different varieties were investigated by available standard procedure. All the varieties of plasma-treated and untreated silk degummed yarns were tied to glass rod. Depending on the denier of the yarn, dead weight was hanged to maintain uniform tension to the silk yarn and markings were made at 6 cm from the glass rod. Color water solutions were prepared by dissolving acid dye in cold water. Then, samples were dipped till the markings. Duration of 20 minutes was allowed to wick the tinted liquid. Then wicked height was noted down for all the samples. Wicking properties (Hmax) are given in Table 8.5.

The wicking property has shown significant improvement in both mulberry and vanya silk fabrics, which was represented in the ANOVA of the results in Table 8.6. The wicking behavior of fibre determines the ability of the fibre to transfer moisture, i.e. perspiration away from the body for increased comfort. The wickability can be explained in the light of capillary action which is the upward movement of water against the gravitational force within the spaces of porous materials. However, in the case of wicking along horizontal fibres, the effect of gravitational force becomes negligible (Slade and Slade, 1997). The capillary action is a function of the force of adhesion (attraction between water molecules and the substrate due to intermolecular forces of attraction), cohesion (attraction between water molecules), and surface tension. Capillary action occurs when the adhesive intermolecular forces between the liquid and the substrate are stronger than the cohesive intermolecular forces within the liquid. Creation of surface irregularity will assist in this capillary action (Periyaswamy et al., 2007).

Table 8.5 Average test results wicking heights (H_{max})

S. no.	Yarn	Wicking heights in cm	
		Control	Plasma treated
1	Mulberry	0.6	1.3
2	Tasar	0.1	0.4
3	Muga	0.1	0.5
4	Eri	1.0	1.5

Table 8.6 ANOVA for wicking height (H_{max}) of Indian silk yarn

Yarn	Groups	Sum of squares	df	Mean square	F	Sig.
Mulberry (*Bombyx mori* L.)	Between groups	0.735	1	0.735	73.5	0.1
	Within groups	0.040	4	0.010		
Tasar (*Antheraea mylitta* D.)	Between groups	0.135	1	0.135	27.0	0.007
	Within groups	0.020	4	0.005		
Muga (*Antheraea assama* Ww.)	Between groups	0.202	1	0.202	121.0	0.001
	Within groups	0.007	4	0.002		
Eri (*Samia (Philosamia) ricini* B.)	Between groups	0.282	1	0.282	84.5	0.001
	Within groups	0.013	4	0.003		

8.6.7   Surface characterization

Surface morphology of the samples was studied using scanning electron microscope (SEM) of LEICA S 40 Cambridge, UK make. The surface morphology of the silk fibres of all four varieties untreated and exposed to plasma has been observed under scanning electron microscope at different magnifications and are shown in Fig. 8.4 to Fig. 8.11. SEM images of untreated mulberry, tasar, muga, and eri fibres show sharp serrations morphology (Figs. 8.4, 8.6, 8.8 and 8.10). The higher and sharp serrations were observed on the morphology of tasar fibres in comparison with muga and eri fibres. The plasma-treated vanya fibres show minimized and blunt edges of the serration to a large extent with micro pitting, i.e. plasma itching causes reduction of sharp edges (Fig. 8.7, 8.9 and 8.11). SEM (Fig. 8.4) of mulberry untreated fibre shows smoother surface. The morphology of the same fibres after plasma treatment show nanopores on the surface and longitudinal flutes (Fig. 8.5), which is a key factor for the color deepness of dyed fibre. Appropriate roughness (etched parts) enables the surface to hold more dyes and causes random scattering of reflecting light, which results in deep color appearance. But the higher roughness on the fibre causes higher friction during the rubbing yielding lower rubbing fastness property.

Figure 8.4 SEM image of untreated Mulberry silk

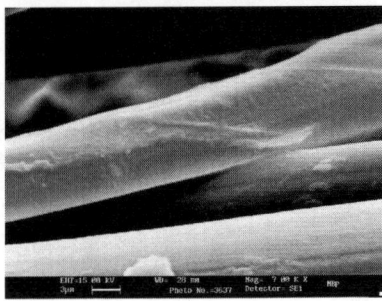

Figure 8.5 SEM image of plasma-treated Mulberry silk

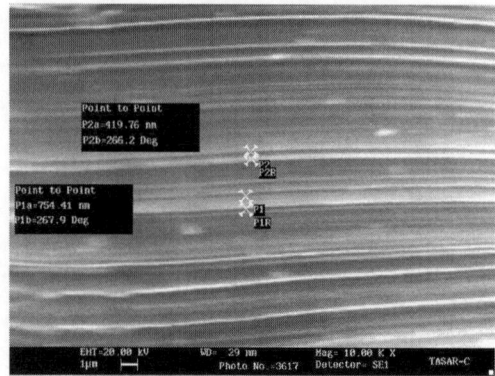

Figure 8.6 SEM image of untreated Tusar silk

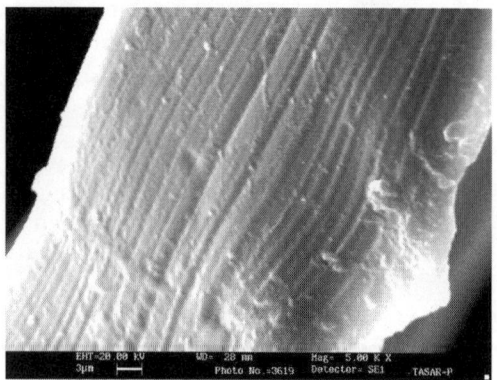

Figure 8.7 SEM image of plasma-treated Tusar silk

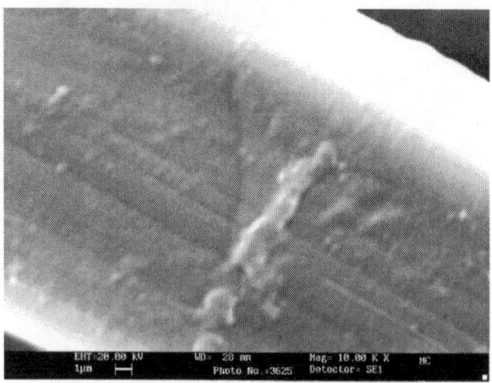

Figure 8.8 SEM image of untreated Muga silk

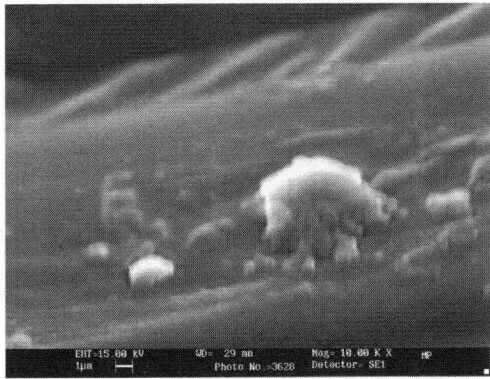

Figure 8.9 SEM image of plasma-treated Muga silk

Figure 8.10 SEM image of untreated Eri silk

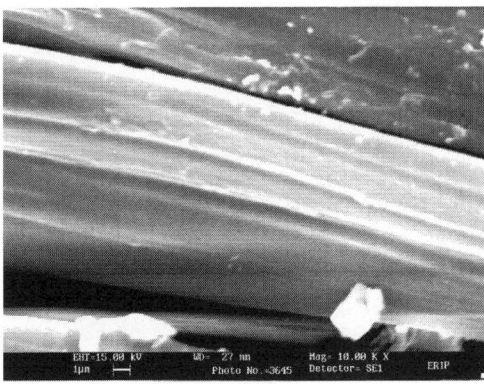

Figure 8.11 SEM image of plasma-treated Eri silk

8.7   Conclusion

Plasma surface modification changes the morphology of the silk fibres significantly thus alters its properties. These morphological changes assist in improving wicking property and the shade depth. However, over etching of silk by plasma results in increase in roughness which reduces its rubbing fastness property. This calls for an extensive study and data base generation on frictional coefficient of yarn or fabric of all the Indian silks, which may help researchers to optimize plasma treatment and improve rubbing fastness properties of silk fabrics. Although the present study has been carried out using low pressure plasma, for scaling up the technology one requires to work either in batch process in vacuum or one can use atmospheric pressure plasma.

8.8   References

Anonymous (2008). *Annual report 2007-2008,* Central Silk Technological Research Institute, Central Silk Board, Bangalore, India.

Anonymous (2009). *Annual report 2008-2009,* Central Silk Technological Research Institute, Central Silk Board, Bangalore, India.

Basu A. (2011). "Research: Present and future trends in textile and apparel industry", *24th NC of Textile Engineers on Textile & Apparel Industry Contemporary Issues to Address in Coming Years"* Institute of Engineers, Bangalore, 19th to 20th August.

Yu-yue C., Hong L., Yu R., Hong-wei W. and Liang-jun Z. (2004). Study on *Bombyx mori* silk treated by oxygen plasma, *Journal of Zhejiang University SCIENCE* 1009-3095, 5(8), pp. 918–922.

FAO Agriculture Services Bulletin (1987). Manuals on sericulture, *Central Silk Board, Gitanjali Printers,* Bangalore, India, pp. 58–60.

Goel R.K. and Krishna Rao J.V. (2004). Oak tasar culture-Aboriginal of Himalayas, *Central Silk Board, APH Publishing House,* New Delhi, India.

Wiener J. and Dejlova P. (2003). "Wicking and wetting in textile". *AUTEX Research Journal,* 3(2), pp. 64–71.

Jhala P.B., Nema S.K. and Mukherjee S. (2008). "Innovative atmospheric plasma technology for improving angora cottage industry's competitiveness" Conference on leveraging innovation and inventions enhancing competitiveness, NRDC, Delhi, October 13.

Samanta K.K., Jassal M. and Agrawal A.K. (2008). "Hydrophobic Finishing of Cellulosic Substrates by Tetrafluoro ethane ($C_2F_4H_2$) Plasma at Atmospheric Pressure" *International Conference on Technical Textile and Nonwovens,* IIT, Delhi.

Sen K. and Murugesh Babu K. (2004). "Studies on Indian silk. I. Macrocharacterization and analysis of amino acid composition, *Journal of Applied Polymer Science,* **92**(2), pp. 1080–1097.

Mahadevappa D., Halliyal V.G., Shankar D.G. and Bhandiwad R. (2000). Mulberry silk reeling technology, *Oxford and IBH Publication,* New Delhi, India.

Hojo N. (2000). Structure of silk yarn, *Oxford and IBH Publication,* New Delhi, India.

Periyaswamy S., Gulrajani M.L. and Gupta D. (2007). Preparation of multifunctional mulberry silk fabric having hydrophobic and hydrophilic surfaces using VUV excimer lamp, *European Polymer Journal,* **43,** pp. 4573–4581.

Riccardi C., Barni R. and Esena P. (2005). Plasma treatment of silk, *Solid State Phenomena* **107,** pp. 125–128.

Sarkar D.C. (1988). Ericulture in India, *Central Silk Board, Grapho Printers,* Bangalore, India.

Sadov F., Korchangin M. and Matetsky (1973). A Chemical technology of fibrous materials, *Mir Publisher,* Moscow.

Satreerat K.H., Supasai T., Paosawatyanyong B., Kamlangkla K., and Pavarajarn V. (2008). "Enhancement of the hydrophobicity of silk fabrics by SF6 plasma" *Applied Surface Science* **254,** pp. 4744–4749.

Singh K.C. and Singh N.I. (1998). Biology and ecology of temperate tasar silkmoths, Antheraea proylei J and Antheraea pernyi Guerin-Meneville (Saturniidae) a review, *Indian J Seri,* **37**(2), pp. 89–100.

Slade S.E. and Slade P.E. (1997). Hand book of fibre finish technology, Marcel Dekker, p. 46.

Suanpoot P., Kueseng K., Ortmann S., Kaufmann R., Umongno C., Nimmanpipug P., Boonyawan D. and Vilaithong T. (2008). Surface analysis of hydrophobicity of Thai silk treated with SF6 plasma, *Surface & Coating Technology* **202,** pp. 5543–5549.

Thangavelu K., Chakraborty A.K., Bhagowati A.K. and Md Isa. (1988). Handbook of Muga culture, *Central Silk Board,* Bangalore, India.

Raaja V. and Ramani S. (2009). Effect of plasma treatment on selected silk fabrics, *The Indian Textile Journal,* pp. 56–63.

Iriyama Y., Mochizuki T., Watanabe M. and Utda M. (2002). Plasma treatment of silk fabrics for better dyeability, *Journal of Photopolymer Science and Technology* **15**(2), pp. 299–306.

9

RF plasma treatment of Muga silk and its characterization

Dolly Gogoi, Arup Jyoti Choudhury, Arup Ratan Pal, Joyanti Chutia

Abstract: Radio frequency (RF) plasma treatment is considered a well-established technique for surface modification of natural silk. In this chapter, RF Ar plasma treatment is carried out to improve physical properties of muga (*Antheraea assama*) silk fibers without affecting their bulk properties. The chapter includes detailed investigation on the plasma discharge characteristics to observe the effect of plasma parameters on surface chemistry and morphology, physical properties and thermal behavior of the plasma-treated fibers. The chapter further discusses the possibility of using plasma-treated muga silk as resourceful clothing material.

Key words: Muga silk, textile, surface modification, radiofrequency plasma, ion energy

9.1   Introduction

Muga (*Antheraea assama*) is the famous, widely cultivated and most prestigious natural silk available in the NE region of India or more specifically in Assam (Ghosh and Ghosh, 1995; Mohanty, 2003). This lustrous natural golden-yellow silk is obtained from a semi-domesticated, multi-voltine and sericigenous silkworm called antheraea assamensis, which is fed on the leaves of primary host plants, namely Som (*Machilus bombycina*) and Sualu (*Litsaea polyantha*) (Ram and Bhatt, 2009). The confined ecological distribution of these food plants and the unique climate condition restrict the cultivation of this species exclusively to Assam. Muga is well known for some of its distinctive properties such as acid resistivity, UV resistance, antibacterial properties, highest tensile strength among all natural silk, hydrophobicity, anti-flammable and high durability (Bora et al., 1993; Das and Saikia, 2000; Das et al., 2009). All such properties have made muga silk a potential candidate to draw a great deal of interest in the fields of textile, biomedical and bioengineering research.

The muga silk fiber is composed of a fibrous protein called fibroin that constitutes the structural core of the silk in the form of two sub-fibers and a glue-like protein, termed sericin, which surrounds the fibroin fibers and cements them together (Altman, 2003). Due to presence of sericin, the silk cocoon is rigid and stiff and lacks in luster (Mishra, 2000). The structural rigidity and stiffness of the cocoon makes it difficult in continuous reeling of fiber. Moreover, the presence of residual sericin on silk fiber makes it vulnerable to bacterial attack and is also major source of severe problems associated with biocompatibility and immunological response to silk that hinders its effective utility in biomedical application (Liu et al., 2007). Considering the above disadvantages of sericin, it is utmost necessary to carry out surface treatment of silk cocoon to remove the adhesive sericin layer from the fiber. The process of removing sericin from cocoon before reeling is known as degumming. Sericin is hydrophilic in nature and can be dissolved in hot water where as fibroin is not soluble in water. Degumming of muga silk cocoons is achieved by treating with acids, alkalis, enzymes, soaps and synthetic detergents (Karmakar, 1999). However, none of these treatments ensures complete and uniform degumming of silk fiber and therefore the quest for developing convenient and efficient degumming process is still continuing.

In textile sector, the preparation of muga yarn from its corresponding fiber to weave the cloths is completely a traditional process depending on manual labour because of its poor tensile strength as compared to artificial fibers. This traditional process is a tedious one and consumes considerable amount of time, energy and manpower. The power looms are not suitable for weaving muga cloths due to lack of control of their weaving speed on the tensile strength of the yarn (i.e. the natural and force vibration of the muga yarn which it has to withstand during weaving). This result in adverse effect on the textile industries in terms of lowering the production of muga cloths made of the corresponding yarns. It is also essential to enhance the hydrophobicity of muga silk fiber to retain its natural golden color by reducing the application of natural/artificial dye on the surface which may sometimes make the muga cloths less comfortable to wear and cause serious skin disease (Gogoi et al., 2011). Moreover, reduction in the application of natural/artificial dye on fiber will save the requirement of chemicals, energy and time as well as the environment. In this regard, it is utmost important to modify the surface properties of muga silk fiber by considering the importance and necessity of increasing its mass production, export quality and accessibility as excellent clothing material for benefit of the textile industries and consumers.

Low temperature radiofrequency (RF) plasma is a convenient and well-known technique to modify the surface properties of natural silk fibers without influencing their bulk properties (Hodak et al., 2008). Extensive research is carried out on the RF plasma treatment of natural silk fibers by modifying their wetting, dying and printing properties, tensile strength, shrink resistance, flame retardance, anti-bacterial activity, biocompatibility, hydrophilicity and hydrophobicity (Demura et al., 1992; Gogoi et al., 2012; Suanpoot et al., 2008). RF plasma treatment is a dry and clean process and does not suffer from any environmental and health concerns (Wang and Wang, 2010). It has the major advantage over the conventional wet chemical treatment in terms of reduction of waste, pollution problems and conservation of chemicals, water, energy and time (Gogoi et al., 2011). Plasma treatment of silk fibers with inert gases only influences their outermost surfaces by increasing or decreasing the surface roughness up to few nanometre skin depths. Removing the contaminants like pollutants, oxide layer, etc., from the surface and also sputtering the weekly bonded chemical components from the fibers can easily be achieved by plasma treatment (Hodak et al., 2008).

This chapter deals with the surface modification of muga silk fibers using RF Ar plasma at various RF powers (10–30 W) and treatment time (5–20 min). The prime objective of this work is to improve the tensile strength and hydrophobicity of plasma-treated fibers without altering their bulk properties. This work also aims at investigating the plasma discharge characteristics to understand the effect of the plasma parameters on the properties of the plasma-treated fibers. The plasma-modified and -untreated silk fibers are characterized using Fourier transform infrared (FTIR) spectroscopy, X-ray photoelectron spectroscopy (XPS), universal tensile machine (UTM), dynamic contact angle (DCA) measurement and scanning electron microscopy (SEM). The observed properties of the plasma-treated and untreated fibers are attempted to correlate with their surface chemistry and morphology.

9.2   Experimental

9.2.1   Materials

Muga silk fibers used in this work are provided by Seri Biotech Laboratory, Institute of Advanced Study in Science and Technology, Guwahati, India. In order to remove sericin from the muga silk cocoons, the degumming process is carried out using a standard process. Briefly, the cocoons are boiled in aqueous solution of 0.03 M sodium carbonate for 30 min at 90°C followed by repeated washing with distilled water (Milli-Q water) to remove the sericin. The fibers are then reeled out of the degummed cocoons and dried in air atmosphere (temperature: 27°C and relative humidity: 65%). Prior to plasma treatment, the fibers are cut into a length of 6 cm and are held stretched on a horizontal stainless steel plate with the aid of holding rings providing a treatment area of 36 cm^2.

9.2.2   Plasma treatment

A schematic diagram of the capacitively coupled RF plasma system used in this work is shown in Fig. 9.1. The plasma treatment is carried out in a vertically placed stainless steel cylindrical chamber of 40 cm in diameter and 45 cm in length. Argon (Ar) gas (purity 99.9 %) is allowed to enter the chamber through a flat stainless steel cylindrical gas shower plate (9.25 cm in diameter). The bottom surface of the shower plate contains 60 tiny holes each having diameter of 5×10^{-2} cm as shown in the inset of Fig. 9.1. The RF electrode (10 cm in diameter) is placed horizontally at the centre of the

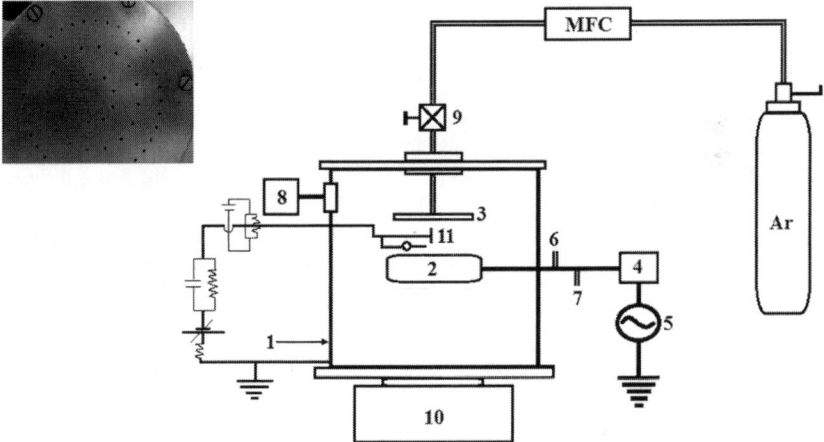

1 = Vacuum chamber 2 = RF electrode   3 = Gas shower plate 4 = Matching network
5 = RF generator   6 = Water inlet   7 = Water outlet   8 = Vacuum gauge
9 = Stop valve   10 = Vacuum system 11 = Self-compensated emissive probe

Figure 9.1 Schematic diagram of the experimental setup of the RF plasma reactor (The self-compensated Langmuir probe is not shown here). The inset shows the image of bottom surface of the gas shower plate.

chamber with proper insulation and connected to an RF (13.56 MHz) source (COMDEL-CPS-500AS, 0–500 W) through an L-type matching network. Water cooling controlling system is attached to the electrode to maintain the temperature of the electrode almost near to inside temperature of the RF plasma chamber (32°C) during Ar plasma treatment at various RF powers (10–30 W) and treatment times (5–20 min). The gas shower plate is placed 5 cm above the RF electrode and the samples are placed on the surface of the electrode. The chamber is evacuated to a base pressure of 1×10^{-3} mbar by a rotary pump (ED-21, Hindhivac, India) with a nominal pumping speed of 21 m^3 h^{-1}. The Ar plasma treatment is carried out at working pressure of 1.2 × 10^{-1} mbar measured with a pirani gauge (DPG-001, Hindhivac, India) and in the RF power and treatment time range of 10–30 W and 5–20 min, respectively. The Ar flow rate is set at 5 sccm using a mass flow controller (MFC) (Aalborg, USA) during whole plasma treatment process. Finally after plasma treatment, the RF power is turned off and the substrates are allowed to cool in an argon flow (35 sccm) for 15 min. The samples are collected from the plasma chamber after 6 h and immediately transferred to vacuum desiccator.

9.2.3   Characterization techniques

The chemical structural investigations of the untreated and Ar plasma-treated muga silk fibers are carried out using FTIR spectroscopy (Bruker Vector 22). For recording FTIR spectra, the fibers are finely cut and mixed to 1% concentration with potassium bromide (KBr) prior to pelletizing. A background spectrum is recorded with pure KBr pellet and is subtracted from the sample spectra. The spectra are obtained in the transmittance mode, within the spectral range of 4000–400 cm^{-1}. All FTIR measurements are performed with 32 scans and at a resolution of 4 cm^{-1}.

The surface chemistry of untreated and Ar plasma-treated muga silk fibers is further investigated by X-ray photoelectron spectroscopy (XPS). The XPS studies are conducted in a UHV chamber (base pressure < 2 × 10^{-8} mbar) using a VG make, CLAM-2 model hemispherical analyzer with a non-monochromatic twin Mg X-ray source. With MgKa line (1253.6 eV), detailed spectra are collected followed by high-resolution scan of relevant core level and valence level photoemission peaks of all the main elements. The XPS curve fitting is performed using "XPSPEAK-4.1" software.

Scanning electron microscopy (SEM) (JEOL-6390 LV) is used to study the surface morphologies of the untreated and Ar plasma-treated muga silk fibers. The samples are coated with platinum in an ion-sputter coater (JFC 100, JEOL) in a low vacuum prior to characterization. The mechanical (tensile) strengths of the untreated and Ar plasma-treated muga silk fibers are evaluated using an Instron tensile tester (INSTRON 4204). The samples are subjected to a constant load of 10 N while maintaining a steady speed of the crosshead at 10 mm/min^{-1} (speed accuracy: ±0.2%). Each measurement is repeated for five times and the average of these measurement values is considered for further analysis.

The dynamic contact angles for untreated and Ar plasma-treated muga silk fibers are measured by tensiometer (Data physics, DCAT-11). The instrument provides a measuring contact angle accuracy of ±0.01° with a position resolution of 0.1 mm. Each sample is immersed into the ionized water up to a depth of 1.5 mm with the help of micro-controlled motor (speed: 0.1 mm/s). The contact angles are determined using Wilhelmy technique (Mennella and Morrow, 1995).

The DSC thermograms are recorded with a Perkin–Elmer thermal analyzer, DSC-6000 (temperature accuracy: 0.25%, weighing precision: 0.01%) coupled with thermo analyzer (TA) processor. Samples of silk fibers (2.1–3.3 mg) are kept in the aluminum sample pans of the DSC cell under nitrogen atmosphere at a rate 20 ml/min. A heating rate of 10°C/min is maintained to get the DSC thermograms with high temperature accuracy within the temperature range of 30–440°C.

The thermal degradation of the muga fibers is studied using a Perkin–Elmer thermal analyzer, TGA-4000 (temperature accuracy: 0.25%, weighing precision: 0.01%). The samples (2–3 mg) are heated in nitrogen atmosphere (flow rate: 20 ml/min), at the heating rate of 10°C/min, over a temperature range of 30–850°C.

9.3   Results and discussion

9.3.1   Plasma parameters

The plasma parameters of the Ar discharge are measured using self-compensated planar Langmuir probe and emissive probe (Schott, 1968; Choudhury et al., 2011a). Self-compensated probe is used to minimize the perturbation effects of the RF voltage across the probe sheath (Chatterton et al., 1991; Sudit and Chen, 1994; Wendt 2001). The Langmuir probe tip consists of high purity (99.99%) stainless steel circular plate having diameter of 6 mm. Prior to each measurement a DC voltage of -100 V is applied to the Langmuir probe so as to clean it by ion impact in Ar plasma. The emissive probe is constructed using a 7 mm long, 0.1 mm diameter thoriated tungsten wire. The probe is heated by a half-wave DC voltage source (0–5 V). During all measurements reported here, both the Langmuir and emissive probe tips are kept 30 mm above the RF electrode. An X-Y recorder (Yokogawa, Japan) is used to plot the I-V characteristics obtained from the Langmuir and emissive probes. The Langmuir probe is used to evaluate electron temperature (T_e) and plasma density (electron density, n_e and ion density, n_i) from the current (I)-voltage (V) characteristic plot. The plasma potential of the Ar plasma discharge is evaluated by a self-compensated emissive probe using inflection point method (Smith et al., 1979). The variation in plasma parameters at various discharge conditions is presented in Table 9.1.

Table 9.1 Ar plasma discharge characteristics studied at various RF powers (10–30 W).

RF power (W)	Ar plasma characteristics					
	T_e (eV)	n_e (× 108 cm^{-3})	n_i (× 108 cm^{-3})	Plasma potential (V_p)	DC self-bias voltage (V_b)	Ion energy (qV)
10	6.4	3.6	3.7	16.0	− 26	q × 42.0
20	6.7	3.9	4.0	16.9	− 50	q × 66.9
30	6.9	4.2	4.1	17.8	− 74	q × 91.8

As revealed from the data given in Table 9.1, T_e, n_e and n_i increase with the increase of RF power. This is simply attributed to the fact that when the RF power is raised more dissipation of power occurs in the plasma and this eventually leads to the increase in electron energy (or T_e). An increase in electron energy will enhance the ionization rate of the Ar atoms and hence ne and ni are also increased with the increase of RF power. During Ar plasma treatment, the DC self-bias voltage (V_b) that is developed on the substrates is observed to be increased from −26 to −74 V with increasing RF power. The DC self-bias voltage is measured using a digital voltmeter connected to the RF electrode through a high impedance (RL = 1.35×10^5 Ω) inductor. However, the plasma potential (V_p) remains almost constant irrespective of the increase in RF power.

With the knowledge of DC self-bias voltage and plasma potential, the maximum energy gained by ions as they travel through the sheath to the substrate is calculated according to the following relation:

$$E_{max} = q (V_p - V_b)$$

[1.1]

where q is the ion charge (Glew et al., 1999). The variation in ion energy at RF power range of 10–30 W is further presented in Table 9.1, which shows an increase in ion energy with higher RF power. In this present work, the ion energy cannot be quantified as the types and charges of the ionized species formed near the plasma sheath are not known (Choudhury et al., 2011a).

9.3.2   Chemical structure and surface chemistry

FTIR spectra of the untreated and Ar plasma-treated muga silk fibers treated at RF power values of 10–30 W and treatment time of 10 min are presented in Fig. 9.2. The assignments of the peaks/bands are listed in Table 9.2 (Das and Saikia, 2000; Dyer, 1978; Hazarika et al., 1998; Pavia et al., 2008). The IR absorption peaks/bands display all the characteristic functional groups of different amino acids that present in the silk fibroin. In the present investigation, the treated muga fibers exhibit similar chemical composition irrespective of the increase in treatment time. However, the increase in RF power seems to affect the chemical composition of the plasma-treated fibers. As observed from Fig. 9.2, the intensity of the band at 1448 and 2928 cm⁻¹ is higher for untreated fiber than all the plasma-treated muga fibers. Interestingly the FT-IR spectra of the plasma-treated fibers show the absence of the band at 800 cm⁻¹ which is observed to be appeared in the spectrum of untreated fiber (Das et al., 2009). This may be attributed to the breakage of the H-bonded amide I and amide V groups due to energetic ion bombardment to the surfaces of the fibers. On the contrary a new and weak absorption peak at 1308 cm⁻¹ appears in the FT-IR spectra of the plasma-treated fibers. It is further observed from Fig. 9.2 that the intensity of the absorption band at 680 cm⁻¹ increase for the fiber treated at lower RF power (10 W) and after that it decreases with increase in RF power. It is assumed that some of the destroyed H-bonded amide I and amide V groups takes part in bond formation process of amide IV group. This possibly results in an increase in the absorption intensity of the band (680 cm⁻¹) at RF power value of 10 W. At RF power >10 W, the energy of the ions is sufficiently high enough to destroy the amide IV groups thereby leading to a decrease in absorption peak intensity. More observable variation in absorption peak intensity is shown by the spectra of the fibers treated at RF power value of 30 W. As seen from Fig. 9.2, the intensities of the absorption bands at 1161, 1231 and 1539 cm⁻¹ considerably decrease as compared to those treated at lower RF powers (10–20 W). It is apparent that at higher RF power value of 30 W, the energy of the impinging ions is sufficiently high enough to destroy the chemical structure of the fibers. The above findings indicate that at lower RF power (10 W) the impinging ions do not possess sufficient energy to influence the chemical composition of the muga fibers whereas at higher RF power (30 W) the energy of ions significantly alters the chemical structure of the fibers.

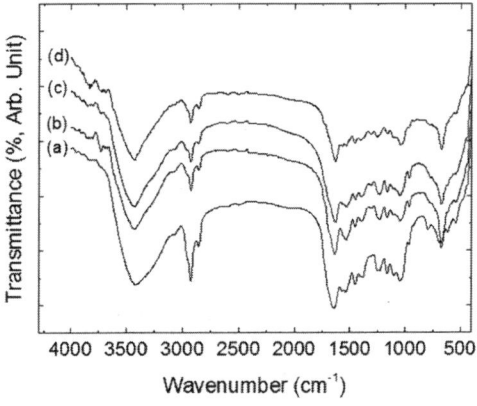

Figure 9.2 FTIR spectra of (a) untreated and Ar plasma-treated muga fibers obtained at RF power value of (b) 10, (c) 20 and (c) 30 W and treatment time of 10 minutes.

Table 9.2 Assignment of the peaks/bands detected in FTIR spectra of untreated and Ar plasma-treated muga fibers obtained in the RF power range of 10–30 W (treatment time: 10 minutes).

Wave number (cm⁻¹)	Assignments
3406	N-H stretching
3070	Amide II
2928	N-H stretching (Amide I)
2852	C-H stretching
1645	Amino acid band 1 (Amide I)
1539	N-H in-plane bending, C-N stretching (amide II)
1448	CH4 group frequency of Ala.
1387	-CH3 deformation
1315	Gem-distributed (=C=CH$_2$) C-H stretching
1308	C-N stretching
1231	C-H stretching, N-H in-plane bending (amide III)
1161	O-H bending
1048	Gly-Ala sequence
964	N-H rocking
800	N-H out of plane bending (amide V)
680	N-C=O in-plane bending (amide IV)
543	N-C=O out of plane bending (amide VI)

Figure 9.3(a) shows the XPS survey spectra of untreated and Ar plasma-treated muga fibers obtained at various RF powers (10–30 W) and treatment time of 10 min. The variation in atomic concentration of the untreated as well as the Ar plasma-treated fibers is shown in Figs. 9.3(b–d). It is revealed from Figs. 9.3 (b–d) that the fiber treated at 10 W exhibits similar variation in atomic composition than that of the untreated fiber, irrespective of the increase in treatment time. At RF power value of 20 W and lower treatment time (5–10 min), the plasma-treated fibers show relatively higher carbon and lower oxygen content as compared to the untreated and other plasma-treated fibers. Above treatment time of 10 min, the fibers show a gradual decrease in carbon content and increase in oxygen content. The variation in nitrogen content follows the similar trend as that of oxygen as the treatment time is increased from 5 to 20 min. From these findings, it is apparent that with increasing treatment time from 5 to 10 min, the impinging ion energy is responsible for the cleavage of peptide chain as well as the breakdown of side chain groups of amino acid, mainly glycine (—NH—CH—CO—) and alanine

$$\begin{array}{c} \text{H} \\ | \\ (\text{—NH—CH—CO—}) \end{array}$$

$$\begin{array}{c} \text{CH}_3 \\ | \\ (\text{—NH—CH—CO—}) \end{array}$$, which are the major constituents of muga fiber

(Rajkhowa et al., 2000; Sen and Babu, 2004). Similar observation has already been made in case of UV/ozone-irradiated Bombyx mori silk fibroin (Shao et al., 2005). The breakage of the side chain groups of amino acid and peptide chain scission may contribute to the removal of loosely bonded fibroin region through the formation of various volatile products (CO, OH, CO_2, etc.), which subsequently lead to a decrease in oxygen and nitrogen contents in the fibers. Above the treatment time of 10 min, the prolonged ion bombardment to the substrates creates severe peptide chain scission as well the breakage of the side chain groups of amino acid, and this possibly lowers the carbon content in the fibers through the formation of more volatile products. Besides this may lead to the formation of free radicals at the surfaces of the fibers and on exposure to atmosphere these free radicals readily react with atmospheric oxygen and nitrogen and/or water vapor thereby increasing oxygen and nitrogen contents in the fibers (Choudhury et al., 2011b). With increase in RF power (30 W) and treatment time (5–20 min), more and more free radicals are generated at the surfaces which contribute to the increase in oxygen and nitrogen contents in the fibers.

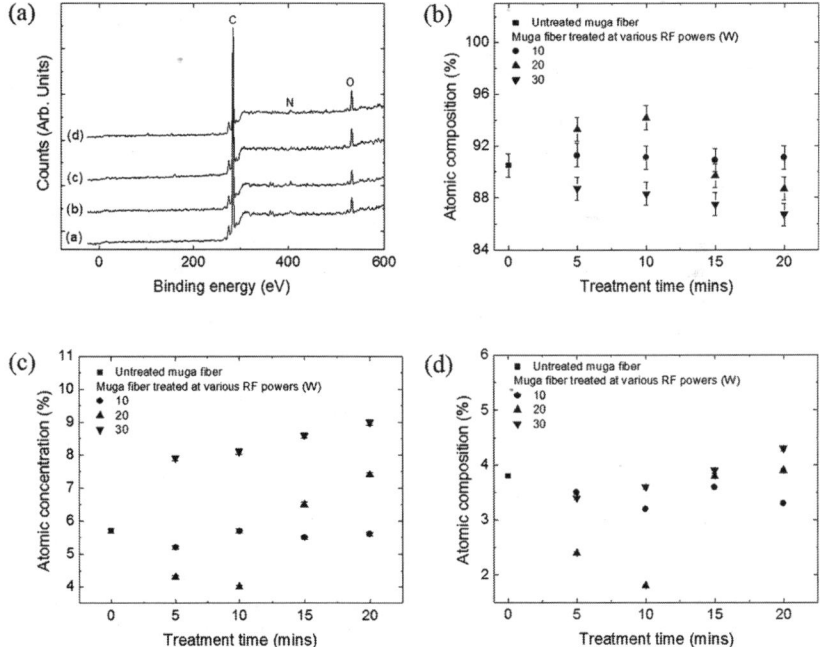

Figure 9.3 (a) XPS survey spectra of untreated and Ar plasma treated muga fibers obtained at various RF powers (10–30 W) and treatment time of 10 minutes. The variation in atomic concentration (%) of (b) carbon (c) oxygen (d) nitrogen content in untreated and plasma-treated muga fiber is also presented as a function of treatment time.

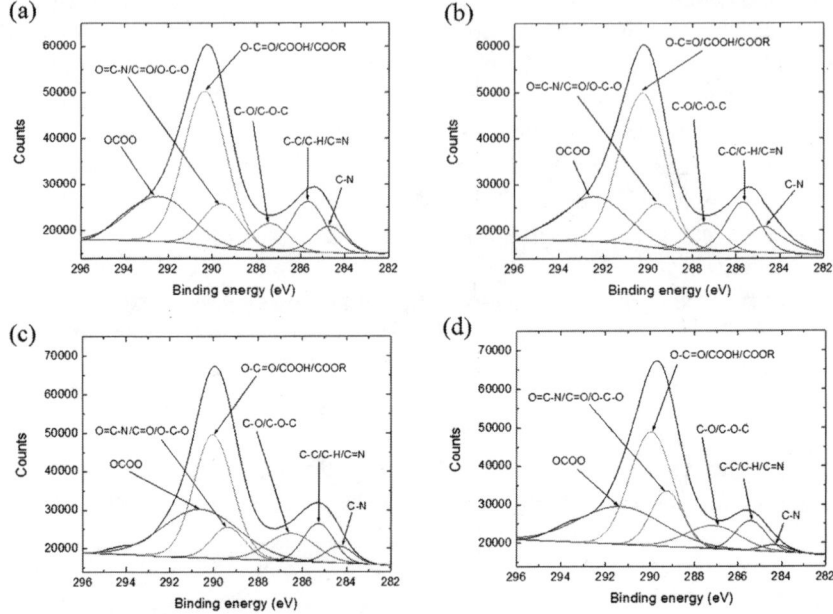

Figure 9.4 Deconvoluted C1s peaks of (a) untreated and plasma-treated muga fibers obtained at RF power value of (b) 10, (c) 20 and (d) 30 W and treatment time of 10 minutes.

In order to have greater insight into the effect of ion energy on the surface chemistry of the plasma-treated muga fibers, the curve fitting of C1 peaks for all the samples is performed. The untreated sample is also analysed and discussed here. The deconvoluted C1s peaks for the untreated and plasma-treated muga fibers are shown in Figs. 9.4(a–d). As observed from Figs. 9.4(a–d), the C1 peak has been fitted with six peaks corresponding to C-N (283.9 eV), C-C/C-H/C=N (284.9 eV), C-O/C-O-C (286.6 eV), O=C-N/C=O/O-C-O (288.7 eV), O-C=O/COOH/COOR (289.5 eV) and OCOO (291.7 eV) units. The details of C1 Peak fitting are given in Table 9.3. It is observed from Table 9.3 that the functional composition (%) in the fibers treated at 10 W and treatment time range of 5–20 min is almost similar to that of the untreated one. This indicates that at RF power value of 10 W, the energy of the impinging ion is insufficient to alter the surface chemistry of the fiber even an increase in treatment time from 5 to 20 min. With increase in RF power (20 W) and treatment time (5–10 min), the functional units (%) C-O/C-O-C, C-C/C-H/ C=N and OCOO increase while the C-N, O=C-N/C=O/O-C-O and O-C=O/ COOH/COOR units (%) decrease as compared to the fibers treated at lower RF power (10 W). With further increase in treatment time (from 15 to 20 min), the O=C-N/C=O/O-C-O, O-C=O/COOH/COOR and OCOO units (%) increase while the C-O/C-O-C and C-C/C-H/C=N units (%) decrease at the fibers. This attributed to the degradation in the chemical structure of the fibers due to prolonged energetic ion bombardment. At higher RF power (30 W) and increase in treatment time, the ion energy is sufficiently high enough to destroy the peptide bond and side chain of amino acid groups through sputtering and thereby leading to a severe loss of C-N, C-O/C-O-C and C-C/C-H/C=N units. It is apparent that ion bombardment to the substrate promotes the chemical interaction of the nearby atoms at the surface of the fiber through momentum transfer which eventually leads to the change in its surface chemistry. Besides at such higher RF power, more and more free radicals are generated at the surfaces of the fibers which possibly contribute to the increase in O=C-N/ C=O/O-C-O and OCOO units in the fibers.

Table 9.3 Summary of the curve fitting of C1s peaks of untreated and plasma-treated muga fibers obtained at various treatment conditions.

Power (W)	Treatment time (min)	Functional composition (%)					
		C-N	C-C/C-H/ C=N	C-O/ C-O-C	O=C-N/ C=O/O- C-O	O-C=O/ COOH/ COOR	OCOO
0	0 (Untreated)	5.7	11.6	6.3	11.4	45.8	19.2
10	5	5.1	10.9	5.8	11.2	46.7	20.3
	10	5.5	11.1	6.0	12.0	45.4	20.0
	15	5.2	11.4	5.5	12.3	45.9	19.7
	20	5.0	10.4	5.6	11.3	46.4	21.3
20	5	3.8	15.1	9.6	7.5	40.6	23.4
	10	3.7	14.3	10.5	8.1	38.7	24.5
	15	3.1	6.5	7.3	9.8	40.5	32.8
	20	3.0	3.2	6.1	11.1	41.8	34.2
30	5	2.0	7.5	8.9	10.2	40.1	31.3
	10	1.5	5.5	6.4	12.6	39.7	34.3
	15	1.3	3.7	3.3	15.3	41.4	35.0
	20	1.2	1.3	2.2	16.3	42.1	36.9

9.3.3   Surface morphology

The typical SEM micrographs of the untreated and Ar plasma-treated muga fibers are shown in Figs. 9.5(a–h). The observed crystals on the surfaces of the fibers are attributed to calcium oxalate $(Ca_2C_2O_4)$ deposits which are generally left by the silkworm during spinning (excrements) (Freddi et al., 1994). Figure 9.5(a) shows smooth surface of untreated muga silk fiber with fewer defects. At RF power value of 10 W, the variation in treatment time does not seem to bring any significant change in surface morphology of the fibers with respect to the untreated one (Figs. 5b–c). This is may be due to the fact that the impinging ion energy on the substrate (at RF power value of 10 W and within treatment time of 5–20 min) is not sufficient enough to alter the surface morphology of the treated fiber. At RF power value of 20 W and treatment time of 5 min, minor surface roughness is introduced on the fiber due to the ion sputtering effect on the substrate (Fig. 9.5d). However, the fiber treated for 10 min shows relatively smooth surface texture as compared to that treated for 5 min (Fig. 9.5e). This is attributed to the efficient removal of fibroin region through the peptide bond scission and breakage of side chain of amino acid residues, which is well revealed from XPS analysis. With further increase in treatment time (15–20 min), more surface degradation in the form of micro pits and voids occurs on the surface of the fiber (Fig. 9.5f). This is attributed to the decrease in carbon content in the fibers through prolonged sputtering effect of the impinging ions. At higher RF power value of 30 W, extensive damage can be observed on the surfaces of the fibers when the treatment time is increased from 5 to 20 min. It is apparent that at such higher RF power (30 W) the ions acquire sufficiently high energy to sputter the surfaces of the fibers thereby causing extensive damage to the peptide bond and side chain of amino acid groups and this results in much more surface degradation to the fibers in the form of micro pits and voids (Figs. 9.5g and h). From these discussions, it is revealed that above a critical plasma discharge parameter value the ions acquire sufficiently high energy to sputter the fiber surface while below that critical value the energy of the ions is too low to produce any observable change in the surface morphologies of the fibers.

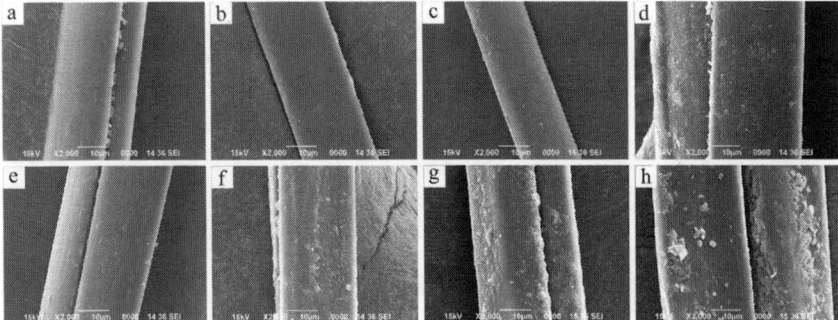

Figure 9.5 Scanning electron micrographs of (a) untreated and Ar plasma-treated muga fibers obtained at (b) 10 W, 5 minutes; (c) 10 W, 20 minutes; (d) 20 W, 5 minutes; (e) 20 W, 10 minutes; (f) 20 W, 20 minutes; (g) 30 W, 5 minutes; and (h) 30 W, 20 minutes.

9.3.4   Physical properties

The variation of tensile strength of the untreated and Ar plasma-treated muga fibers as a function of treatment time is shown in Fig. 9.6. The tensile strength of untreated muga fiber is evaluated to be 3.81 gm/den. At lower RF power value of 10 W, no significant variation in tensile strength of the treated fibers can be observed. This is possibly attributed to similar surface chemistry of the treated fibers as compared to the untreated one which is well revealed from XPS analysis. For muga fibers treated at 20 W, tenacity reaches a maximum value of 4.5 gm/den at treatment time of 10 min and after that it decreases with further increase in treatment time (from 15 to 20 min). The higher value of tenacity at 10 min is attributed to the presence of highest percentage of carbon content and also the increase in more C-O/C-O-C and C-C/C-H/C=N units (%) at the surfaces of the fibers. The decrease in carbon content and C-O/C-O-C and C-C/C-H/C=N units (%) are the possible reasons for lowering the tensile strength of the fibers treated for longer times (15–20 min). Moreover increase in oxygen content (%) and O=C-N/C=O/O-C-O, O-C=O/COOH/ COOR and OCOO units (%) may also contribute to a decrease in tensile strength of the fibers. At higher RF power (30 W) tensile strength of the fibers gradually decreases with the increase in treatment time. This may be caused by an increase in oxygen content (%) and decrease in C-N, C-O/C-O-C and C-C/C-H/C=N units (%) in the fibers. The tensile strength of the fibers treated at 30 W and treatment time of 5–20 min lies well below that of the untreated one.

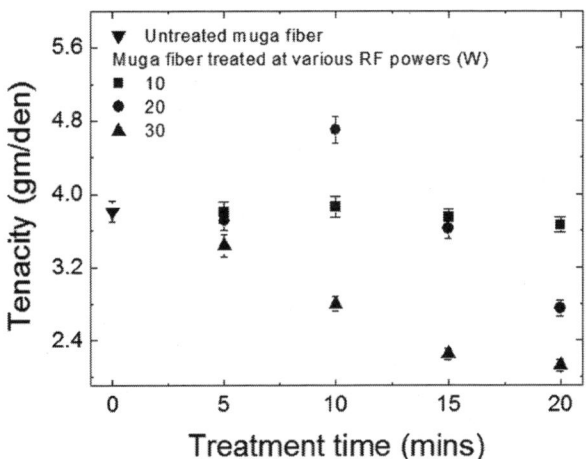

Figure 9.6 Variation in tensile strength of the untreated and Ar plasma-treated muga fibers as a function of treatment time.

The observed tensile strength behavior of treated muga fibers can further be explained from stress concentration effect induced by Ar plasma treatment (Cioffi et al., 2002). As observed from SEM micrographs (Figs. 9.5b and 9.5c), the energy of the impinging ions on the substrate does not seem to introduce significant surface roughness or defect on the fibers thereby making the stress concentration effect almost independent of the treatment time. These result in nearly similar variation of tensile strength of the muga fibers treated at RF power value of 10 W and within the treatment time range of 5–20 min. At RF power value of 20 W and in the treatment time of 5 min, the impinging ions on the fibers produce defect in the form of irregularities (micro pits and voids) through the cleavage of peptide bond and side chain of amino acid residues. This may have concentration effect on the fibers thus leading to a decrease in their tensile strength (Fig. 9.5d). At treatment time of 10 min, the removal of more weakly bonded fibroin region possibly smoothen the surface of the fiber, thereby inducing less defects than that of the fiber treated for 5 min (Fig. 9.5e). The minimization of the stress concentration effect due to less defects thus probably leads to a higher value of tensile strength (maximum in this case) in the plasma-treated fibers. With higher treatment time, much more destruction in the chemical structure and subsequent increase in surface defects take place due to prolonged energetic ion bombardment and this can be associated with lowering in tensile strength in the fibers (Fig. 9.5f). At RF power value of 30 W, the ions acquire sufficiently high energy to bombard the substrate leading to a severe damage in the surface chemical structure of the fibers even at lower treatment time which is well revealed from XPS analysis. This incorporates more surface defect in the fiber (Figs. 9.5g and 9.5h) with higher treatment time and consequently leads to the decrease in their tensile strength.

The variation in water contact angle on the untreated and plasma-treated muga fibers as a function of treatment time is shown in Fig. 9.7. The water contact angle for untreated muga fiber is measured to be 100°. No significant variation in water contact angle on the fibers, treated at 10 W, can be observed within the treatment time range of 5–20 min. Maximum water contact angle of 115° is observed for the muga fiber treated for 5 min and at RF power value of 20 W. This is due to the increase in carbon content (%) as well as the C-C/C-H/C=N and C-O/C-O-C units (%) in the fibers. Besides, lowering in oxygen content (%) and O=C-N/C=O/O-C-O and O-C=O/COOH/COOR units (%) may also contribute to the increase in water contact angle on the fibers. The water contact angle on the treated fibers remains almost same as the treatment time is increased from 5 to 10 min. This is attributed to the almost similar variation in atomic concentration and functional composition of the fibers as revealed from XPS analysis. With further increase in treatment time (from

15 to 20 min) the water contact angle decreases by ~30° thereby indicating more increase in surface roughness due to energetic ion bombardment to the substrates for longer period of time. The decrease in water contact angle on the fiber surface above 10 min of treatment time can be correlated with the decrease in carbon content (%) and C-C/C-H/C=N units (%) in the fibers. An increase in oxygen content (%) and the O=C-N/C=O/O-C-O and OCOO units (%) also apparently contribute to the decrease in water contact angle on the fibers. At higher RF power value of 30 W, the water contact angle is observed to be further decreased with increase in treatment time, and this may be attributed to the growth of more surface roughness on the surfaces of the fibers caused by energetic ion bombardment to the substrates. The variation in water contact angle of the fibers treated at 30 W can be well corroborated with the XPS results.

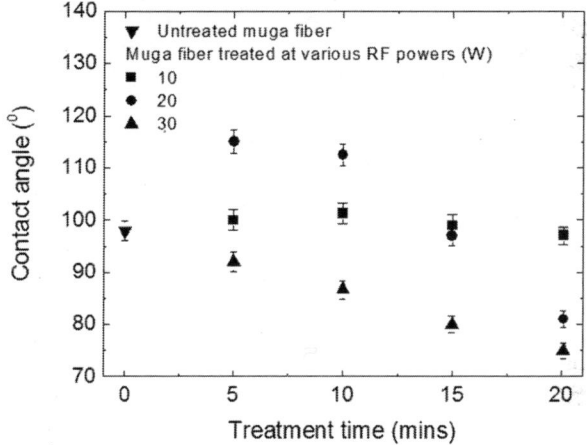

Figure 9.7 Variation in water contact angles on the untreated and Ar plasma-treated muga fibers as a function of treatment time.

9.3.5   Thermal behavior

Figure 9.8 shows the DSC thermograms of the untreated and Ar plasma-treated fibers obtained at various RF powers (10–30 W) and treatment time of 10 min. All the plots presented in Fig. 9.8 shows similar DSC thermograms thereby indicating that the variation in plasma discharge conditions have less significant effect on the thermal behavior of the treated fibers. The first broad endotherm below 100°C is due to the evaporation of water. The thermal stability of the untreated and Ar plasma-treated fibers remains almost unchanged up to 206°C. As revealed from Fig. 9.7, two minor and broad shoulder peaks appear at 231°C and 297°C, respectively, and this can be related to the molecular conformation of the fibroin chains in the fibers (Hazarika et al., 1998). It is assumed that the water molecules adsorbed in the fibers probably restrict the alignment of the fibroin chain molecules. On removal of these water molecules from the fibers above 100°C (Fig. 9.7), the restricted force being withdrawn, the molecular chains are free to rearrange within the fibers. This explanation may find good agreement with the results obtained for some plant fibers (Ray, 1969). The prominent endothermic peak at 362°C is attributed to the thermal decomposition of muga fibers.

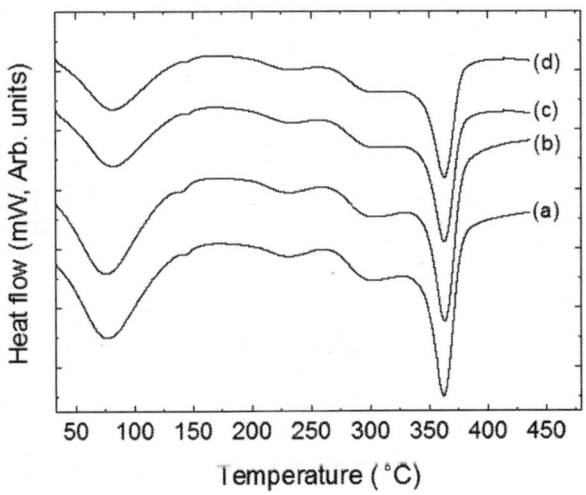

Figure 9.8 DSC thermograms of (a) untreated and Ar plasma-treated muga fibers obtained at RF power value of (b) 10, (c) 20 and (d) 30 W and treatment time of 10 minutes.

The thermal behaviour of the untreated and Ar plasma-treated muga fibers are further studied by TGA. The thermograms of the untreated and plasma-treated fibers obtained at various RF powers (10–30 W) and treatment time of 10 min and the corresponding differential thermograms (DTGA) are shown in Figs. 9.9(a and b). All the Ar plasma-treated fibers exhibit almost similar thermal stability irrespective of the change in RF power and treatment time. From Figs. 9.9(a and b) no significant variation in the weight loss of the treated fibers from the untreated one can be observed within the temperature range of 33–850°C. As observed from the plot presented in Fig. 9.9(b), each of the thermogram is accompanied with four steps of weight loss. The thermal behaviour of the untreated and Ar plasma-treated fibers are summarized in Table 9.4. The results obtained from Table 9.4 indicate that all the endotherms attributed in the DSC records are accompanied by the weight loss steps as shown in Fig. 9.9b. The occurrence of the first weight loss step is attributed to the removal of adsorbed water while the fourth step weight loss is due to decomposition of the untreated and plasma-treated fibers as observed earlier from the DSC results. The DTGA curves reveal that in the temperature region (333–846°C) for decomposition, the untreated muga fiber show slightly more thermal stability (370°C) than all the plasma-treated fibers (366°C). From the above findings, it is apparent that Ar plasma treatment within the RF power range of 10–30 W and treatment time of 5–20 min does not much affect the thermal stability of the fibers.

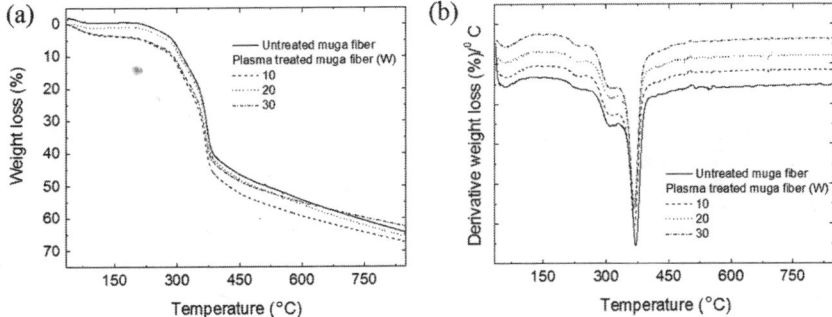

Figure 9.9 (a) TGA and (b) DTGA curves of untreated and Ar plasma-treated muga fibers obtained in the RF power range of 10–30 W and treatment time of 10 minutes.

Table 9.4 Thermal behavior of untreated and Ar plasma-treated muga fibers obtained in the RF power range of 10–30 W (treatment time: 10 minutes).

RF power (W)	First step	Second step	Third step	Fourth step
	Weight loss (%)	Weight loss (%)	Weight loss (%)	Weight loss (%)
	Temperature range (° C)	Temperature range (° C)	Temperature range (° C)	Temperature range (° C)
0 (Untreated)	2.9	2.7	15.2	44.8
	(39–117)	(117–269)	(269–333)	(333–846)
10	3.6	3.6	14.9	45.7
	(35–130)	(130–261)	(261–329)	(329–846)
20	3.0	3.5	15.1	46.2
	(37–128)	(128–263)	(263–330)	(330–846)
30	3.6	3.1	15.9	45.3
	(36–121)	(121–260)	(260–334)	(334–846)

9.4   Industrial scale-up and marketing

The correlation between in-situ plasma discharge characteristics and ex-situ surface analysis of plasma-treated muga silk fibers demonstrates that the impinging energy of ions can induce significant alteration in surface chemistry and morphology of the fibers prior to their reeling into yarns. The transfer of both mass and energy of the impinging ions to the substrates is apparently governed by the sheath voltage, working pressure and the plasma power density (power per area) controlling the degree of ionization. With proper knowledge of external (working pressure, plasma power density, etc.) and internal (plasma density, sheath voltage, DC self-bias voltage, etc.) plasma parameters coupled with material surface analysis, one can assure in-depth understanding of fundamental processes occurring in the plasma and at the plasma material interfaces, process reproducibility and the feasibility to scale-up plasma reactor and processes to industrial throughputs.

For the need of industrial adaptability and consumer satisfaction, the scale-up of plasma reactors must be associated with the low cost effective production of plasma-treated muga silk fibers. Based on the present research investigation, similar work can be carried out at atmospheric pressure plasma environment so that the need of expensive and high power vacuum system can be eliminated. Such atmospheric pressure plasma system will have the crucial advantage of treating large quantity of muga silk fibers in low processing time. However, considerable research and development work must be carried out before industrialization of such plasma treatment process is practically realized.

Improvement in the quality of muga silk fabric using low cost and high efficiency driving plasma treatment system is very challenging but vital one as these clothing materials have huge demand in domestic as well as international market. In this regard, both state and central governments should take promising measures to initiate and sustain extensive research on developing and setting up of large scale plasma treatment systems in north-eastern region, particularly in Assam where muga silk cocoon is available and widely cultivated. State and central governments along with non-government organizations should organize exhibitions, seminars, workshops, awareness camps, etc., with a sole aim of promoting this environment-friendly surface treatment technique and its usefulness to the common people and industrialists. Successful development and implementation of such plasma treatment system will essentially help the industries to explore the potential market value of this natural silk fabric, and this will also have positive impact on the economic development of the north-east region.

9.5   Conclusion

Studies of Ar plasma discharge characteristics reveal an increase in electron temperature and plasma density with increasing RF power. Plasma potential is observed to remain almost same irrespective of the increase in RF power. The thermal behavior of the treated fibers remains unaffected by the variation in treatment time and RF power. Significant alteration in the chemical structure of the plasma-treated fibers can be observed at higher RF power (30 W). As revealed from XPS analysis, the peptide bond scission and the breakage of side chain groups of amino acid caused by impinging energy of ion contribute to the variation in atomic concentration and functional composition in the plasma-treated fibers and subsequently affect their tensile strength and hydrophobicity. At critical RF power value of 20 W and lower treatment time (5–10 min), the plasma-treated fibers exhibit enhanced tensile strength and hydrophobicity, and this may be attributed to the higher carbon content (%) in the fiber and low stress concentration effect on the fibers induced by smoother surface structure. Besides, the presence of more carbon containing functional (C-O/C-O-C and C-C/C-H/C=N) units (%) at the surfaces may also enhance the observed tensile strength and hydrophobicity of the fibers. Higher RF power and treatment time result in severe destruction of the peptide bond and side chain groups of amino acid that contributes incorporation of more surface roughness and stress concentration effect thereby deteriorating the observed properties of the plasma-treated fibers. Moreover, incorporation of more oxygen content (%) and oxygen containing functional (O=C-N/ C=O/O-C-O, O-C=O/COOH/COOR and OCOO) units (%) is also observed to be responsible for decrease in the tensile strength and hydrophobicity of the fibers. Encouraged by the development of this work, it is believed that the Ar plasma-treated muga silk fiber may find potential application in textile industries in terms of weaving and decorative purposes.

9.6   References

Altman G.H., Diaz F., Jakuba C., Calabro T., Horan R.L., Chen J., Lu H., Richmond J. and Kaplan D.L. (2003). 'Silk-based biomaterials', *Biomaterials*, **24**, pp. 401–416.

Bora M.N., Baruah G.C. and Talukdar C.L. (1993). 'Investigation on the thermodynamical properties of some natural silk fibers with various physical methods', *Thermochim Acta*, **218**, pp. 425–434.

Cioffi M.O.H, Voorwald H.J.C., Ambrogi V., Monetta T., Bellucci F. and Nicolais N. (2002). 'Tensile strength of radio frequency cold plasma treated PET fibers-Part I: influence of environment and treatment time', *J Mater Eng Perform*, **11**, pp. 659–666.

Chatterton P.A., Ress J.A., Wu W.L. and Assadi K. (1991). 'A self-compensating Langmuir probe for use in rf (13.56 MHz) plasma systems', *Vacuum*, **42**, pp. 489–493.

Choudhury A.J., Barve S.A., Chutia J., Pal A.R., Chowdhury D., Kishore R., Jagannath, Mithal N., Pandey M. and Patil D.S. (2011). 'RF-PACVD of water repellent and protective HMDSO coatings on bell metal surfaces: correlation between discharge parameters and film properties', *Appl Surf Sci,* **257**, pp. 4211–4218.

Choudhury A.J., Barve S.A., Chutia J., Kakati H., Pal A.R., Jagannath, Mithal N., Kishore R., Pandey M. and Patil D.S. (2011). 'Effect of impinging ion energy on the substrates during deposition of SiOx films by radiofrequency plasma enhanced chemical vapor deposition process', *Thin Solid Films,* **519**, pp. 7864–7870.

Das A. and Saikia C.N. (2000). 'Graft copolymerization of methylmethacrylate onto non-mulberry silk-Antheraea assama using potassium permanganate-oxalic acid redox system', *Bioresour Technol,* **74**, pp. 213–216.

Das A.M., Chowdhury P.K., Saikia C.N. and Rao P.G. (2009). 'Some physical properties and structure determination of vinyl monomer-grafted antheraea assama silk fiber', *Ind Eng Chem Res,* **48**, pp. 9338–9345.

Demura M., Takekawa T., Asakura T. and Nishikawa A. (1992). 'Characterization of low-temperatureplasma treated silk fibroin fabrics by ESCA and the use of the fabrics as an enzyme-immobilization support', *Biomaterials,* **13**, pp. 276–280.

Dyer J.R. (1978). *Applications of absorption spectroscopy of organic compounds,* New Jersey, Prentice-Hall.

Freddi G., Gotoh Y., Mori T., Tsutsu I. and Tsukada M. (1994). 'Chemical structure and physical properties of antheraea assama silk', *J Appl Polym Sci,* **52**, pp. 775–781.

Ghosh G.K. and Ghosh S. (1995). *Indian textiles: past and present,* New Delhi, APH Publishers, p. 73.

Glew A.D., Saha R., Kim J.S. and Cappelli M.A. (1999). 'Ion energy and momentum flux dependence of diamond-like carbon film synthesis in radio frequency discharges', *Surf Coat Tachnol,* **114**, pp. 224–229.

Gogoi D., Choudhury A.J., Chutia J., Pal A.R., Dass N.N., and Patil D.S. (2011). 'Enhancement of hydrophobicity and tensile strength of muga silk fiber by radiofrequency Ar plasma discharge', *Appl Surf Sci,* **258**, pp. 126–135.

Gogoi D., Chutia J., Choudhury A.J., Pal A.R., Dass N.N., Devi and Patil D.S. (2012). 'Effect of radiofrequency plasma assisted grafting of polypropylene on the properties of muga silk yarn', *Plasma Chem Plasma Process,* **32**, pp. 1293–1306.

Hazarika L.K., Saikia C.N., Kataky A., Bordoloi S. and Hazarika J. (1998). 'Evaluation of physico chemical characteristics of silk fibres of Antheraea assama reared on different host plants', *Bioresour Technol,* **64**, pp. 67–70.

Hodak S.K., Supasai T., Pasawatyanyong P., Kamlangkla K. and Pavarajan V. (2008). 'Enhancement of the hydrophobicity of silk fabrics by SF_6 plasma', *Appl Surf Sci,* **254**, pp. 4744–4749.

Karmakar S.R. (1999). *Chemical technology in the pre-treatment processes of textiles,* **12**, Amsterdam, Elsevier, pp. 115–118.

Liu H., Wang Y., Toh S.L., Sutthikhum V. and Goh J.C.H. (2007). *J Biomed Mater Res Part B,* **82B**, pp. 129–138.

Mennella A. and Morrow N.R. (1995). 'Point-by-point method of determining contact angles from dynamic Wilhelmy plate data for oil/brine/solid systems', *J Colloid Interface Sci,* **172,** pp. 48–55.

Mishra S.P. (2000). *A textbook of fiber science and technology,* New Delhi, New Age International Publishers, p. 128.

Mohanty P.K. (2003). *Tropical wild silk cocoons of India,* New Delhi, Daya Publishers, p. 51.

Pavia D.L., Lampman G.M., Kriz G.S. and Vyvyan J.R. (2008). *Introduction to spectroscopy,* 4th Edition, Kentucky, Brooks Cole Publisher.

Rajkhowa R., Gupta V.B. and Kothari V.K. (2000). 'Tensile stress–strain and recovery behavior of Indian silk fibers and their structural dependence', *J Appl Polym Sci,* **77,** pp. 2418–2429.

Ram R. and Bhatt M.M. (2009). 'Prospects of muga culture in North-West India', in Pandey B.N., Trivedi S.P., Jaiswal K., Karnatak A.K., *Silk in new millennium,* Sarup Book Publishers, New Delhi, p. 84.

Ray P.K. (1969). 'On the degree of crystallinity in jute and mesta fibers in different states of purifications and moisture conditions', *J Appl Polym Sci,* **13,** pp. 2593–2600.

Schott L. (1968). 'Electrical probes', in Lochte-Holtgreven W, *Plasma Diagnostics,* New York, John Wiley & Sons Inc., pp. 668–731.

Sen K. and Babu M.K. (2004). 'Studies on Indian silk. I: macrocharacterization and analysis of amino acid composition', *J Appl Polym Sci,* **92,** pp. 1080–1097.

Shao J., Zheng J., Liu J. and Carr C.M. (2005). 'Fourier transform Raman and Fourier transform infrared spectroscopy studies of silk fibroin', *J Appl Polym Sci,* **96,** pp. 1999–2004.

Smith J.R., Hershkowitz N. and Coakley P. (1979). 'Inflection-point method of interpreting emissive probe characteristics', *Rev Sci Instrum,* **50,** pp. 210–218.

Suanpoot P., Kueseng K., Ortmann S., Kaufmann R., Umongno C., Nimmanpipug P., Boonyawan D. and Vilaithong T. (2008). 'Surface analysis of hydrophobicity of Thai silk treated by SF6 plasma', *Surf Coat Technol,* **202,** pp. 5543–5549.

Sudit I.D. and Chen F.F. (1994). 'RF compensated probes for high-density discharges', *Plasma Sources Sci Technol,* **3,** pp. 162–168.

Wang C. and Wang C. (2010). 'Surface pretreatment of polyester fabric for ink jet printing with radio frequency O2 plasma', *Fibers Polym,* **11,** pp. 223–228.

Wendt A.E. (2001). 'Passive external radio frequency filter for Langmuir probes', *Rev Sci Instrum,* **72,** pp. 2926–2931.

10

Cold plasma for dry processing, functionalization and finishing of textile materials

G. Thilagavathi and S. Periyasamy

Abstract: Cold plasma treatment of textile materials has gained importance over the past few decades, though the use of the plasma technology in electronic industries dates long back. One of the unique advantages of the cold plasma treatment is that it only alters the surface characteristics of the textile materials without affecting its bulk properties. Hence many novel surface functionalities, improvements in the surface-related applications such as inter-fibre frictions in fibre processing, enhanced surface adhesion properties in composite applications, improvements in dyeing affinity, etc., can be achieved. Further cold plasma treatments have number of advantages over the conventional wet chemical processing of textile materials such as dry processing, no or little use of chemicals and hence no effluent problems, applicability to all kinds of substrates, etc. In this chapter, principle of cold plasma treatment of textiles is first discussed and then the cold plasma treatments specific to various natural and synthetic textile materials have been reviewed.

Key words: Cold plasma, textiles, surface modification, dry processing, dyeing and finishing

10.1   Introduction

Textile materials, the second primary need of human beings, are characterized by its fineness, flexibility and comfort. Basically, there are three forms of textile materials, i.e. fibre, yarn and fabric. Of these, the fibres are considered to be the building blocks and the basic elements of the final textile products such as apparels, home textiles and technical textiles. According to the textile institute terms and definition, a textile fibre may be defined as units of matter characterized by fineness, flexibility and a high ratio of length to width. The fibre must also have sufficiently high temperature stability and a certain minimum strength, extensibility and elasticity. Additionally, the textile fibres are expected to have certain minimum moisture regain particularly for apparel end uses. Yarn is produced from such fibres through the process known as spinning, and fabric is produced from the yarns through a process known as weaving.

As far as properties of textiles are concerned, it can be classified as physical, mechanical, thermal, chemical, optical, electrical and frictional properties. The arrangement of (i) fibre-forming polymer molecules inside a fibre, (ii) fibres in yarn, (iii) yarns in fabric and the dimensions of the fibre, yarn and fabrics can be regarded as physical properties. Strength, modulus, elongation, elasticity, flexibility, bending rigidity, torsional rigidity of fibres, yarns and fabrics; additionally impact, bursting and tearing strengths, low stress mechanical properties such as shearing, bending, handle properties such as drape, bending length, crease recovery of fabrics can be classified under mechanical properties. Melting temperature, softening temperature, degradation temperature, crystallization temperature, heat setting temperature, scorching temperature are the properties under thermal properties. Resistance to chemicals such as organic and inorganic chemicals, resistance to sunlight, resistance to UV, moisture properties and dyeability can be classified as chemical properties. Refractive Index, birefringence, absorption, dichroism, reflection and lustre are the optical properties. Electrical resistance/conductivity and static charge accumulation are the electrical properties of textile fibres.

Of all these properties, the macro morphological properties such as the fibre surface smoothness and roughness; chemical properties such as surface functional groups; reflection and lustre properties; static charge accumulations and frictional properties are classified under surface-specific properties. These surface properties have high practical significance both in the processing stages where the fibres are processed into final products and in the final products as such for the specified end uses. For example, fibre-to-fibre cohesion which depends on the macro morphological properties and fibre friction is one of the important requirements in spinning process (manufacture of yarn from fibres). If the fibre is too smooth as in the case of angora wool, it is very difficult to spin a yarn and hence it is necessary to introduce artificial surface irregularities to increase the fibre-to-fibre cohesion. In the case of synthetic fibres manufacture, crimping is introduced to increase the fibre-to-fibre cohesion for better spinnability. Similarly the surface chemical groups and the static charge present on the fibre surface are interdependent and influence both the processing stages and the final product properties. If the surface of the fibres contains hydrophobic groups, then the static charge accumulation will be more, and this will result in roller lapping, clogging in the machines in the spinning lines, fibre-to-fibre, yarn-to-yarn entanglements, and it will also attract oppositely charged dirt particles and cause series of problems. It also affects the final products in the case of apparels causing discomfort to the wearer, quick and deep soiling, etc. The surface functional groups also influence the rate of dyeing in the textile chemical processing operations.

The major textile fibres can be broadly classified into hydrophilic fibres and hydrophobic fibres. Almost all the natural fibres such as cotton, silk and wool are hydrophilic in nature, while the synthetic fibres such as polyethylene terephthalate (PET), polyamides, cellulose triacetate, polypropylene and polyethylene are hydrophobic in nature. These synthetic fibres have static charge accumulation problems, dyeing problems and wearing comfort problems. Hence these problems can be addressed through surface modification process.

Conventional chemical treatment methods are used for modification of surface properties. However, they not only alter the surface properties but alter the bulk properties particularly the mechanical properties and hence the fibre becomes relatively weak. For example, the weight reduction treatment of polyester with NaOH improves its hydrophilic characteristics and the static charge dissipation properties, but reduces the strength notably. Therefore, treatment techniques without affecting the bulk properties would be quite attractive.

Moreover, one of the serious problems for both the textile industry (to run its business) and for the society is the effluents produced out of series of textile operations particularly from the textile chemical processing units. The problem can be looked into two-fold, one being the use of large amount of water as a medium of treatment and the other one being the removal of various chemicals present in the water from various processes such as desizing, scouring, bleaching, dyeing and finishing. If the textile chemical processing operations could be carried out with low or no use of water, both the problems of using large quantity of water and the subsequent recovery of the chemicals could be done easily with no or less effluent discharges.

Apart from the above-mentioned problems, the use of water in the textile chemical processing adds to energy consumption in the hydro-extraction and drying process. Drying of textile materials particularly the hydrophilic fibres, due to its strong bonding with water, would require high amount of energy to evaporate to its standard moisture regain. So, if there can be a process with no use of water, it can avoid the energy required for drying of textile materials and can effectively bring down the processing cost.

Low temperature plasma (LTP), by virtue, is a highly reactive phase which can initiate various kinds of physicochemical reactions on the materials being exposed to it. Such reactions develop various types of functional groups on the materials being treated based on the type of feed gas. For example exposing hydrophobic fibres like PET to oxygen plasma, hydrophilic groups such as –COOH, –OH, –COH can be developed and exposing hydrophilic fibres like cotton to fluorine containing gas, hydrophobic groups like $-CF_2$, $-CF_3$, $-CFH-$, can be developed on the fibre surfaces. As plasma is produced using only gas, treatment of textile with LTP is purely considered as dry process and hence addresses the problem of effluent discharge/treatment, and it also avoids the intermediate drying processes. Moreover, LTP treatment modifies the surface of fibres to submicron level and so, it does not alter the bulk properties (Rakowski, 1982; Mirjalili, 2004 and Yuyue, 2004). Hence low temperature plasma treatment satisfies all the above-mentioned requirements of surface modification of textile materials. Further, surface modification, as a sole process or as a pre-treatment, can effectively be used for functional finishing of textile materials. Special finishes such as antistatic and antisoiling finish for synthetics, flame retardant and water-repellent finishes for cotton can also be imparted using this treatment (Hegemann, 2007; Yamada, 2008; Badyal, 2005 and Shahidi, 2005). LTP treatment has been used for other fibre-specific treatments such as anti-shrinking of wool, desizing of cotton and degumming of silk, etc. (Prabaharan, 2005).

Originally low temperature plasma treatment has been used in electronic industry for preparing the printed circuit boards, magnetic coated data storage devices. Later the LTP treatment to textile has been explored for textile treatments many decades (at least 60 years) before and from then various research works have been carried out and are reported in the literature. Although some companies like Diener Plasma-Surface-Technology, Sigma Technologies, Porton Plasma Innovations Limited (P2i) and Europlasma are supplying plasma systems, LTP for commercial level continuous treatment of textiles has not yet become common in industry which may be due to problems in operating the plasma system under ambient conditions. It requires development of commercially available, reliable, and large plasma systems which can operate at atmospheric conditions in a continuous mode. Plasma systems are now available (mostly in research laboratories), and the application of plasma to industrial problems has been increasing rapidly for the past 10 years.

In this chapter, the principle of cold plasma treatment is first discussed. Subsequently, a brief review on plasma treatment of textile materials has been presented. Finally, specific research works on hydrophilization of silk and development of antimicrobial cotton fabric using RF plasma and neem is presented.

10.2   Principle of cold plasma treatment of textiles

10.2.1   Definition of cold plasma

Plasma can be defined as the ionized state of gas which is considered to be the fourth state of matter. Based on the operating system temperature, the plasma can be classified into thermal plasma and non-thermal plasma or cold plasma. In thermal plasma, essentially equal density of positive and negative charges would be present and can exist over wide range of temperature and pressure.

Hence, normally the operating temperature of thermal plasma would be above 10,000°C, while in the cold plasma, the temperature would be near ambient conditions. The solar corona, a lightning bolt, a flame and a "neon" sign are all examples of plasma of which the first three belongs to thermal plasma category while the last one belongs to cold plasma category.

The difference between the two types of plasmas (thermal and non-thermal) is primarily based on non-equilibrium conditions of electrons, ions and gas temperature as well as due to difference in degree of ionization. In the case of thermal plasma, electron, ion and gas temperature are used to be in local thermodynamic equilibrium (LTE) and the degree of ionization of gaseous molecules is substantially high (≥10%), whereas in the case of cold plasma electron temperature is used to be much higher in comparison to ion or gas temperature and this plasma has very low degree of ionization ≤1% of the total gas molecules. The cold plasma comprises of electrons, ions, radicals, metastables and UV radiations with *degree of ionization is ≤ 1% and the overall plasma used to be at ambient temperature. Therefore, the cold plasma environment is extremely energetic and highly reactive for surface reactions.*

10.2.2   Description of the general plasma setup

Principle schematic diagram of a typical low temperature and low pressure plasma reactor is shown in Fig. 10.1 (Periyasamy, 2007). It consists of a closed chamber wherein two electrodes, separated at a distance of few centimeter, are present. The top electrode is connected to high frequency and high voltage supply (HF) and the bottom electrode is connected to ground (G). It has the provision for gas inlet (I) and gas outlet (O) for passing the working gas. Further the plasma setup also has separate arrangements for creating vacuum which contains a control valve (V) and a suction pump (P). When high voltage is applied to the electrodes, the electric field accelerates small number of ions present in the gas. These collide with other gas molecules, knocking electrons off them and creating more positive ions in a chain reaction called a Townsend discharge (Glenn, 2000; Von Engel, 1957) and thus forming plasma.

10.2.3   Plasma treatment of textiles

Textile products are available in any of the three basic forms such as fibres, yarns and fabrics. Fibres, being the smallest form of the textiles with the approximate diameter of 10 μm, are the building blocks of the yarn whose diameter would be around 0.1 mm (100 μm). Figure 10.2 shows the micrographs of typical fibre and yarn. Apparel grade fabrics are produced through interlacing (weaving process) or interloping (knitting process) of yarn. Fabrics can also be made directly from fibres through some special techniques, and the fabrics are known as nonwovens which are generally used for technical applications. Generally fabrics would have thickness of less than 1 mm (1000 μm) except those fabrics meant for certain technical applications which might have more thickness. Microscopic view of a typical textile fabric is shown in Fig. 10.3a, and its schematic side view is shown in Fig. 10.3b.

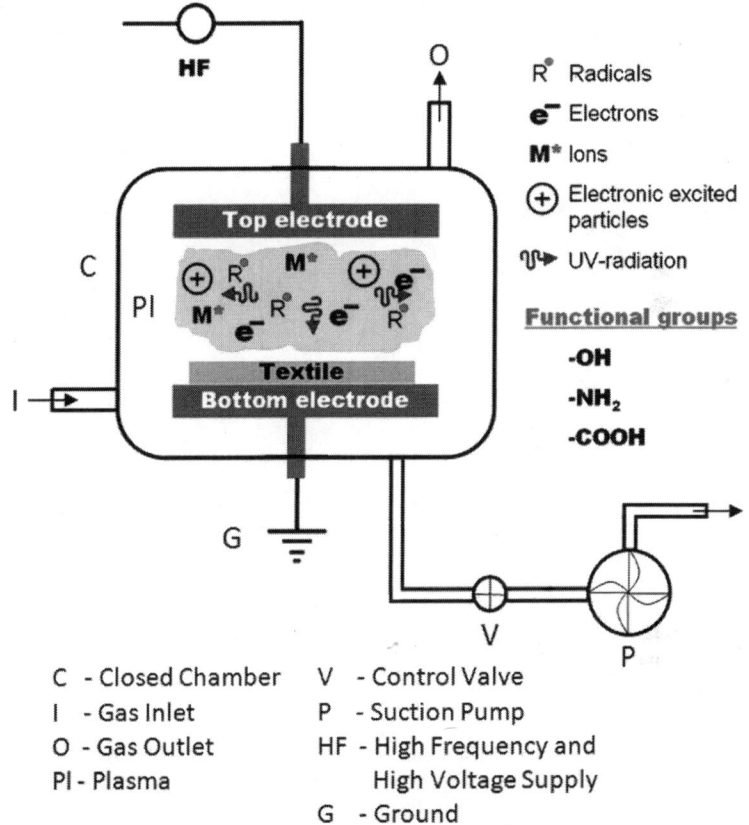

C - Closed Chamber   V - Control Valve
I - Gas Inlet   P - Suction Pump
O - Gas Outlet   HF - High Frequency and
Pl - Plasma   High Voltage Supply
                        G - Ground

Figure 10.1 Schematic diagram of a plasma chamber.

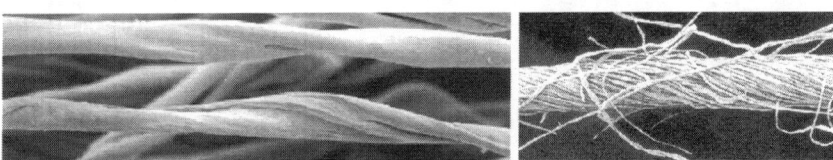

Figure 10.2 Micrographs of (a) a typical fibre, (b) a typical yarn.

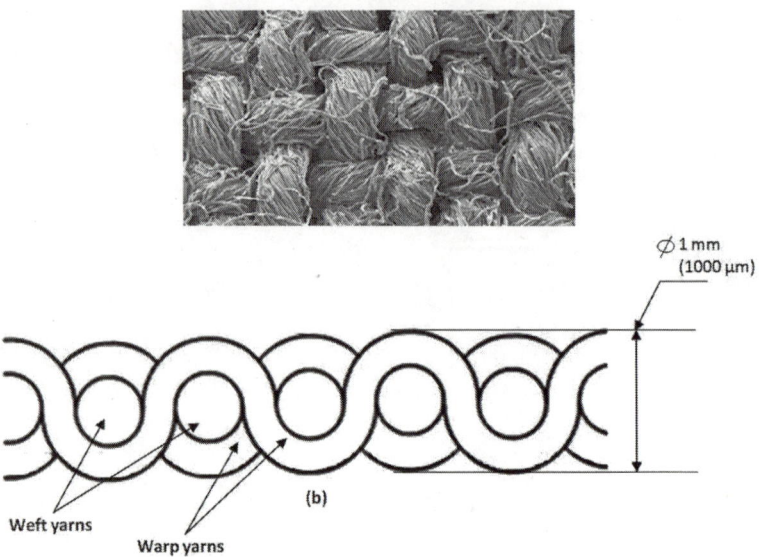

$\varnothing$ 1 mm
(1000 μm)

(b)

Weft yarns

Warp yarns

Figure 10.3 Typical textile fabric, (a) micrographic view, (b) schematic of the side view.

Plasma treatment of textile involves placing textile materials between the two electrodes as shown in Fig. 10.1. When the electric filed is applied to the electrodes at above the break down voltage, the plasma is formed. Depending upon the type of feed gas used for the generation of plasma, the type of plasma species present in the reactive environment would vary.

The specific reactive species, thus formed, act on the textile substrates and initiate various physicochemical reactions leading to the formation of various functional groups of any type as shown in Tables 10.1 and 10.2, which would depend upon the type of feed gas. For example, the use of oxygen (Wakdia, 1996; Kan, 2004) as the feed gas for the generation of plasma would generate hydrophilic groups over the textile surface while hydrophobic groups (Oktem, 2005) will be generated if CF_4 is used as feed gas (Chaivan, 2005). In the case of oxygen plasma, various reactive oxygen species (ROS) such as diatomic oxygen, superoxide anion, peroxide, hydrogen peroxide, hydroxyl radical, hydroxyl ion, etc., would be present. Apart from these species, it would also contain ozone (O_3), 10–20% atomic oxygen and a similar percentage of excited oxygen molecules. Both the oxygen atoms and excited oxygen molecules convert the organic products into their oxidation products. Similarly, when nitrogen is used as a feed gas, various reactive nitrogen species (RNS), such as nitrous oxide, peroxynitrite, peroxynitrous acid, nitroxylanion, nitryl chloride, nitrosylcation, nitrogen dioxide, dinitrogen trioxide and nitrous acid, would be created which react with the textile substrates.

Table 10.1 Functional groups containing oxygen and nitrogen.

Functional groups containing oxygen			Functional groups containing nitrogen		
Chemical class	Group	Structural formula	Chemical class	Group	Structural formula
Alcohol	Hydroxyl	R—O—H		Carboxamide	R—C(=O)—N(R')—R''
Ketone	Carbonyl	R—C(=O)—R'		Primary amine	R—N—H, H
Aldehyde	Aldehyde	R—C(=O)—H	Amines	Secondary amine	R—N(H)—R'
Acyl halide	Haloformyl	R—C(=O)—X		Tertiary amine	R—N(R'')—R'
Carbonate	Carbonate ester	R_1—O—C(=O)—O—R_2		4° ammonium ion	R_1—N$^+$(R_2)(R_3)(R_4)
Carboxylate	Carboxylate	R—C(=O)—O$^-$		Primary ketimine	R—C(=N—N)—R'
Carboxylic acid	Carboxyl	R—C(=O)—OH	Imine	Secondary ketimine	R—C(=N—R'')—R'
Ester	Ester	R—C(=N)—OR'		Primary aldimine	R—C(=N—N)—H
Methoxy	Methoxy	R—O—CH_3		Secondary aldimine	R—C(=N—R')—H
Hydroperoxide	Hydroperoxy	R,H O—O	Imide	Imide	O=C(R)—N(R)—C(R)=O
Peroxide	Peroxy	R O—O R'	Azide	Azide	R—N=N$^+$=N$^-$
Ether	Ether	R—O—R'	Azo compound	Azo (Diimide)	R(N=N)—R'

Functional groups containing oxygen			Functional groups containing nitrogen		
Chemical class	Group	Structural formula	Chemical class	Group	Structural formula
Hemiacetal	Hemiacetal		Cyanates	Cyanate	
Hemiketal	Hemiketal			Isocyanate	
Acetal	Acetal		Nitrate	Nitrate	
Ketal (or Acetal)	Ketal (or Acetal)		Nitrile	Nitrile	$R\!-\!\!\equiv\!N$
Orthoester	Orthoester			Isonitrile	$R\!-\!\overset{..}{N}\!\equiv\!C^{-}$
–	–	–	Nitrite	Nitrosooxy	
–	–	–	Nitro compound	Nitro	
–	–	–	Nitroso compound	Nitroso	
–	–	–	Pyridine derivative	Pyridyl	

Table 10.2 Functional groups based on hydrocarbons and halogens.

R_1 $\diagdown$ R_3 / R_2 $\diagup$ R_4 Alkenyl functional group	R$\equiv$R' Alkynyl functional group	R$+$$)_n$ Alkyl functional group	R$\diagup$◯ Benzyl functional group
R$-$◯ Phenyl functional group	R$-$Cl Chloro functional group	R$-$F Fluoro functional group	R$-$X Halo functional group
O‖ R$\diagup$$\diagdown$X Haloformyl functional group	R$-$I Iodo functional group	R$-$Br Bromo functional group	O‖ R$\diagup$$\diagdown$H Aldehyde functional group

Surface chemical changes

Textile materials are made of linear polymer molecules with a high degree of molecular weight in order that the fibre formed out of it has certain minimum mechanical, thermal and chemical resistance. Cellulosics and proteins are the two primary categories of the natural and regenerated type man-made fibres. Synthetic-type man-made fibres can be classified based on the functional groups linking the monomeric repeat units. Accordingly, some of the major synthetic fibres are polyester (–COO–, ester – the linking group), polyamide (–NHCO–, amide, the linking group), polyurethane (–CONHCO–, urethane, the linking groups), etc. There are other polyvinyl-based synthetic fibres such as poly (acrylonitrile), poly (ethylene) and poly (propylene) fibres. All these fibres' polymer chains primarily contain carbon, hydrogen, oxygen and nitrogen with the carbon as the backbone atoms. So the reactive species can act upon either on the functional group and cleave the link resulting in depolymerisation and forming terminal functional groups as shown in Fig. 10.4.

It can be seen from Fig. 10.4 that, in the first step, the UV photon present in the plasma chamber first attacks the functional groups and cleaves the chain into two and forms radicals at the ends. Subsequently, in the second step, these radicals reacts with the hydrogen and hydroxyl radicals, formed as a result of complex reactions, present in the plasma and hence form carboxyl and amine terminal groups. Thus for each chain scission action, two functional groups would form.

1. Radical formation

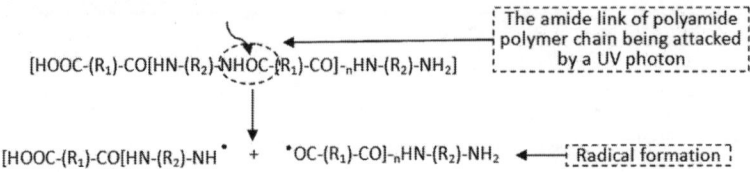

2. Functional groups formation

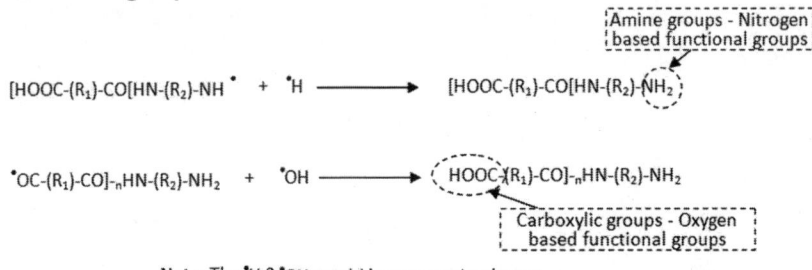

Note: The °H & °OH would be present in plasma

Figure 10.4 Radical reaction scheme showing polymer chain scission and end functional groups formation.

Apart from chain scission, any atom from the main backbone carbon atoms can also be removed by the action of photons or by the action of ions and thus form radicals. Such radicals would further react with the atomic or molecular radicals present in the plasma environment and form functional groups. For example, a photon or ion would attack and cleave the hydrogen atom attached to the main backbone carbon atom and form radical which react further with reactive oxygen species (ROS) or with reactive nitrogen species (RNS) or with both ROS and RNS in the plasma and form functional groups as shown in Fig. 10.5. The presence of various types of atomic and molecular radicals is due to etching of the polymers, i.e. removal of the atoms or molecules from the polymer chain. Such etched-out atoms or molecules in turn take part in further reactions with the large number of reactive species formed in plasma and help in the formation of various functional groups.

Surface morphology changes

The UV photon in the plasma environment is highly responsible species for chain scisson. Such chain scisson action would fragment the surface polymer long chains into short chains and ultimately converting them into small molecules and atoms which would be volatilized. When many such polymer chains are fragmented locally, it forms pores. This process is technically known as etching process. The size and depth of pore depends upon the photon energy, intensity and the time of irradiation. For example, in a research work irradiation of silk with photons of 172 nm wave length, photon energy of 7.2 eV, irradiation power of 50 mW/cm^2 has been reported (Periyasamy, 2007). It has been reported that due to such irradiation pores on the silk fibres are formed, and they increase with irradiation time i.e. pores with average diameter of 95 nm for 1 min irradiation and pores with average diameter of 190 nm for 30 min irradiation (Fig. 10.6).

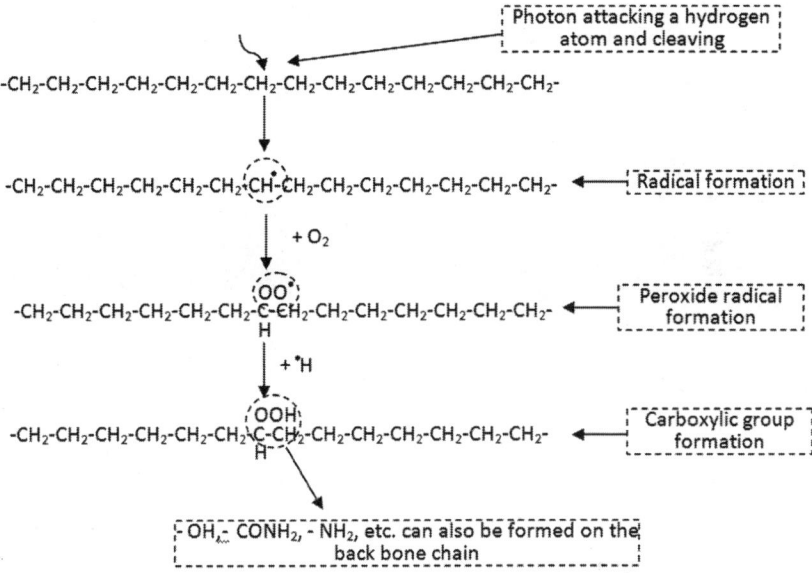

Figure 10.5 Reaction scheme showing formation of functional groups on the backbone polymer chain.

Depth effect

The depth of modification in plasma treatment is primarily restricted to the submicron level because of the nature of the species present in the plasma. The depth effect is primarily due to the UV photon. It is known that the UV and higher energy photons are strongly absorbed by the molecules; therefore, the higher energy photons do not penetrate deep inside the structure but they are absorbed strongly on the surface. Moreover the other types of reactive species like ions, electrons, excited atoms and molecules only participate in the chemical reactions in forming the new functional groups on the immediate surface. Hence the surface modification is restricted to submicron level. For instance, in the same research work where silk fibre has been irradiated with strong VUV light, the depth of surface modification has been restricted to only 10 nm (Periyasamy, 2007).

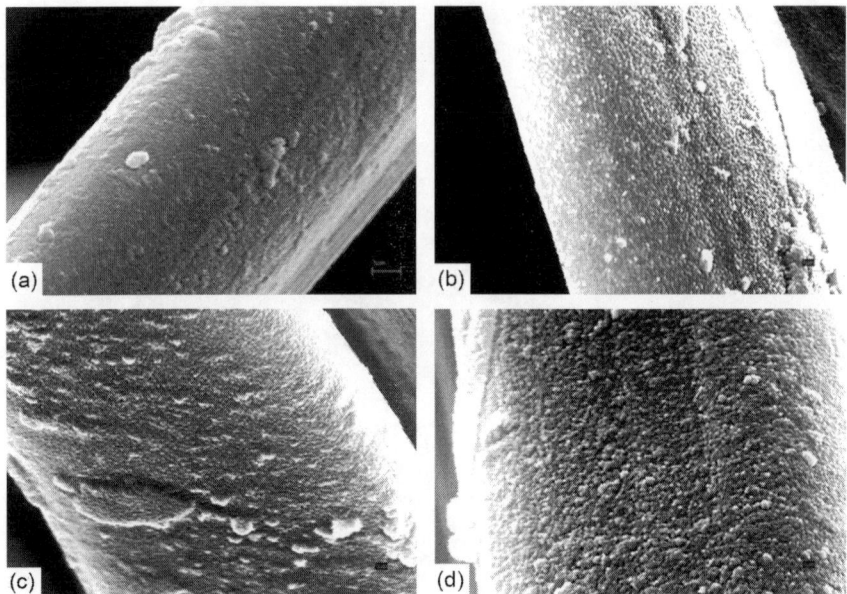

Figure 10.6 SEM micrographs of silk fibre irradiated with excimer lamp (a) untreated,
(b) 10 min, (c) 15 min, (d) 30 min (Periyasamy, 2007a).

10.3   Low temperature plasma (LTP) treatment of textile materials

Whether natural or synthetic, each fibre has some special property. For example cotton is known for its hydrophilicity, while polyester is popular for its wash and wear property. However, very few fibres have all the essential properties and almost always a fibre would lack some essential properties. For example, although polyester fibre has excellent mechanical and aesthetic properties, it suffers severely from its lack of hydrophilic nature. Therefore, it is always used with other natural fibres as a blended yarn like polyester/cotton, polyester/wool, etc. Through surface modification, it is possible to engineer the fibre surface according to the specific need and applications. For example, through low temperature plasma treatment, it is possible to make the polyester surface hydrophilic, and it is possible to make the cotton fabric repellent to hydrophilic stains. The subsequent section is a brief review on modification of major textile fibres by low temperature plasma treatments.

10.3.1   LTP treatment of cotton

Cotton is the single-most fibre whose volume of production and consumption (approx. 45%) is very high compared to any other fibers of natural and synthetic origin. Most of the process effluents result from cotton processing industry. The chemical processing of cotton textiles involves pretreatment processes, dyeing and printing, post treatments and the finishing processes. In all these processes LTP has almost always been used as pretreatment in order to enhance the process efficiency with minimal use of chemicals and water. For example, plasma-assisted desizing, scouring, dyeing and printing processes have been tried and some improvements have been reported. In the case of finishing, LTP treatment has been used as an alternative process particularly in imparting hydrophobic finishes like water and stain repellent finishes. In such studies, effects of various parameters such as feed gases, treatment time and power intensity on fibre surface have been reported (Abidi, 2004; Chen, 1996 and Malek, 2003). Malek et al. (2003) have reported that the plasma treatment leads to surface erosion of the cotton fibres, which generates a weight loss, accompanied by an increase in the fibre carboxyl group and carbonyl group contents (Malek, 2003). The increase in fibre carboxyl group content leads to a more wettable fibre and the rate of fabric vertical wicking is increased. Researchers have attempted to develop water-repellent surface on cotton fabric by plasma-initiated polymerization (Tsafack, 2007), plasma thin film deposition (Zhang, 2003), direct plasma treatment using fluorine-containing feed gas (Höcker, 2002 and Selli, 2001a). In plasma-initiated polymerization, the active centers of plasma-treated cotton fabric are used for initiating copolymerization reactions with vinyl monomers to impart hydrophobic character to the fabric. Although this method yields excellent water repellency in cotton the fabrics (Tsafack, 2007), it involves two different stages. However, the fabric can be made hydrophobic, directly by treating with the plasma generated by fluorine-containing feed gas such as hexafluoroethane plasma. Apart from hydrophobization there are studies on removal of size from the cotton using plasma treatment. Cai et al. (2002, 2003) have reported that atmospheric plasma treatment of sized cotton followed by cold washing removes the polyvinyl alcohol (PVA) size considerably. They have further added that the surface chemical changes such as chain scission and formation of polar groups promote the solubility of PVA in cold water. Although the cotton fibre is quite dyeable with various classes of dyes, the possibilities of increasing its dyeability using plasma technique has been explored like improving colour yield and fastness properties (Ozdogan, 2002; Jahagirdar, 2001) and improving rate of dyeing (Yoon, 1999).

Antimicrobial property of cotton has been investigated using low temperature plasma (RF oxygen and air plasma) (Vaideki, 2008; Vaideki, 2007). RF magnetron sputtering unit was used to produce oxygen plasma. The schematic diagram of the system is shown in Fig. 10.7. The chamber and the electrodes are made of stainless steel. The sample is placed on the lower electrode which is grounded. The frequency of the RF system is 13.56 MHz. The authors have first treated the cotton fabrics under RF Oxygen Plasma and have found that such treatment increases the hydrophilicity of cotton as per the static immersion test. Further, the antimicrobial activity has been imparted to the RF oxygen plasma-treated samples using methanolic extract of neem leaves containing Azadirachtin. The antimicrobial activity of these samples has been analysed and compared with the activity of the cotton fabric treated with neem extract alone. The authors have reported that such surface modification due to RF oxygen plasma increases the hydrophilicity and hence the antimicrobial activity of the cotton fabric when treated with Azadirachtin (Vaideki, 2007). When such trials were carried out using low temperature RF Air plasma, the authors have found similar results (Vaideki, 2008).

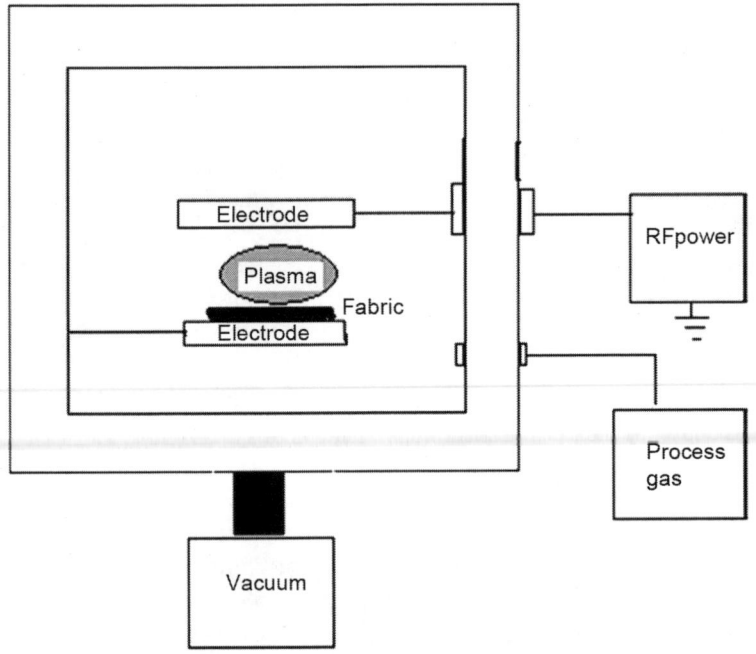

Figure 10.7 Schematic diagram of RF plasma system

A mechanism for increasing the hydrophilicity of the cotton fibre was proposed, which is attributed to the production of a high concentration of chemically active species such as excited oxygen molecules, oxygen radicals, free electrons and oxygen ions for the above-mentioned process parameters. These chemically active species impinge on the sample surface which results in the generation of cellulosic radicals on the fabric surface. These radicals are a result of any one of the following mechanisms (Fig. 10.8) (McCord, 2003;

Ward, 1987):

1. Bond breakage between C1 and glycosidic bond oxygen.
2. Dehydrogenation and dehydroxylation between C2 and C3 after the ring opening of anhydroglucose.
3. Dehydrogenation at C6.
4. Dehydroxylation at C6.
5. Bond breakage between C1 and ring oxygen.

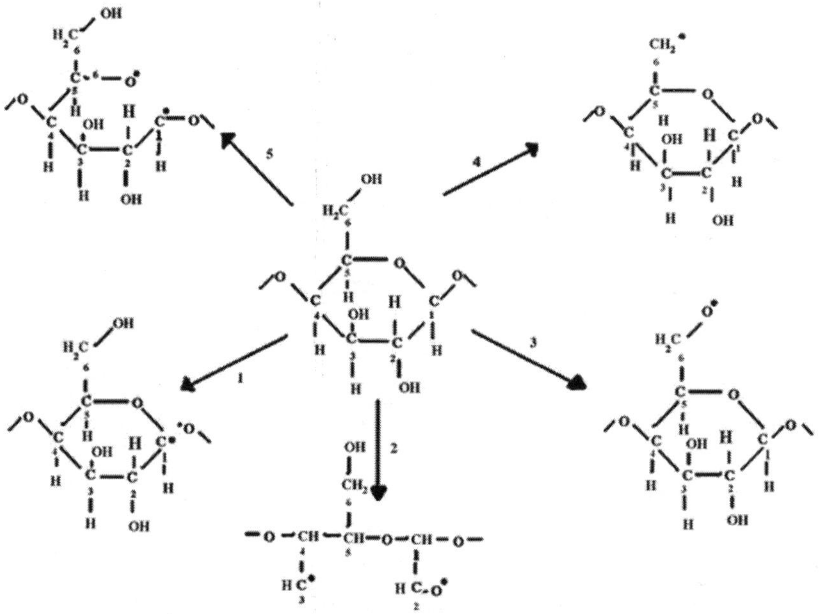

Figure 10.8 Cellulosic radical formation (McCord, 2003; Ward, 1987)

They have reported that of the five mechanisms, those that would result in the formation of carbonyl (aldehyde) group which has an increased polarity than the hydroxyl group are bond breakage between C1 and glycosidic bond oxygen, dehydrogenation at C6 and bond breakage between C1 and ring oxygen.

Table 10.3 shows the comparison of absorption percentage, percentage reduction and zone of bacteriostasis of untreated sample, sample treated with neem extract alone and a sample treated with oxygen plasma and neem extract. Tables 10.4 and 10.5 show the antimicrobial activity and antifungal activity of the LTP (air plasma) treated and untreated cotton fabrics, respectively. From these results, it can be concluded that the samples underwent plasma treatment prior to the application of the neem extract show an improved activity.

Table 10.3 Comparison of absorption percentage, percentage reduction and zone of bacteriostasis of untreated sample, sample treated with neem extract alone and a sample treated with oxygen plasma and neem extract (Vaideki, 2007).

Sample	Absorption percentage (%)	Percentage bacteria reduction (%)		Zone of bacteriostasis (mm)	
		Staphylococcus aureus	Escherichia Coli	Staphylococcus aureus	Escherichia Coli
Untreated sample	53.7	0	0	0	0
Sample treated with azadirachtin alone	53.7	86	86	37	37
Sample treated with oxygen plasma (using optimized process parameters) and Azadirachtin	61.5	100	100	50	49

Table 10.4 Antimicrobial activity by modified Hohenstein test method – challenge test (Vaideki, 2008).

Specimen	Name of the organism	Percentage of bacteria reduction (%)
Cotton fabric treated with antimicrobial finish	S. aureus	97.98
	E. Coli	96
Cotton fabric treated with RF air plasma using optimized system parameters and antimicrobial finish	S. aureus	100
	E. Coli	99.05

Table 10.5 Agar diffusion assay (antifungal activity) (Vaideki, 2008).

Specimen	Zone of inhibition (mm)	
	Penicilium funiculosum	Trichoderma riridea
Cotton fabric treated with antimicrobial finish	5.0	3.9
Cotton fabric treated with RF air plasma using optimized system parameters and antimicrobial finish	5.9	4.2

In another study, Thilagavathi et al. (2008) have developed dual antimicrobial and blood repellent hospital fabrics using plasma deposition techniques. They have applied antimicrobial and blood-repellant finish to cotton fabrics used for surgical gowns, bed linens and drapes to reduce the surgical site infections. They have applied the extract of neem to the fabric for imparting antimicrobial activity by pad-dry-cure method. The neem-treated fabric has then been imparted blood repellency through two different techniques, namely by treatment with fluoropolymer (3%, 4%, and 5% owf) using pad-dry-cure method and by 'sputter deposition of teflon' technique using argon plasma. The antimicrobial activity has been found to be higher for Teflon-deposited fabric than for the fluoropolymer finished fabric. Blood repellency increases with the higher concentration of fluoropolymer and the highest repellency for the Teflon-deposited fabric has been observed at 80 W power and 20 min exposure in the plasma chamber.

10.3.2   LTP treatment of silk

Surface modification of silk has mainly been carried out by low temperature plasma for improving dyeing and printing characteristics, imparting hydrophobicity for enhanced fabric care, improving fabric handle, degumming and so on.

It has been reported that plasma treatment of silk leads to formation of microcraters, evident cracks and flutes (Nakano, 1994). SF6 plasma treatment of silk increases its crystallinity, making the surface properties more durable (Selli, 2001b). Nadiger et al. (1985) have reported that the nitrogen plasma treatment of mulberry silk leads to surface decrystallization while the crystallinity increases in case of tasar silk. As the plasma treatment increases hydrophilicity of textiles in general, except when fluorine containing gases are used, the plasma treatment of silk also leads to increase in wettability (Bhat, 1978). This increase is attributed to the formation of oxygen-containing functional groups in the case of air, oxygen or other feed gases. The action of plasma on the fibre surface creates free radicals by surface polymer molecule bond cleavages (Kuzuya, 1993). These radicals react with the atmospheric oxygen present in the air and form oxygen-containing functional groups such as –COOH, –COO– and –CO. It has been reported that the plasma treatment does not alter the mechanical properties of the silk fibre such as tensile strength and so on (Yuyue, 2004). However the plasma treatment may cause a change in the properties such as stiffness, crease recovery angle, etc., that are related with the fabric handle. The presence of ozone in the plasma discharge may cause the silk fabric slightly crisp (Wakida 2004). The plasma treatment alone decreases the crease recovery angle (Nadiger, 1985; Bhat, 1978) while grafting of acrylamide (AAm) followed by plasma pretreatment increases the crease recovery by 20% with the grafting weight of 1%. The grafting brings the AAm branch polymer into the main chain of silk by the initiation of some kind of O and N groups formed by the plasma and thus the grafted silk fabric also has high elastic recovery angle (Zhang, 1997).

As the raw silk fibre is made of fibroin, a highly crystalline fibrous protein, and a gummy protein, the sericin, which constitutes about 20–30% in weight, experiments using LTP for the removal of sericin has also been carried out. Riccardi et al. (2005) observed that air and oxygen plasma remove selectively the sericin up to a weight loss corresponding to more than half of the sericin content. Increasing the weight loss beyond a point starts to remove fibroin as well, because of the non-uniform etching rate or probably due to the difference in the sericin distribution along the fibres. These investigators further observed that the etching rate on silk depends on RF power and pressure inside the chamber.

Silk the most lustrous of all natural fibres has always been surrounded by an aura of glamour. However, the increased lustre of the fibre may also reduce the depth of shade. Unlike wool, silk is a compact and crystalline fibre. Hence it may inhibit the penetration of dye molecule inside the fibre structure. LTP has been effectively used to improve its dyeing characteristics. Iriyama et al. (2002) treated silk with O_2, N_2, and H_2 plasmas for deep dyeing and improved colour fastness to rubbing with a reactive black dye (Iriyama, 2002). They found that plasma-treated silk showed higher K/S value with 6% shade as compared to untreated ones with 10% shade. Moreover they have reported an increase in shade depth regardless of the kind of gas used.

In a recent study, Periyasamy et al. (2008) have compared the plasma-treated silk samples against VUV irradiated silk samples. In this work, the authors have used dielectric barrier discharge-based atmospheric plasma reactor system which is shown in Fig. 10.9. The system consists of a plasma chamber, power supply and a HT transformer (Fig. 10.9a). The plasma chamber consists of a dielectric material covered bottom roller which is connected to ground and is driven by a servomotor (Fig. 10.9b). The plasma chamber contains a top electrode plate which is permanently fixed and is connected to HT supply. The silk fabric runs over the bottom roller and is treated in the plasma formed between the top electrode and the bottom roller.

The authors have found that the LTP treatments are on par with the results of UV-irradiated samples (Fig. 10.10). It has been found that the wicking property of plasma-treated sample irradiated for 1 min is better than the excimer-irradiated sample for the same time, i.e. 1 min. This indicates that the surface modification in plasma is more effective due to presence of UV photon as well as active species, i.e. high energy electrons, ions and radicals which are absent in excimer system. However, the increase in wetting and wicking stabilizes beyond 5 min of treatment probably due to the simultaneous development and etching of the hydrophilic groups.

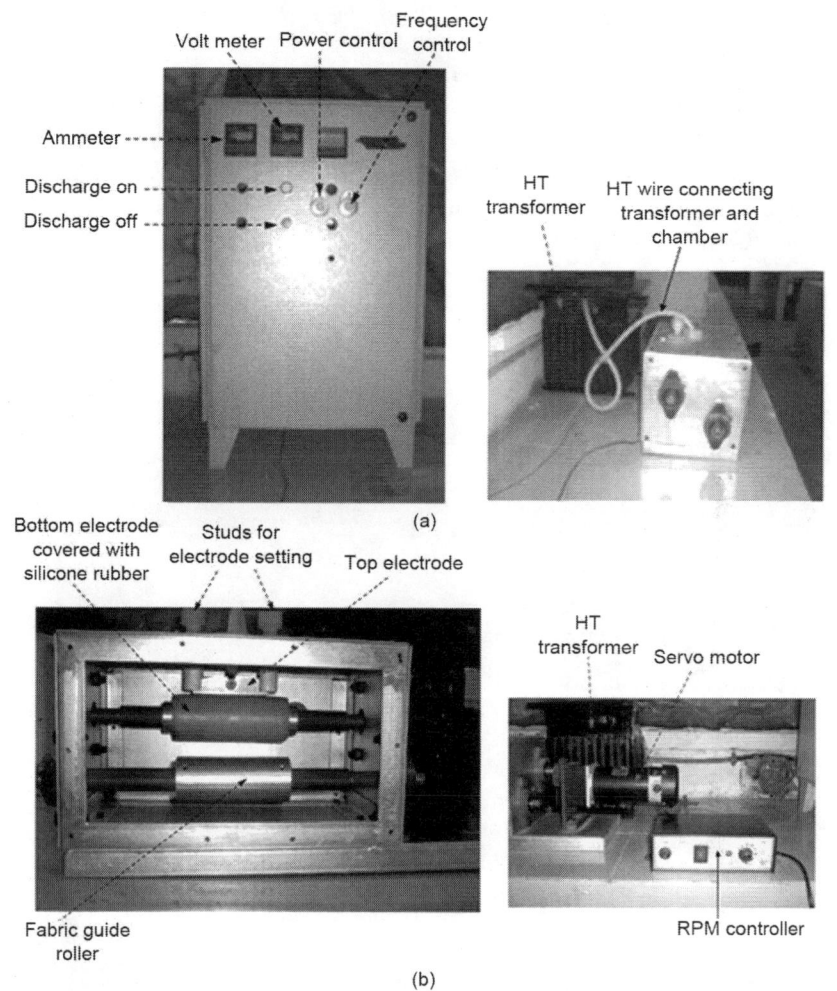

Figure 10.9 Dielectric Barrier Discharge (DBD) system works at atmospheric pressure: (a) power supply, ht transformer and the plasma chamber; (b) front view of the plasma chamber and the servo motor connected to the plasma chamber

10.4   Current situation of plasma technology in textiles

The application of plasma technology on textiles started in Russia in the sixties. It got a revival in the West from the eighties onwards, where numerous studies have been published as a result of experiments in vacuum reactors designed for the treatment of inorganic (micro) electronic materials. In the meantime – from the seventies onwards – Russian researchers developed a full industrial scale roll-to-roll vacuum reactor. Also in the West such reactors have been built, but the Russian one remains the only large scale (up to 3, 4 m fabric width) textile related plasma reactor with a significant industrial "experience". Such reactors can – in principle – treat any type of fabric and are only limited in the amount of material that can be charged in one batch, i.e. the fabric roll diameter is limited to e.g. 70 cm diameter.

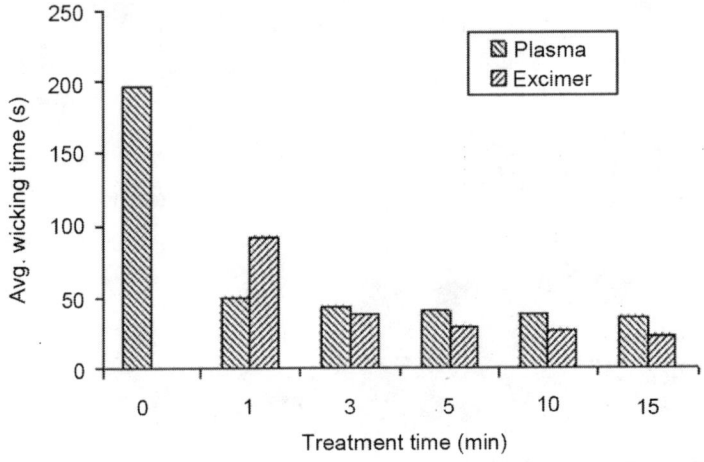

Figure 10.10 Effect of plasma treatment versus excimer irradiation on wicking (Periyasamy, 2007b).

While the development of vacuum plasma technology for textile surface modification has come to a virtual standstill – it is still the most perfected plasma technology available – literature regularly reports new designs of plasma sources working at atmospheric pressure. Most of these are based on, or are a combination of the corona and dielectric barrier discharges. Their main aim is to further improve treatment uniformity and increase size and energy density of the discharge. These improvements are needed to give atmospheric plasmas a competitive industrial usability. Another current limitation of atmospheric reactors is their apparent ability to successfully treat only thin, light weight textiles with an open structure.

More inventive plasma sources try to form truly uniform glow discharges at atmospheric pressure. In this type of reactor, the material is treated at a distance away from the electrodes instead of in between them. Such treatments are said to be done in the plasma "afterglow". The driving force behind the development of atmospheric plasma sources is the complaint from the textile industry that vacuum plasma technology is non-continuous. Much hope was raised from the availability of corona discharges (in the mid eighties) and from dielectric barrier discharge reactors (in the mid nineties) at atmospheric pressure, for in-line continuous application. However, in spite of increasing treatment uniformity and energy density, plasma technology for the treatment of textiles remains "promising" - as it has been for decades - without the textile industry picking in on its really enormous potential. This may be because of some of the hurdles as stated in the following sections. Hence, further industry and institute collaborative research works are required to work on overcoming such hurdles and transfer the eco friendly dry processing LTP technology for textile processing industry.

10.5   Characterization

As it has been stated in the previous sections, the LTP treatment modifies the textile substrates to the submicron level. In order to assess the modifications objectively, sophisticated characterization techniques need to be adopted. The LTP treatments can be broadly classified into physical – the surface morphological changes and chemical – surface functionality changes. Both of them can be characterized using the following instruments, whose descriptions can be found in the standard literatures.

Instruments and methods used to characterize the LTP treated textile materials are:

- Surface Analysis
  - XPS (X-ray Photoelectron Spectroscopy)/ESCA (Electron Spectroscopy for Chemical Analysis)
  - AFM (Atomic Force Microscope)
- Functional Group Analysis
  - ATR-FTIR
  - Carbonyl groups analysis
  - Dyeing and Staining
- Contact Angle
- Wetting Time
- Wicking Time (Vertical and Horizontal)

Since the surface modification in the cold plasma treatment is restricted to the submicron level, though, as mentioned above, many methods can be used for surface characterization, XPS (X-ray Photoelectron Spectroscopy) and atomic force microscope (AFM) are considered to be precise in detecting the chemical and physical surface changes, respectively.

10.5.1   X-ray Photoelectron Spectroscope (XPS)

X-ray Photoelectron Spectroscopy detects accurately the change in atomic composition at the submicron surface level. The principle of operation of the XPS instrument is by measuring the characteristics binding energies of the core electrons of the atoms. Hence the principle of operation of XPS instrument is first to irradiate the samples with X-rays which hits the core electrons (e⁻) of the atoms. The ejected core electrons are carefully detected and thus the atomic composition is precisely measured to a depth of around 10–100 Å on the surface. A typical XPS spectrum of plasma treated and untreated sample is shown in Fig. 10.11 (Samad, 2010). It is obvious from the figure that the surface modifications by plasma treatments are detected to atomic precisions, i.e. the C content decreases while the oxygen content increases in the plasma-treated samples which indicate the incorporation of oxygen-containing functional groups.

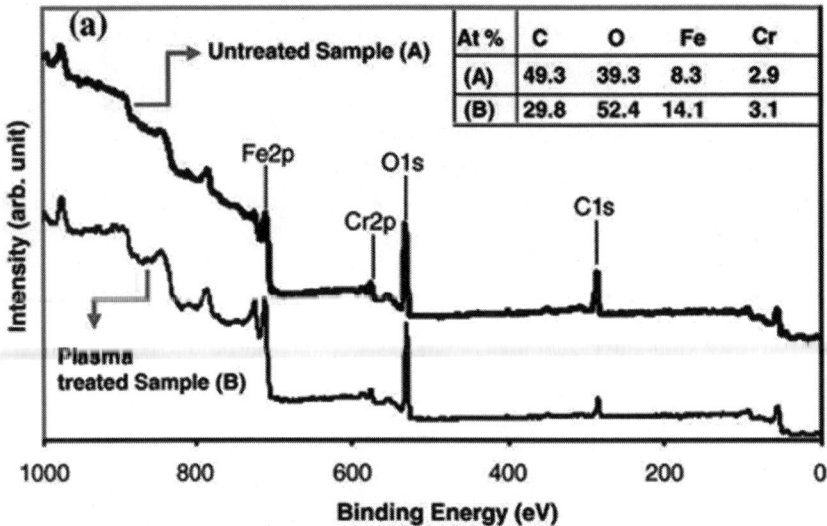

Figure 10.11 Comparison of the XPS spectra for the (A) untreated and the (B) plasma treated samples (Samad, 2010).

10.5.2   Atomic Force Microscope

AFM works on the principle of measuring the atomic force that is developed when a probe is closely brought to the surface of the samples. The force that exists between the probe and the sample is sensitive to the submicron level roughness on the surface, and hence the changes at the submicron level i.e. the surface roughness with average size of less than even 100 nm can be detected. The typical results of AFM images of the untreated and air plasma-treated PET textile, after an exposure time to air plasma of 60 s and 120 s, are shown in Fig. 10.12 (Polettia, 2003). The surface of the untreated PET textile appears flatter in comparison to the air plasma-treated sample and shows a topography which resembles rolling hills with a well-oriented pattern, which is lost by the air plasma treatment. In fact AFM images of treated samples show that the air plasma treatment creates pits and micropores whose density, depth and size increase as a function of the exposure time (Polettia, 2003).

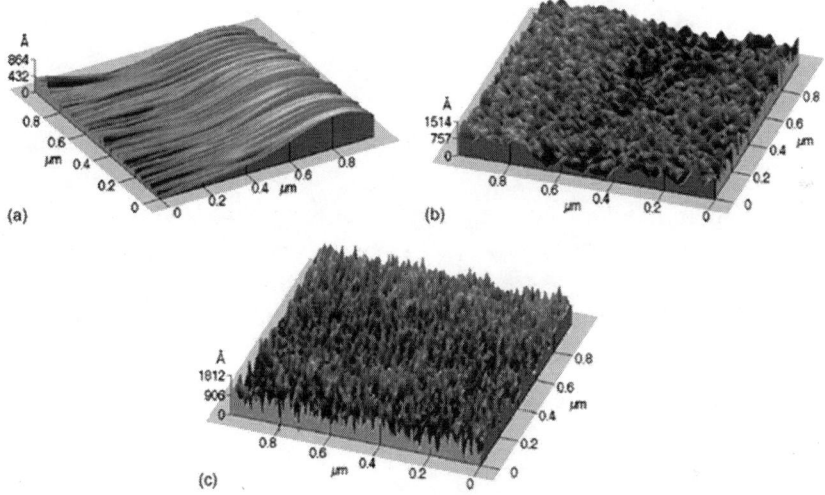

Figure 10.12 3D views of non-contact mode AFM images of PET textile surface. Scan area 1 μm × 1 μm. (a) untreated surface; (b) 60 s treated surface; (c) 120 s treated surface (Polettia, 2003).

Since the changes are very sensitive to contaminations, the samples should be extremely handled properly in order to avoid the experimental errors. Therefore the following points should be observed during characterization.

- Sample preparation
- Free from contamination
- Instrument sensitivity
- Instrumental error
- Experimental error
- Operator error
- Reproducibility

10.6 Advantages and shortcomings of plasma technology

Cold plasma treatment is much explored for treating materials for surface functionalization. In order to make use of such cold plasma for effective treatment of textiles, its advantages and shortcomings in terms of the plasma system itself and the modifications initiated by the cold plasma treatment are to be considered. In the previous sections they have been elaborately discussed whose summary is presented below.

10.6.1 Advantages of plasma treatment

• The plasma treatment of polymeric material has a lot of benefits compared with classical wet chemistry finishing. The main advantages are as follows:

- Applicable to all substrates suitable for vacuum processes, i.e. almost free choice of substrate materials
- Optimization of surface properties of materials without alteration of bulk characteristics
- On polymers, which are unable or very difficult to modify with wet chemicals, the surface properties can also easily be changed
- The consumption of chemicals is very low due to the physical process
- The process is performed in a dry, closed system, and excels in high reliability and safety
- Environmentally friendly

10.6.2 Aging of plasma-treated surfaces

In general, the concentration of functional groups introduced on a polymer surface by plasma treatment may change as a function of time depending on the environment and temperature. This is because polymer chains have much greater mobility at the surface than in the bulk, allowing the surface to reorient in response to different environments. Surface orientation can be accomplished by the diffusion of low-molecular-weight oxidized materials into the bulk and the migration of polar function groups away from the surface. Aging of plasma-treated polymer surfaces can be minimized in a number of ways. An increase in the crystallinity and orientation of a polymer surface increases the degree of order and thus reduces mobility of polymer chains, resulting in slower aging. A highly cross-linked surface also restricts mobility of polymer chains and helps to reduce the rate of aging.

10.7   Conclusions

One of the main applications of the low-temperature plasma (LTP) in textiles is its use for modifying the surfaces of the textile substrates. The hydrophobic nature of synthetic fibres like polyester (MR-0.4%) makes discomfort to the wearer. Also the synthetic fibres are rounder and regular in shape, this makes the fibres more lustrous making it difficult to obtain deep shades in it. The LTP treatment with different gases, particularly with oxygen gas introduces functional groups like carboxylic groups, hydroxyl groups, and so on. This improves the wettability of the fabric. In addition the LTP treatment results in surface etching of the fibres producing microcraters on the surface. This reduces the lustre of the fibre and increases surface area and hence deep shades are possible to obtain. The LTP surface treatment does not alter the bulk properties of the fibres because it alters only sub-micron levels of the fibre surface. The LTP, with fluorine-containing gases, treatment is given to cotton, PET and wool fabrics to introduce water repellency. Unfortunately the use of low pressure makes the process discontinuous and is a costly process. So the atmospheric pressure plasma treatments have an edge. The R&D work has been intensively initiated to modify the surfaces of the polymers and fibres with atmospheric pressure plasma, which will make the process continuous and cheaper. Companies like Europlasma, Dow Corning, Diener, Surfx Technologies, Polyplas GmbH, etc., have developed various cold plasma systems meeting the requirements of the industry for treating various materials including textiles. Hence the cold plasma treatment of textiles is an industrially upcoming process. Moreover, since there is no/little chemical used in LTP treatment, there is absolutely no environmental pollution and the LTP treatment is considered to be an eco-friendly and energy efficient dry process.

10.8   References

Samad A.M., Satyanarayana N. and Sinha S.K. (2010). 'Tribology of UHMWPE film on air-plasma treated tool steel and the effect of PFPE overcoat', *Surf Coat Technol,* **204**, pp. 1330–1338.

Abidi N. and Hequet E. (2004). 'Cotton fabric graft copolymerization using microwave plasma. Universal attenuated total reflectance–FTIR study', *J Appl Sci,* **93**, pp. 145–154.

Badyal J.P., Ward L., William Y.D., Brooker A.T., Summers S., Crowther J.M., Roberts N.P.S., Yates A.T. and Saswati D. (2005). *PCT Int Appl,* WO 2005028741 29 pp.

Bhat N.V. and Nadiger G.S. (1978). 'Effect of Nitrogen Plasma on the Morphology and Allied Textile Properties of Tasar Silk Fibers and Fabrics', *Text Res J,* **48**(12), pp. 685–691.

Cai Z.S., Qiu Y.P., Zhang C.Y., Hwang Y.J. and McCord M. (2003). 'Effect of Atmospheric Plasma Treatment on Desizing of PVA on Cotton' *Text Res J,* **73**(8), pp. 670–674.

Cai Z.S., Hwang Y.J., Park Y.C., Zhang C.Y., McCord M. and Qiu Y.P. (2002). 'Preliminary investigation of atmospheric pressure plasma-aided desizing for cotton fabrics', *AATCC Review,* **2**(12), pp. 18–21.

Chaivan P., Pasaja N., Boonyawan D., Suanpoot P. and Vilaithong T. (2005). 'Low-temperature plasma treatment for hydrophobicity improvement of silk', *Surf Coat Technol,* **193**(1–3), pp. 356–360.

Chen J.R. (1996). 'Study on free radicals of cotton and wool fibers treated with low-temperature plasma', *J Appl Sci,* **62**(9), pp. 1325–1329.

Knoll G.F. (2000). *Radiation Detection and Measurement,* 3rd edn. New York, USA. Hegemann D. and Dawn B. (2007). 'Nanostructured textile surfaces: plasma deposition shows several advantages' *NanoS,* **1**, pp. 8–13.

Höcker H. (2002). 'Plasma treatment of textile fibers' *Pure Appl Chem,* **74**(3), pp. 423–427.

Iriyama Y., Mochizuki T., Watanabe M. and Utada M. (2002). 'Plasma treatment of silk fabrics for better dyability', *J Photopolym Sci Technol,* **15**(2), pp. 299–306.

Jahagirdar C.J. and Srivastava Y. (2001). 'Effects of plasma treatment and metal-ion chelation onlight fastness of dyed polyester/cotton fabric' *J Appl Polm Sci,* **82**(2), pp. 292–299.

Von Engel A. (1957). 'John Sealy Edward Townsend. 1868-1957', *Biographical Memoirs of Fellows of the Royal Society,* **3**, pp. 256–272.

Kan C.W., Chan K. and Yuen C.W. (2004). 'A study of the oxygen plasma treatment on the serviceability of a wool fabric', *Fibers and Polymers* **5**(3), pp. 213–218.

Kuzuya M., Kamiya K., Yanagihara Y. and Matsuno Y. (1993). 'Nature of plasma-induced free radical formation of several fibrous polypeptides', *Plasma Sources Sci Technol* **1**(2), p. 251.

Leshkov (2007). 'Plasma Jet Systems for Technological Applications', WDS'07 Proceedings of contributed papers, Part II, MATFYZPRESS, pp. 202–206.

Malek R.M.A. and Holme I. (2003). 'The effect of plasma treatment on some properties of cotton' *Iranian Polymer Journal* **12**(4), pp. 271–280.

McCord M.G., Hwang Y.J., Qiu Y., Hughes L.K. and Bourham M.A. (2003). 'Surface analysis of cotton fabrics fluorinated in radio-frequency plasma'. *J Appl Polym Sci* **88**, pp. 2038–2047.

Mirjalili M. and Golshan-Tafti A.R. (2004). 'Effect of irradiation on physical and chemical properties of polyester yarn' Yazd Islamic Azad University, Iran. International Conference on Nuclear Science and Technology in Iran, 2nd, Shiraz, Islamic Republic of Iran, Apr. 27–30, 215/1-215/5.

Laroussi M., Liu C., Roth J.R., Spence P.D., Tsai P.P. (1995). 'One atmosphere, uniform glow discharge plasma', US5414324 A.

Nadiger G.S. and Bhat N.V. (1985). 'Effect of plasma treatment on the structure and allied textile properties of mulberry silk', *J Appl Polym Sci,* **30**(10), pp. 4127–4135.

Nakano S., Isono T., Furutani M., Senzaki T. and Suzuki M. (1994). 'Changes in Specefic Surface Area of Silk by Treatment with Low-Temperature Oxygen Plasma', *Sen'i Gakkaishi* **50**(3), pp. 136–141.

Oktem T., Seventekin N., Ayhan H. and Piskin E. (2000). 'Modification of polyester and polyamide fabrics by different in situ plasma polymerization methods', *Turk J Chem,* **24**(3), pp. 275–285.

Ozdogan E., Saber R., Ayhan H. and Seventekin N. (2002). 'New approach for dyeability of cotton fabrics by different plasma polymerisation methods' *Color Technol,* **118**(3), pp. 100–103.

Periyasamy S., Deepti G. and Gulrajani M.L. (2007). 'Nanoscale Surface Roughening of Mulberry Silk by Monochromatic VUV Excimer Lamp' *J Appl Polym Sci,* **103**, pp. 4102–4106.

Periyasamy S. (2008). 'Studies on modification of silk', Ph.D. Thesis, Department of Textile Technology, Indian Institute of Technology Delhi, India.

Polettia G., Orsini F., Raffaele-Addamo A., Riccardi C. and Selli E. (2003), 'Cold plasma treatment of PET fabrics: AFM surface morphology characterization', *Appl Surf Sci,* **219**(3–4), pp. 311–316.

Prabaharan M. and Carneiro N. (2005). 'Effect of low-temperature plasma on cotton fabric and its application to bleaching and dyeing', *Indian J Fibre Text Res,* **30**(1), pp. 68–74.

Rakowski W., Okoniewski M., Bartos K. and Zawadzki J. (1982). 'Plasma treatment of textiles – potential applications and future prospects, *Melliand Textilberichte,* **11**, pp. 301–308.

Riccardi C., Barni R., Esena P. (2005). 'Plasma treatment of silk', Diffusion and Defect Data- Solid State Data, Pt. B: Solid State Phenomena 107 (Particle Beams & Plasma Interaction on Materials and Ion & Plasma Surface Finishing 2004), pp. 125–128.

Selli E., Mazzone G., Oliva C., Martini F., Riccardi C., Barni R., Marcandalli B. and Massafra M.R. (2001a). 'Characterisation of poly(ethylene terephthalate) and cotton fibres after cold SF6 plasma treatment', *J Mater Chem,* **11**(8), pp. 1985–1991.

Selli E., Riccardi C., Massafra A.R., Marcandalli B. (2001b). 'Surface Modifications of Silk by Cold SF6 Plasma Treatment', *Macromol Chem Phys,* **202**(9), pp. 1672–1678.

Shahidi S., Ghoranneviss M., Moazzenchi B., Dorranian D. and Rashidi A. (2005). 'Water repellent properties of cotton and PET fabrics using low temperature plasma of argon', International Conference on Phenomena in Ionized Gases, Proceedings, 27th, Eindhoven, Netherlands, July 18–22.

Thilagavathi G. and Kannaian T. (2008). 'Dual antimicrobial and blood repellent finishes for cotton hospital fabrics', *Indian J Fibre Text Res,* **33**, pp. 23–29.

Tsafack M.J. and Levalois-Gruetzmacher J. (2007). 'Towards multifunctional surfaces using the plasma-induced graft-polymerization (PIGP) process: Flame and waterproof cotton textiles', *Surf Coat Technol,* **201**(12), pp. 5789–5795.

Vaideki K., Jayakumar S., Thilagavathi G. and Rajendran R. (2007). "A study on the antimicrobial efficacy of RF oxygen plasma and neem extract treated cotton fabrics", *Appl Surf Sci,* **253**, pp. 7323–7329.

Vaideki K., Jayakumar S., Rajendran R. and Thilagavathi G. (2008). "Investigation on the effect of RF air plasma and neem leaf extract treatment on the surface modification and antimicrobial", *Appl Surf Sci*, **254**, pp. 2472–2478.

Wakida T., Lee M., Jeon J.H., Tokuyama T., Kuriyama H. and Ishida S. (2004). Ozone-gas treatment of wool and silk fabrics, *Sen-i Gakkaishi*, **60**(7), pp. 213–219.

Wakdia T., Muncheul L., Yukihiro S., Shinji O., Yi G. and Shouhua N. (1996). 'Dyeing properties of oxygen low-temperature plasma -treated wool and nylon 6 fibers with acid and basic dyes', *J Soc Dyers Colour* **112**(9), pp. 233–236.

Ward T.L., Jung H.Z., Hinojosa O.R. and Benerito R.J. (1979). 'Characterization and use of radio frequency plasma-activated natural polymers', *J Appl Polym Sci* **23**(7), pp. 1987–2003.

Yamada H. and Shimizu T. (2006). 'Fiber cloths with excellent antisoiling properties after sterilization and their manufacture', *Jpn. Kokai Tokkyo Koho*, JP 2006316360.

Yoon N.S., Lim Y.J. and Takagishi T.M. (1996). 'Mechanical and dyeing properties of wool and cotton fabrics treated with low temperature plasma and enzymes', *Text Res J*, **66**(5), pp. 329–336.

Yuyue C., Hong L., Yu R., Hongwei W. and Liangjun Z. (2004). 'Study on Bombyx mori silk treated by oxygen plasma', *Journal of Zhejiang University*, **5**(8), pp. 918–922.

Zhang J., France P., Radomyselskiy A., Datta S., Zhao J.A. and Ooij W. (2003). 'Hydrophobic cotton fabric coated by a thin nanoparticulate plasma film', *J Appl Sci*, **88**(6), pp. 1473–1481.

Zhang J. (1997). 'The surface characterization of mulberry silk grafted with acrylamide by plasma copolymerization', *J Appl Polym Sci*, **64**(9), pp. 1713–1717.

11

Nanotitania synthesis and its integration in textiles using plasma technology

Balasubramanian C.

Abstract: Nanotechnology has made inroads into many areas of applications ranging from medical to metallurgy and optoelectronics to textiles. Prominent among the nanotechnology applications in textile are imparting self-cleaning and anti-bacterial properties to the fabric by using nonmaterial. These two applications have not only been found feasible but also hold a high market potential.

In this chapter, use of nanoparticles of titanium dioxide for imparting self-cleaning effect to cotton and silk fabrics has been described. This includes various methods of generation of nanoparticles including plasma route of synthesis. It also covers introduction to both nanoscience as well as plasma processes and efforts made by various research groups to produce self-cleaning fabrics with nano titania. It primarily focuses on the work carried out at FCIPT on nanotitania synthesis and its application to silk and cotton fabrics and the various tests conducted and results obtained therein. Finally, the potential hazards of nanoparticles of titania which can pose risk to human are discussed briefly.

Key words: Nanotitania, self-cleaning fabrics, thermal plasma

11.1   Introduction

Technological advances have been the root for all industrial revolution, leading to a mechanised or automated process for bulk production leading to cost reduction. The revolution in textile industry moved from a cottage scale to industrial arena in the eighteenth century with the invention of '*flying shuttle*' – a mechanism to weave cotton in a mechanised way. Textile industry has been keeping abreast and adapting the technological advances in various fields; for example, advances in chemistry adapted to produce new synthetic fibres, dying processes, etc., advances in automation was adapted for mass production of fabrics and to meet the ever increasing demand for fabrics. In India, textile industry continues to be the second largest employment generating sector.

The basic use of fabrics for covering modesty has undergone a sea change with the incorporation of a number of functionality to the apparel – fire-retardant apparels, chemical resistant apparels, apparels for varying climatic conditions, etc. Both R&D organizations as well as industrial houses have been working towards bringing out fabrics with specific properties and purposes.

Of the scientific and technological breakthroughs achieved at the fag end of last century has been the use of nanomaterials/nanotechnology in various areas including medical and electronic equipments. The textile industry too was not be left behind, and got on to the nanotechnology band wagon with research on incorporating nanomaterials in fabrics to impart certain functionalities. The thrust has been on the study and use of nanomaterials in textiles during manufacturing (Harholdt et al., 2003) as well as finishing.

Some of the most widely studied nanomaterials for textile applications include nano silver and nanotitania. The former incorporates anti-microbial properties and the later self-cleaning and hydrophobic properties. Titania, even in its bulk form, is frequently used in the production of paints, paper, plastics, cosmetics, food colouring agent, etc. Apart from these two materials, zinc oxide in nano form has also been studied and is found to incorporate UV blocking property to the fabric.

Other areas of research wherein nanomaterials have potential use in textile industry is imparting nano-based chemical finishes – use of nanoscale emulsification for an even, thorough and stronger bonding of the chemical to the fabric material. The chemical finishes are incorporated with the objective of imparting wrinkle resistance, hydrophilicity, stain resistance, etc.

On the manufacturing side, use of nanofibers like carbon nanofibers along with traditional synthetic fibers like nylon, polyester, polyethylene have been studied and are reported to impart a greatly enhanced mechanical strength. Carbon nanoparticles as well as clay nanoparticles in a synthetic matrix like nylon are reported to have not only enhanced mechanical properties but also other functionalities like chemical resistance, flame retardant and UV blocking properties (Lei Qian, 2004).

11.1.1   Nanoscience and nanotechnology

Nanoscience is the study of processes, properties and behavioral pattern of materials when their sizes are in nanometers (one billionth of a meter: 10^{-9} m). It is the study of control of materials at atomic and molecular scale. Nanotechnology, on the other hand, is the application of these materials and their properties in devices.

Nanotechnology is an interdisciplinary technology with applicability as wide ranging as electronics, optics, medicine, materials science and mechanics. It is projected to bring in a large societal impact and is most widely stated to bring in a second 'industrial revolution'.

Nanomaterials are broadly defined as those materials whose size lies in the range of 1–100 nm at least in one of the dimensions. However, this range varies depending on the material type, property that is to be employed, etc. For example, for semiconductor materials like CdS, CdSe, ZnO, etc., wherein enhancement of optical properties is the focus, the size range typically is less than 10 nm. On the other hand, ceramic materials like alumina, etc., whose mechanical properties are the focus, a size range of up to 100 nm has also applicability and can be called as nanomaterial.

The basics of nanotechnology lie in the fact that the properties of materials change dramatically when their size is reduced to nanoscale. For example, ceramics that are brittle in bulk form become deformable at nanoscale. Opaque substances like copper turn transparent at nanoscale. Aluminium which is stable in bulk form becomes highly reactive at nanoscale.

11.1.2   Titanium dioxide (titania)

Titanium dioxide (TiO_2) is a semi-conductor metal oxide with photocatalytic and anti-microbial properties. Titanium dioxide is commercially very important as a white pigment and is non-toxic, chemically inert and a dielectric ceramic. When exposed to UV light in the sub 400 nm range, TiO_2 becomes a photo catalyst oxidizer (PCO) as well, which creates hydroxyl radicals and super oxide ions which are two times stronger disinfectants than chlorine and 1.5 times stronger disinfectant than ozone. TiO_2 is safe and widely used in many applications (Othmer, 1997; Quan et al., 2005; Fujishima and Rao, 1997) and houschold products such as toothpaste, food and teeth-whitening solutions. It can also be used for waste water remediation.

Titania exists in different mineral forms – *anatase, rutile* and *brookite*. Anatase-type TiO_2 with a band gap of 3.3 eV has a crystalline structure that corresponds to the tetragonal system (with dipyramidal habit) and is used mainly as a photocatalyst under UV irradiation. Rutile-type TiO_2, with a band gap of 3.1 eV, also has a tetragonal crystal structure but with prismatic habit. This type of titania is mainly used as whitening agent/pigment in paints. Brookite-type TiO_2, with a varying band gap of 3.2–3.8 eV, has an orthorhombic crystalline structure. Table 11.1 summarizes the various properties of titania polymorphs.

Table 11.1 Physical properties of the Titania polymorphs

Crystalline phase	Anatase	Rutile	Brookite
Band gap energy (eV)	3.2	3.0	3.2 to 3.8
Refractive index	2.49	2.61	2.58
Unit cell	Tetragonal	Tetragonal	Orthorhombic
No. of TiO_2/unit cell	4	2	8
Density (g/cm³)	4.26	3.84	4.11

Titanium dioxide, in the presence of ultraviolet light, will oxidize a wide range of organic materials. When activated by *UV-A* irradiation, titania exhibits photo catalytic properties. This photo catalytic property is greatly enhanced when titania used is in the nanosize. Nanotitania-coated fabrics, tiles, etc., are already in the market, and nanotitania gives a functional edge to the product – acting as a self-cleaning agent in the presence of *UV-A* radiation. The *UV* component, present in the sunlight, is sufficient enough to carry out the photocatalytic activity and clean the stained surfaces. Figure 11.1 depicts the process of stain removal from a fabric surface which has been coated with nanotitania. When light hits the nanotitania, it frees up electrons within the crystal and these react with oxygen from the air. This generates free-radical oxygen, a powerful oxidizing agent that can break down grime into smaller particles such as carbon dioxide and water. Because the catalyst does not get used up, it can keep on working as long as it is exposed to sunlight.

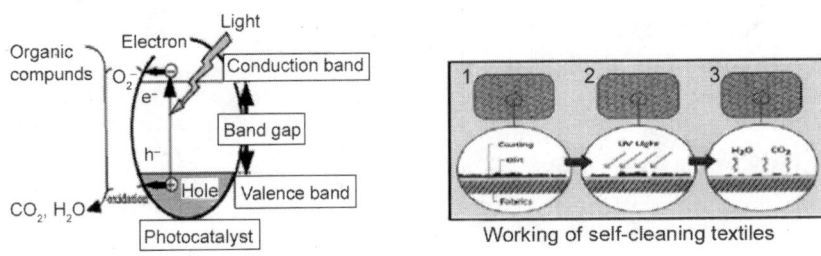

Figure 11.1 Schematic of the photocatalytic activity of titania (left) and the self-cleaning effect in textiles (right)

Daoud et al. (2008) from Monash University and the Hong Kong Polytechnic University were the first to report the use of nanotitania for self-cleaning of cotton fabric. They dipped cotton fabric pieces into a liquid slurry of titanium dioxide nanoparticles (of 20 nm size) for half a minute before removing them, padding them dry, and heating them to 97°C in an oven for 15 minutes. This was followed by 3 hours in boiling water to complete the coating process. These coated cotton fabrics were found to effectively remove wine stains and coffee stains.

11.1.3   Plasma processes

Plasma consists of negatively charged electrons, positively charged ions radicals, metastable and forms the fourth state of matter. Though more than 99% of the universe contains plasma, natural terrestrial plasma is limited to lightening. Man-made plasma includes fluorescent bulbs, neon signs, etc.

Plasmas can be broadly classified into "thermal or hot plasmas" and "non-thermal or cold plasmas". Both the types have great applicability in various fields. In thermal plasma, the electron energy approximately equals the ion energy which approximately equals the neutral atoms' energy (present in small percentage) leading to an ambient gas temperature of thousands or tens of thousands of Kelvin depending on the input power. On the other hand, in non-thermal plasmas, the electron energy is much higher than either the ion or neutral atom energy resulting in a low gas temperature. Therefore, non-thermal plasmas are used for surface modification of polymers, textiles, metals, and alloys.

11.2   Fabric pre-treatment by plasma

Since adhesion is a surface-dependent property, plasma can be effectively used to achieve modification of this near-surface region without affecting the bulk properties of the material. Researchers in the past have successfully pre-treated fabric surface prior to the nanoparticle deposition to bind the nanoparticle and fibre. Surface pre-treatment of fabrics and polymers have been carried out using atmospheric pressure plasma by Szabová et al. (2009).

Szabová et al. (2009) show an incorporation of both oxygen and nitrogen atoms into polypropylene surface modified by N_2 plasma. These modified samples were dipped in three concentrations of TiO_2 dispersed in water and ultrasonicated to prevent agglomeration. AFM analysis showed the surface has increased on treated PPNW (polypropylene nonwoven). SEM images recorded after coating of titania show smooth surface samples in contrast to rough surface obtained with water dispersion. Surface analysis by XPS confirms the incorporation of both oxygen and nitrogen atoms into polypropylene surface.

Cotton and polyester fibres treated in atmospheric plasma by Nasadil and Benesovsky (2008) increased the hydrophilization of cotton thus improving the performance of the fabric particularly in sorption of inks on sample. Anatase nano titania in colloidal form was applied on plasma-treated samples, which inhibited Klebsiella pneumonia bacterial generation even after 55 hours of exposure. It is shown that plasma treatment has effectively improved the binding of the nanoparticles on cotton and thus enhancing the functionality.

11.3   Methods of nanomaterial synthesis

Nanomaterials can be synthesised by many physical and chemical methods. Both dry as well as wet processes can be used to prepare nanomaterials; high temperature as well as low temperature can be employed to synthesize nanomaterials. There are two broad approaches for the synthesis of nanomaterials: (a) Top–down approach wherein the bulk material is broken down mechanically or by electrochemical etching to nanodimensions. (b) Bottom–up approach wherein the nanosized clusters are built from atoms or molecules. The later approach is more predominantly used for the synthesis of nanomaterials. Laser ablation, chemical vapour deposition, plasma processes, sol-gel process, inert gas phase condensation are some of the synthesis techniques based on bottom–up approach. Figure 11.2 shows the schematic of techniques for synthesis of nanomaterials.

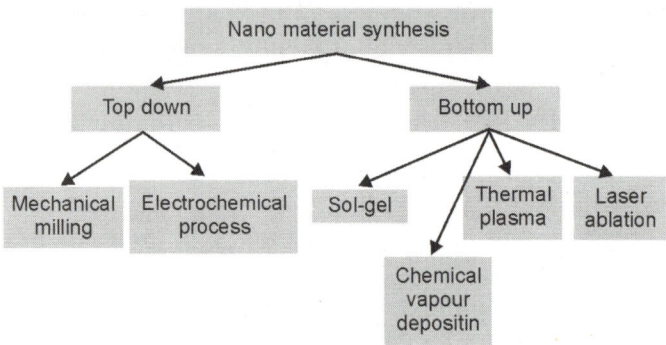

Figure 11.2 Flow chart of the various processes for synthesis of nanomaterials

11.3.1   Synthesis of nanomaterials by thermal plasma process

Thermal plasma has been widely used for synthesis of nanomaterials. As described earlier and also as the name suggests, thermal plasma has within it a large amount of heat energy. This heat content can be used to melt and evaporate materials. The materials in the ionised or gas phase can be suddenly cooled and condensed to form very fine particles whose size range would be in nanometer region.

Thermal plasma process could be either a non-transferred plasma torch or transferred arc plasma. In both the cases the material is evaporated and condensed to form nanomaterials. The materials synthesized by these high temperature process generally yields a highly crystalline form of product. However, it is to be noted that the high temperature gradients that exist in plasma also results in a size distribution which is much wider than nanomaterials formed by any other technique. This is due to the fact that nucleation and growth of particles happening at slightly different locations within the plasma chamber have undergone varying thermal histories due to which the cluster sizes will be different. Whereas by low temperature (sol-gel) process one could get narrow size distribution (Daoud and Xin, 2004).

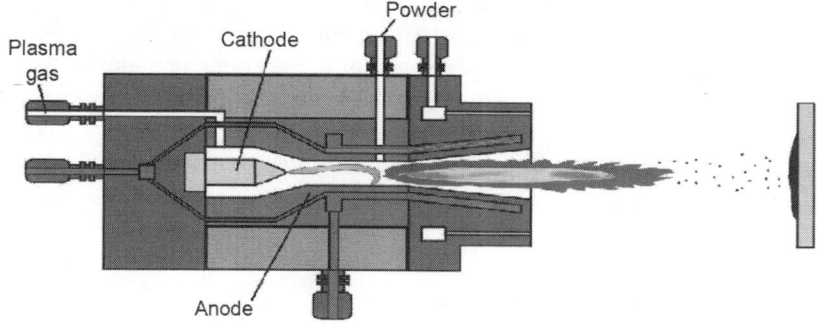

Figure 11.3 An illustration of plasma torch (image taken from Centre for Advanced Coating Technologies, University of Toronto)

In a plasma torch process, as shown in Fig. 11.3, the electrodes are arranged in a coaxial geometry with the cathode at the centre surrounded by a coaxial tube of the anode. A gas, which typically could be argon or nitrogen, is made to flow between the electrodes pushing the plasma plume through the tapered nozzle at the end of the torch. Depending on the gas flow rate and the voltage applied, the plasma plume can extend much further away from the electrodes. Materials whose nanoparticles are of interest could either (a) be fed between the electrodes which gets evaporated in the plasma plume and condensing the vapour at a cooler region downstream resulting in nanoparticle formation or (b) the material kept at a distance from the torch but well within the reach of the plume resulting in heating of the material and subsequent evaporation and cooling. Kakati et al. (2008) have used plasma torch for synthesis of nanomaterials.

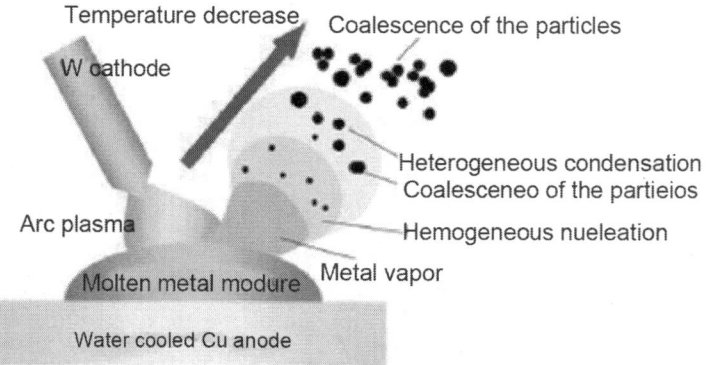

Figure 11.4 A schematic of a typical transferred arc plasma process for generation of nanomaterialsA few of the available literature on the use of nanotitania to impart self-cleaning functionality to textiles are elaborated here.

In an arc plasma process, as shown in Fig. 11.4, the material of interest (material to be evaporated) itself serves as one of the electrodes and so the plasma zone is created between the material to be evaporated and the other electrode. It is to be noted that the material to be evaporated has to be conducting, unlike the non-transferred plasma torch process where the material could be non-conducting as well. In the arc plasma process, one of the electrodes is thus a "*consuming*" electrode – as the size gets diminished due to the evaporation.

11.3.2   Review of other synthesis methods and application of nano titania in fabrics

Several other methods for synthesis of nano TiO_2 have been reported: Chemical method (Mejia et al., 2009; Kim et al., 2004; Okuyama et al., 1998), sol–gel and other wet chemical techniques (Bozzi et al., 2005; Walid et al., 2008, Uddin et al., 2008; Ramakrishna et al., 2004), synthesis of TiO_2 using argon plasma (Vijay et al., 2009), synthesis of TiO_2 using thermal plasma (Kakati et al., 2008). Preparation of titania thin films by MOCVD has also been reported by Babelon et al. (1998).

Yuranova (2006) reports the structure-forming function of cotton textile at low temperatures on a mixture of TiO_2–SiO_2 colloids effective in photo-discoloration of red wine stains under solar simulated radiation. TiO_2–SiO_2 was produced by sol-gel process and it was observed that the most suitable Ti-content of the coating was found to be 5.8% and for SiO_2, the content was 3.9% (w/w). The discoloration of red wine led to CO_2 evolution that was more efficient for TiO_2–SiO_2-coated cotton samples than of TiO_2-coated ones. The TiO_2–SiO_2 layer thickness on the cotton fibres was detected to be 20–30 nm. The TiO_2 and SiO_2 were both observed to have particle sizes between 4 and 8 nm. EDS analysis showed that the Ti-particles were always surrounded by amorphous silica.

Mejia et al. (2009) reports on the synthesis of nanocrystalline anatase TiO_2 with particles size ranging between 8 and 18 nm and subsequent loading on cotton fabrics by two innovative approaches: (a) fabrics were pre-treated in air under RF-plasma at atmospheric pressure as well as 0.1 mbar were able to bind TiO_2 due to the drastic localized heating of the cotton. This heating effect is reported to break the intermolecular H-bonding between the cellulose surface-OH groups of adjacent molecules and (b) by UV-C light (185 nm) pre-treatment at atmospheric pressure led to the formation of atomic O and excited O* in the gas phase. It also introduced functionalities into the cotton surface enabling the binding of TiO_2 to the cotton fabric.

Good reproducibility for the discoloration was obtained in both cases for red wine stains under simulated solar light. A more uniform coating of TiO_2 on cotton was obtained by pre-treating with UV-C as compared to RF-plasma.

Tung and Daoud (2009) successfully synthesised nanocrystalline anatase in 1% N-sol and 1.4% H-sol along with titanium isopropoxide and galcial acetic acid by hydrolysis and condensation method at 60°C under vigorous stirring for 16 hours. The XRD pattern of titanium dioxide extracted from N-sol and H-sol demonstrates a single phase anatase with an average particle size of 8 nm for Nsol and 10–80 nm of H-sol. The photo-degradation on protein samples showed a serious yellowing with uneven and rough deposition of N-sol, compared to evenly deposited H-sol giving strong photo-resistance with no peel-off effect. With similar ultraviolet absorption ability and photocatalytic properties, the degradation of coffee stain was possible after 8–20 hours of exposure to UV light by both H-sol and N-sol coated fibres. H-sol has led to comparatively less deterioration in fibre-tearing strength and improvement in air permeability compared to N-sol confirming H-sol to have better affinity to fibres with improved self-cleaning functionalization. The same authors (Tung and Daoud 2009) also show the self cleaning effect of nanotitania on keratin type wool fibers.

Oh and Ishigaki (2004) reports on the synthesis of TiO_2 nanopowder with anatase crystal structure from titanium tetrachloride by thermal plasma process. To improve photocatalytic activity, silicon tetrachloride and iron (III) acetylacetonate were added to the plasma reactor. The photocatalytic activity of prepared pure TiO_2 (T-powder), Si-doped TiO_2 (ST) and Fe-doped TiO_2 (FT) powder was evaluated by photo-degradation of acetaldehyde.

Decomposition efficiency of T-powder under UV-light increased with the content of anatase titania. Authors report improvement in the photocatalysis with the presence of small amount of Si-dopant. Excessive Si-dopant of over 2% is reported to decrease the photocatalytic activity of ST-powder because of the reduction of active sites on the catalyst. FT-powder was tested for the photocatalytic activity under visible light and UV-light. Decomposition efficiency increased with the addition of Fe-dopant because of suppression of electron–holes recombination. However, excessive Fe-dopant by over 15% is reported to inhibit the crystallization of anatase and acted as recombination center, leading to decrease in the decomposition efficiency.

11.4 Nanotitania synthesis and its application in textiles at FCIPT

The Facilitation Centre for Industrial Plasma Technologies (FCIPT) Division of the Institute for Plasma Research (IPR), Gandhinagar, India, is primarily engaged in development of various plasma-based (both thermal as well as non-thermal) technologies which have potential industrial uses. One of the activities undertaken is generation of nanotitania particles by thermal arc plasma process, which was then tested for its self-cleaning activity in silk and cotton fabrics. Also a system has been designed and fabricated that can generate nanotitania in large quantities in an automated way. The details of the study – both synthesis as well as the properties of nanotitania generated – are elaborated in the following sections.

11.4.1   Synthesis of nanotitania by thermal arc plasma

Nanotitania was prepared by thermal plasma processing in air ambient at atmospheric pressure. An electric arc was struck between two electrodes, both consisting of titanium rod/blocks, with the help of a high current DC power source. The applied voltage was varied (depending on the distance between the electrodes) between 25 and 50 V, and the arc current was maintained at 80 A. The high thermal energy content in the arc plasma melts the anode material (titanium) and evaporates it. In the arc column the titanium is in the ionised state and as it moves away from the arc zone, the temperature falls and neutral titanium atoms form, which in turn react with the oxygen present in ambient air to form molecules of titania. Clusters of titania molecules form this clustering and particle formation continues till sufficient energy exists for such action. The energy availability is drastically reduced by cooling the synthesis chamber so the cluster formation is arrested when the particle size reaches nano dimension. These particles condense on the inner walls of the reactor chamber (substrates can also be placed inside the chamber on which deposition could take place), and these particles, in the form of powder, are then collected for analysis. The particles were analysed for their phase, morphology and for its self-cleaning activity.

After the analysis of the nanoparticles, they were then coated onto the silk and cotton fabrics by ultrasonic spray system. One can find a number of reports on the coating and binding of nanoparticles onto the fabrics and most of them report on the dip padding followed by drying process. This random-coating process has been replaced with a controlled coating process of ultrasonic spray system. The fabric pieces thus coated were then studied for its properties especially related to removal coffee, tea and turmeric stains, which are the commonly unrelenting stains encountered in India.

11.4.2   Analysis of nanotitania

The as-synthesised nanotitania powder was collected and analyzed using various characterization techniques to ascertain its morphology, composition, etc.

X-ray diffraction analysis
The powder of titanium dioxide was characterized by powder X-ray diffraction technique to verify its phase as well as crystal size of nanoparticles. The diffraction spectra was recorded using Seifert make XRD 3000 PTS model and CuK_α radiation.

The powder XRD spectra, as shown in Fig. 11.5, clearly indicate the titania product to be crystalline in nature, with prominent diffraction peaks at 25.28°, 48°, 54° and 55.04° – all corresponding to Anatase phase. However, one can observe the presence of Rutile phase as well (peaks at 27.44°, 36.08° and 54.38°), though to a lesser concentration. The Selected Area Electron Diffraction (SAED) pattern recorded, as shown in Fig. 11.6, from TEM apparatus also confirms these 'd' values. The SAED pattern also confirms the product to consist of both polycrystalline as well as single crystal particles. The average particle size (approximated from the crystallite size) as calculated from the FWHM of the prominent diffraction peaks in XRD is found to be ~30 nm.

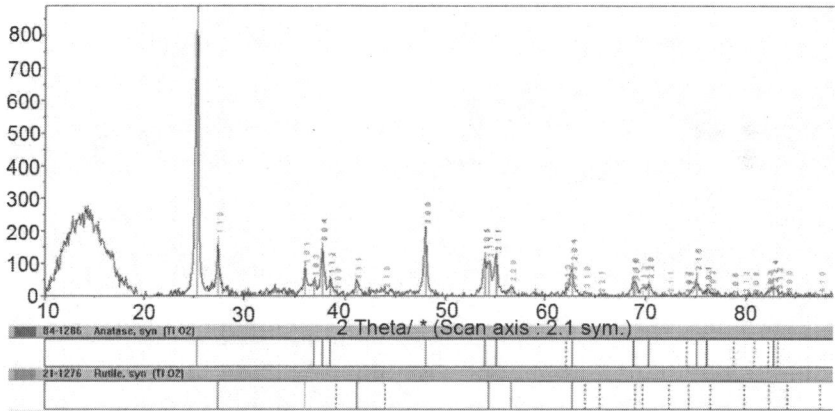

Figure 11.5 XRD spectra of nanotitania. The hkl planes are indicated adjacent to the diffraction peaks

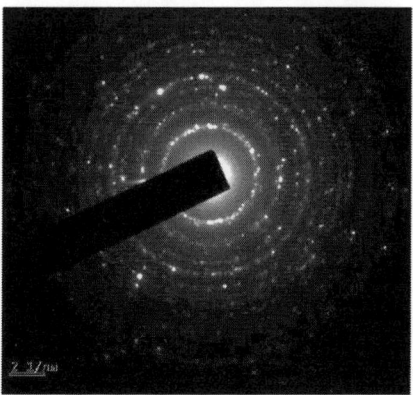

Figure 11.6 Selected Area Electron Diffraction (SAED) pattern of the nanotitania

Transmission Electron Microscopy Analysis

A high resolution Transmission Electron Microscope (beam energy 200 keV) was used to study the morphology of the particles. Images shown in Fig. 11.7(a) and (b) are the TEM images of the nanotitania particles. The particles are clearly seen to be spherical in shape and with size ranging from 16–40 nm with a few scattered particles of size as low as 7 nm and some as high as 60 nm. The TEM image shown in Fig. 11.7 clearly brings out the single crystal nature of the particles.

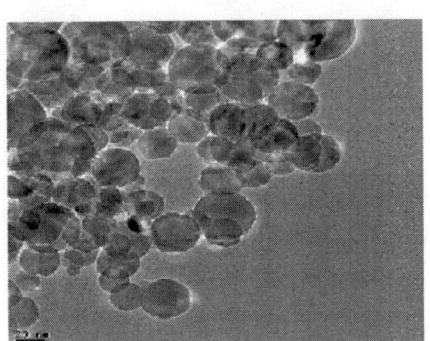 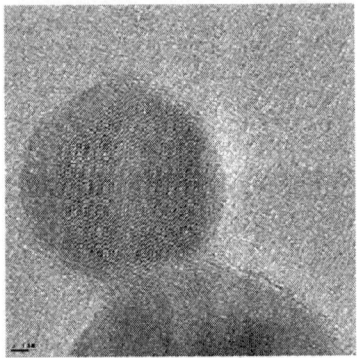

Figure 11.7(a and b) Transmission electron microscope image of the nano-particles indicating the spherical shape of the particles

An analysis of the particle size distribution, as shown in Fig. 11.8, from a number of TEM images indicate that the maximum number of particles to lie in the size range between 26–30 nm.

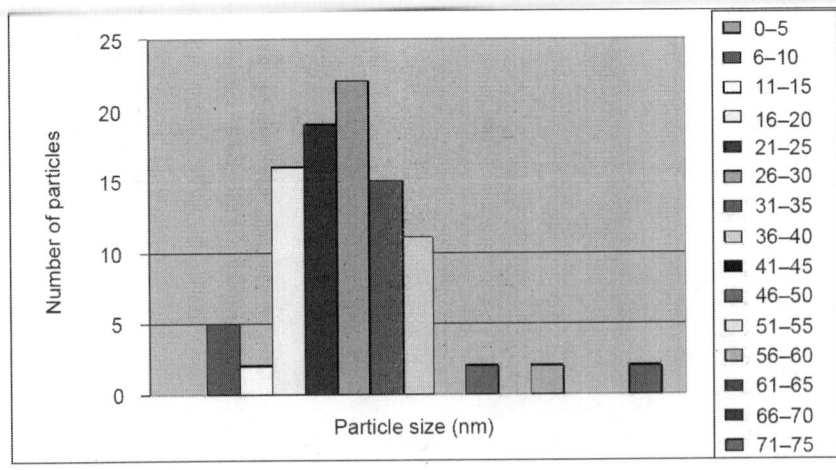

Figure 11.8 Particle size distribution, analyzed using the TEM images recorded

Fourier Transform Infra Red (FTIR) Analysis

The nanotitania powder was mixed with KBr powder and analysed for its
molecular spectroscopy to reconfirm its composition. FTIR spectra given in
Fig. 11.9 show same behaviour with the spectra of standard titania sample for
most of transmission peaks. The absorption peaks for the synthesised TiO_2 are
at wave number of 1384 cm⁻¹, 2424 cm⁻¹, 2364 cm⁻¹ which corresponds to
the wavenumber 1350 cm⁻¹, 2320 cm⁻¹, 2320 cm⁻¹ respectively from standard
FTIR spectra of TiO_2 showing a shift of 34 cm⁻¹, 104 cm⁻¹, 44 cm⁻¹.

The above analyses clearly indicate the nanosize of the synthesised
products as well as its high crystallinity.

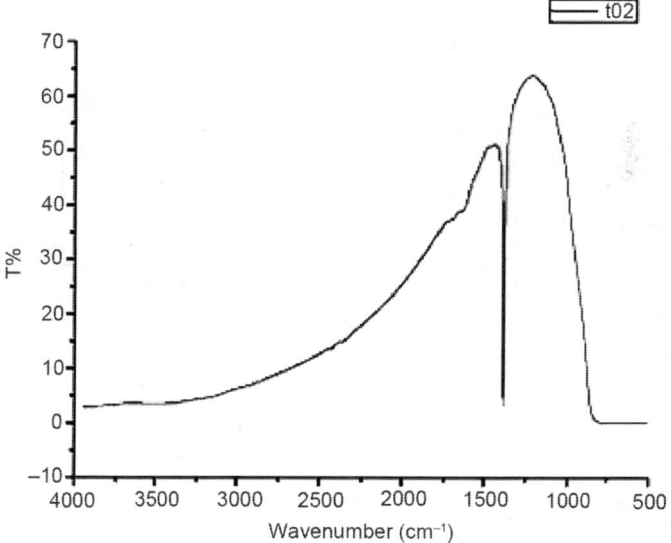

Figure 11.9 FTIR spectra of nanotitania powder sample

11.4.3   Fabric pre-treatment by DBD plasma

Nanotitania particles were then coated onto cotton and silk fabric pieces
separately with an automated ultrasonic spray nozzle system as shown in
Fig. 11.10. Prior to the deposition of the titania nanoparticles, the fabric pieces
were pre-treated with Dielectric Barrier Discharge (DBD).

The DBD system used here is an AC discharge. It works under
atmospheric pressure with air as the ambient gas used for plasma generation.
The electrodes are separated by Teflon of 5 mm diameter. The samples are
placed in between the electrodes and a voltage of 6 kV is applied between
them. The plasma treatment is done on the fabric for duration of 60 seconds.

This pre-treatment greatly enhanced the surface roughness of the fabric at the nanoscale and improved the adhesion of the nanoparticles on to the fabric.

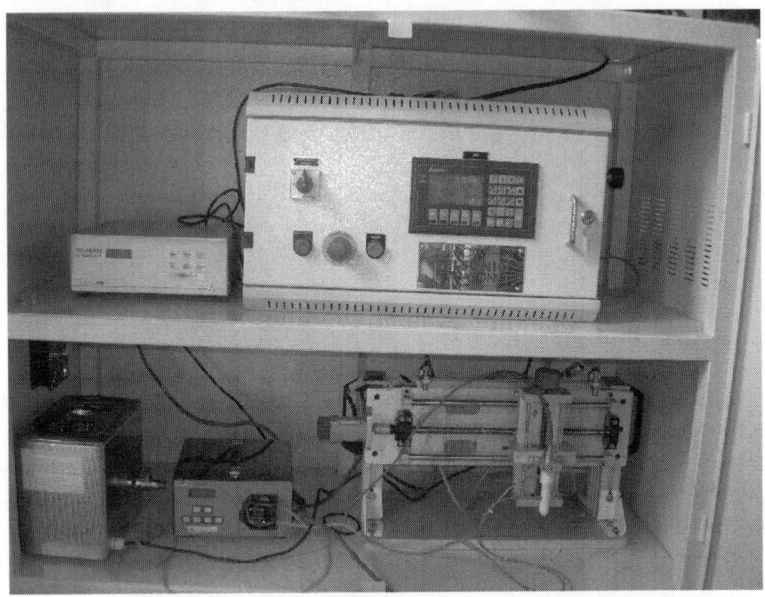

Figure 11.10 Photograph of ultrasonic spray nozzle system with plc control for controlled spraying of nanotitania on fabric surfaces

The nanoparticles of known quantity was dispersed in measured isopropyl alcohol and were ultrasonicated continuously while it was passed onto the ultrasonic spray nozzle through a peristaltic pump as shown in Fig. 11.10. The spray nozzle also has a transducer generating 35 kHz ultrasonic waves to ensure no agglomeration of the particles occur prior to depositing on the fabric.

11.4.4   Fiber morphology changes due to plasma pre-treatment and nanotitania coating

The extent of morphological changes occurring due to the air plasma treatment is shown in Figs. 11.11(a) and (b). Figure 11.11(a) shows typical micrographs of smooth surface of cotton fiber with characteristic parallel ridges and grooves as expressed by Svetlana et al. (2008). Air plasma-treated cotton fibre images shown in Fig. 11.11(b) shows rough surface with minimum parallel grooves without affecting the characteristic parallel ridges. Treating of cotton fibres with air plasma induces changes in morphology superficially but not in the bulk properties of the fibre.

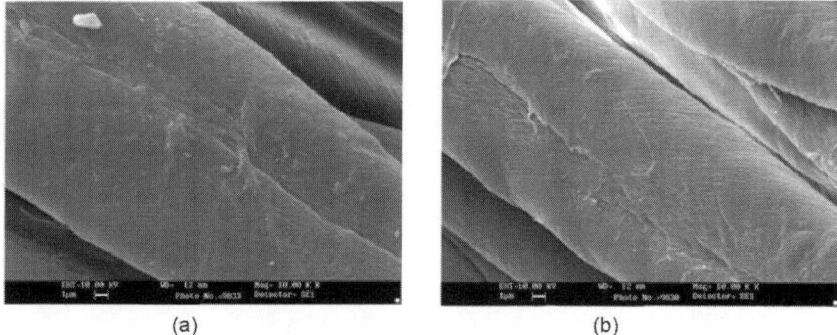

(a)   (b)

Figure 11.11 Surface morphology of the cotton fibre (a) untreated and
(b) plasma treated

Figure 11.12(a) and (b) shows the SEM image of nano titanium dioxide coating on plasma untreated as well as plasma-treated cotton fabric. Both the samples were coated with equal amounts of nanotitania powder. The images clearly show the differences in the presence of nano titanium dioxide particle on the surface of the cotton. In the untreated sample, Fig. 11.12(a), the surface being relatively smooth results in the nanotitania being present over the smooth surface. On the other hand, in the treated sample, the surface being more rough takes in the nanotitania within its roughness and the surface after coating with nanotitania appears smoother. Similar results are also reported by Szabová et al. (2009). These results indicate that the plasma pre-treatment helps in stronger adhesion of the particles on the fabric surface.

(a)   (b)

Figure 11.12 Surface morphology of the cotton fibre (a) without plasma treatment
deposited with TiO_2 (b) plasma treated deposited with TiO_2

Figure 11.13(a) and (b) shows the SEM image of silk fibre deposited with TiO$_2$. Figure 11.13(b) is the image of silk fiber treated with plasma followed by TiO$_2$ coating shows without lumps on the fibre surface as compared to the images of untreated shown in Fig. 11.13(a).

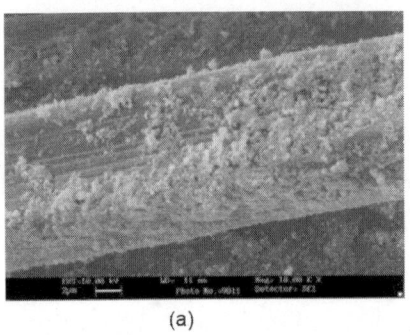
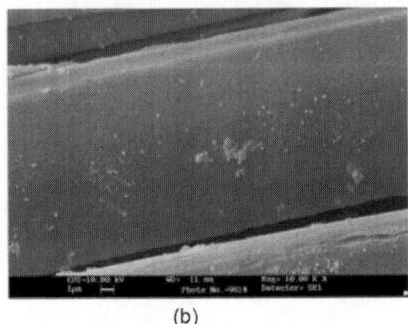

(a)   (b)

Figure 11.13 Surface morphology of the mulberry silk fibre (a) without plasma treatment deposited with TiO$_2$ (b) plasma treated deposited with TiO$_2$

11.4.5   Self-cleaning effect of nanotitania on cotton fabric

Further to the fabric (cotton/ silk) pre-treatment, the nanotitania particles were coated on the fabric pieces, measuring 10 cm ′ 10 cm in size, and then these coated fabric were studied for self-cleaning property. To carry out these studies, a predetermined quantity of nanotitania powder was dispersed in a 50 ml of iso-propyl alcohol and kept in an ultrasonic bath. This solution, while still being kept in the sonicator, was passed through an ultrasonic spray nozzle and was uniformly sprayed over the fabric material. The fabric samples, after getting dried, were treated with tea, coffee and turmeric stains (a drop of concentrated solution of tea, coffee and turmeric were placed on the fabric material). The fabric material was then exposed to sunlight and UV-A light source separately. A fabric sample without any nanotitania coating but stained by the above three compounds served as the control sample. This was also tested under sunlight and UV light separately. The decolouration of the stains over various intervals of time was recorded with the help of a camera. The experiments were performed for three different weight percentages of the nanotitania viz. 0.5 mg/m^2; 1.0 mg/m^2 and 1.5 mg/m2. The results of this study are shown in Figs. 11.14–11.16.

As can be seen in Figs. 11.14–11.16, the coffee and turmeric stains disappear within 4 hours when kept under sunlight, whereas the same decolouration occurs in less than 2 hours when exposed to UV light. The control sample did not show any decolouration indicating that the observed decolouration is purely due to the presence of nanotitania. The stain due to tea, however, did not show much response either under UV or sunlight exposure. It is found that the maximum decolouration occurs for fabric samples coated with 1.0 mg/m² weight percentage of nanotitania.

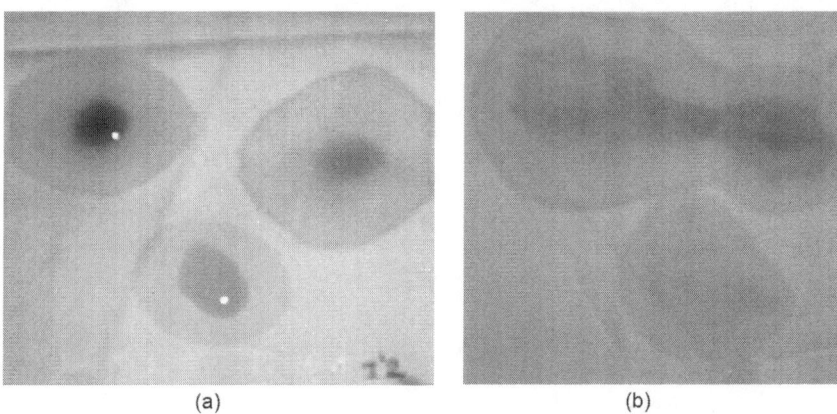

(a)   (b)

Figure 11.14 Virgin cotton fabric sample with stains – top left is tea stain; top right is coffee stain and bottom yellow is turmeric stain (a) before exposure and (b) after exposure to UV over a period of 1 hour (no change was recorded even after 8 hours exposure)

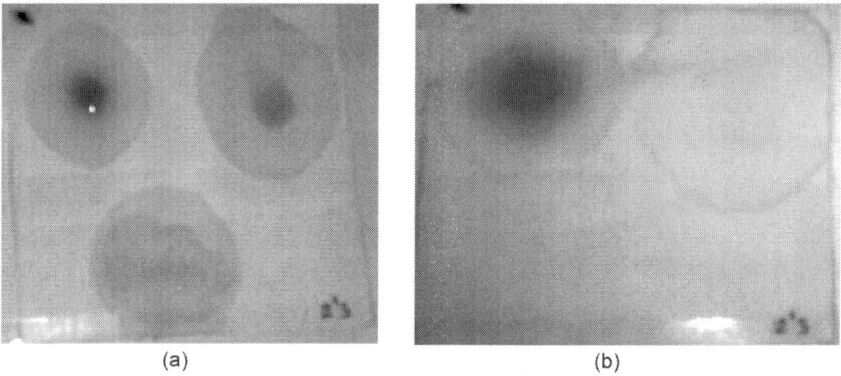

(a)   (b)

Figure 11.15 Fabric sample coated with Nanotitania (concentration 1 gm per m²). Top left is tea stain; top right is coffee stain and bottom yellow is turmeric stain (a) before exposure and (b) after exposure under UV for 2 hours.

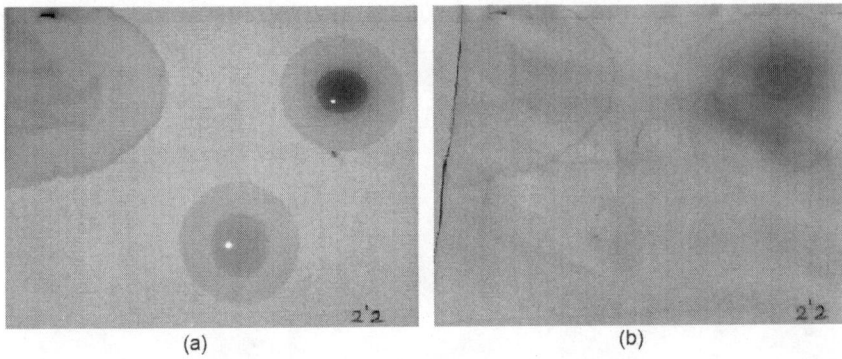

Figure 11.16 Fabric sample coated with Nanotitania (concentration 1 g per m²). Top left is coffee stain; top right is t.ea stain and bottom yellow is turmeric stain (a) before exposure and (b) after exposure under sunlight for 4 hours.

11.4.6   Analysis of Ultra-Violet Protection Factor (UPF)

The UPF rating and transmittance of plasma-untreated and -treated silk and cotton fabric samples are shown in Table 11.2.

Table 11.2 UPF values, %blocking UVA and UVB of silk and cotton fabric sample

Fabrics	Sample types	UPF value	UV% blocking	
			UVA	UVB
Silk	Control	45	97.95	97.7
	Plasma untreated 10 mg TiO$_2$/100 cm²	54	98.14	98.13
	Plasma untreated 30 mg TiO$_2$/100 cm²	41	97.62	97.56
	Plasma treated 10 mg TiO$_2$/100 cm²	49	98.12	98.06
	Plasma treated 30 mg TiO$_2$/100 cm²	37	97.37	97.30
Cotton	Control	397	99.7	99.7
	Plasma untreated 10 mg TiO$_2$/100 cm²	762	99.8	99.8
	Plasma untreated 30 mg TiO$_2$/100 cm²	129	99.2	99.2
	Plasma treated 10 mg TiO$_2$/100 cm²	188	99.9	99.4
	Plasma treated 30 mg TiO$_2$/100 cm²	758	99.8	99.8

The sample show an increase in UPF value of 49 to 54 and increased UV blocking of 98% in both UVA and UVB with, lower concentration of TiO_2 in both plasma-untreated and -treated silk fabric samples which may be due to the even deposition of TiO_2 providing a good shielding effect compared to the control sample.

11.4.7   Air permeability

The air flow through the fabrics tested for silk and cotton fabric samples are discussed below.

From Table 11.3, the air permeability of silk and cotton are found to be high with the treated samples compared to control. The maximum air permeability value of 30.8 for silk and 21.64 for cotton were found in case of plasma untreated followed by 10 mg coating.

Table 11.3 Average results of air permeability for silk and cotton fabrics samples

Parameter	Sample types	Silk	Cotton
Air permeability $(m^3/min/m^2)$	Control	25.8	15.81
	Plasma treated	28.3	15.81
	Plasma untreated 10 mg/100 cm² TiO_2	30.8	21.64
	Plasma untreated 30 mg/100 cm² TiO_2	27.47	19.14
	Plasma treated 10 mg/100 cm² TiO_2	25.8	16.65
	Plasma treated 30 mg/100 cm² TiO_2	26.64	17.48

With respect to air resistance of the fibers, it was found the air permeability was enhanced after the coating treatment; this is possibly due to the change of the surface structure of the coated fibers. Air penetrates fibrous materials through random tiny cavities distributed on its surface; the improvement in air permeability denotes that a uniform nanoparticle coating on the whole fiber's surface may have reduced the level of extruding surface fuzz that blocks the air movement. Fabrics coated with plasma untreated titanium dioxide increased the air permeability of silk and cotton fabrics compared to plasma-treated titanium dioxide coated fabrics, confirming that titanium dioxide has better affinity and compatibility to the fibres leading to a uniform deposition, thus enhancement in air permeability occurred as a result.

11.5   Potential hazards of nanotitania for human skin

In spite of the wide ranging applications of nanomaterials including cancer treatment, there exists serious concern about the hazards on human beings from the usage or coming into physical contact with nanomaterials. Due to the sheer small size of nanomaterials, the probability of them entering the body through pores in the skin or through breathing is quite high and currently available filters are incapable of stopping them from entering the human body. There are also concerns about nanoparticles clogging the vascular system. These serious and need-to-address concerns have resulted in various governmental safety agencies working on the possible hazards of various nanomaterials. This chapter would not be complete without providing a review of various tests conducted to estimate the hazards to humans from using titania (and zinc oxide) nanoparticles.

Cloths being in constant contact with human skin can have potential health hazard if it has nano titania on its surface – which can in principle enter the human body through the skin pores. Studies have been undertaken to establish its hazards to humans.

Nakissa et al. (2010) carried out a dermal penetration study with three TiO_2 particles (uncoated submicron sized, uncoated nano-sized, and dimethicone or methicone copolymer-coated nanosized) applied 5% by weight in a sunscreen. These control formulations were topically applied to minipigs at 2 mg cream/ cm^2 skin (4 applications per day, 5 days per week, 4 weeks). Skin (multiple sites), lymph nodes, liver, spleen, and kidneys were removed, and the TiO_2 content was determined (as titanium) using inductively coupled plasma mass spectroscopy. Titanium levels in lymph nodes and liver from treated animals were not increased over the values in control animals. The epidermis from minipigs treated with sunscreens containing TiO_2 showed elevated titanium. Increased titanium was detected in abdominal and neck dermis of minipigs treated with uncoated and coated nanoscale TiO_2. Using electron microscopy-energy dispersive X-ray analysis, all three types of TiO_2 particles were found in the stratum corneum and upper follicular lumens in all treated skin samples (more particles visible with coated nanoscale TiO_2). Isolated titanium particles were also present at various locations in the dermis of animals treated with all three types of TiO_2-containing sunscreens; however, there was no pattern of distribution or pathology suggesting the particles could be the result of contamination. At most, the few isolated particles represent a tiny fraction of the total amount of applied TiO_2. These findings indicate that there is no significant penetration of TiO_2 nanoparticles through the intact normal epidermis.

Similar studies were also carried out by Lademann et al. (1999) on titania particles and the results are in agreement with those of Nakissa et al. (2010).

Toxicity (pathological changes) that could occur following exposure to titanium dioxide (UV filters) has been linked to free radical generation. In the limited number of studies, there is evidence that TiO_2 can induce the production of free radicals (likely hydroxyl radicals through oxidation) and cause adverse effects in isolated cell experiments.

There were three unpublished studies investigating photo-mutagenicity (possibly linked to free radical formation) that had been discussed by the SCCNFP (Scientific Committee on Cosmetic Products and Non-Food Products, European Commission) in their review of micronised ZnO. The overall conclusion was that micronised material (ZnO) has been found to be clastogenic, possibly aneugenic and inducing DNA damage in cultured mammalian cells *in vitro*, under the influence of UV light. A fourth article looking at photo-mutagenicity of micronised ZnO in cultured bacteria (Ames assay) was negative for mutagenic activity.

Photo-toxicity was assessed by the SCCNFP on intact skin of human volunteers, with no evidence of any reactions in 2 photo-irritation studies and 2 photosensitisation studies. Toxicokinetic assessment (*in vivo*) using human volunteers with healthy and diseased skin (psoriasis) found no evidence of an increase in systemic Zn levels after dermal application of ZnO. An *in vitro* assay using human skin (with stratum corneum stripped away) indicated 0.34% absorption of an applied ZnO dose based on recovery from receptor fluid.

Dermal penetration of TiO_2 and ZnO through the outer layer of the skin was investigated in 8 studies, with 7 of the 8 indicating an inability of TiO_2 and ZnO to reach viable cells.

It is possible that TiO_2 and ZnO in the presence of sunlight could catalyse the generation of free radicals; however, the potential toxicity associated with this event would be nullified if it did not take place in viable cells. The available evidence suggests that the likelihood of penetration beyond the stratum corneum into viable cells is very low.

11.6   Conclusion

Nano-sized titania particles have been synthesised by thermal plasma process at FCIPT. The nanoparticles that were synthesized by plasma process were found to be spherical in shape with an average size of approximately 25–30 nm. These particles were found to be highly crystalline with anatase being the predominant phase.

The tests conducted to study the self-cleaning ability of nanotitania-coated fabric samples gave positive results in respect of turmeric and coffee stains when exposed to UV source as well as sunlight. The exposure time needed for decolouration under sunlight and UV was 6 hours and 4 hours, respectively.

There is evidence from isolated cell experiments that ZnO and TiO_2 can induce free radical formation in the presence of light, and that this may damage these cells (photo-mutagenicity with ZnO). However, this would only be of concern in people using sunscreens if the ZnO and TiO_2 penetrated into viable skin cells. The weight of current evidence is that they remain on the surface of the skin and in the outer dead layer (stratum corneum) of the skin.

11.7   Acknowledgement

The author is grateful to Mr. Shillin Sangappa, CSTRI, Bengaluru for providing various analytical facilities in carrying out the tests on nanotitania coated fabrics. Thanks are also due to Ms. Anusha R for carrying out many of the tests as part of her Master's degree project work

11.8   References

Babelon P., Dequidt A.S., Mostefa-Sba H., Bourgeois S., Sibillot P., Sacilotti M., (1998). *Thin Solid Films* **322**, p. 63.

Bozzi A., Yuranova T., Kiwi J. (2005). *J. Photochem. Photobiol. A: Chem.* **172**, p. 27.

Daoud W.A. and Xin J.H. (2004). *J. Am. Ceram. Soc.,* **87**, pp. 953–955.

Daoud W.A., Leung S.K., Tung W.S., Xin J.H., Cheuk K., Qi K. (2008), *Chem. Mater.* **20**, p.1242.

Fujishima A. and Rao T.N. (1997). *Proc. Indian Acad. Sci. (Chem. Sci.),* **109**, p. 471.

Harholdt K. (2003). Carbon Fiber, Past and Future, *Industrial Fabric Products Review,* **88**(4), p.14.

Kakati M., Bora B., Sarma S., Saikia B.J., Shripathi T., Despande U., Dubey A., Ghosh G., and Das A.K. (2008). *Vacuum* **82**(8), p. 833.

Kim B.-H., Lee J.-Y., Choa Y.-H., Higuchi M., Mizutani N. (2004). *Mater. Sci. Eng.* **B107**, p. 289.

Lademann J., Weigmann H., Rickmeyer C., Barthelmes H., Schaefer H., Mueller G., Sterry W. (1999). *Skin Pharmacol Appl Skin Physiol,* **12**, pp. 247–256.

Lei Qian (2004). *J. Textile and Apparel Technology Management,* **4**(1), pp. 1–7.

Mejia M.I., Marı J.M., Restrepo G., Pulgarı C., Mielczarski E., Mielczarski J., Arroyo Y., Lavanchy J.-C., Kiwi J. (2009). *J. Appl. Catal. B: Environmental* **91**, p. 481.

Nakissa Sadrieh et al. (2010). *Toxicol Sci.* **115**(1), pp. 156–166.

Oh S.M. and Ishigaki T. (2004). *Thin Solid Films* **457**, p. 186.

Okuyama K., Shimada M., Fujimoto T., Maekawa T., Nakaso K., Seto T. (1998). *J. Aerosol Sci.,* **29**(1), pp. S907–S908.

Othmer K. (1997). Encyclopedia of Chemical Technology, 4th ed. Interscience, New York, pp. 186–349.

Nasadil P. and Benesovsky P. (2008). "Plasma in Textile Treatment", *Chem. Listy,* **102**, pp. s1486–s1489.

Quan X., Yang S., Ruan X., Zhao H. (2005). *Environ. Sci. Technol.* **39**, pp. 37–70.

Ramakrishna G., Singh A.K., Palit D.P., Ghosh H.N. (2004). *J. Phys.Chem. B***108**, pp. 47–75.

Svetlana V., Karthik A., Eunkyoung S., Behnam P. (2008). *Textile Res. J.,* **78**(6), pp. 540–548.

Szabova R., Cernakova L., Wolfova M., and Černáka M. (2009). *Acta Chimica Slovaca,* **2**(1), pp. 70–76.

Tung W.S. and Daoud W.A. (2008), *J. Colloid and Interface Sci.* **326**, p.283

Tung W.S. and Daoud W.A. (2009). *J Acta Biomaterialia,* **5**, pp. 50–56.

Uddin M.J., Cesano F., Scarano D., Bonino F., Agostini G., Spoto G., Bordiga S., Zecchina A. (2008). *J. Photochem. Photobiol. A: Chem.* **199**, p. 64.

Vijay M., Selvarajan V., Sreekumar K.P., Yu J., Liu S., Ananthapadmanabhan P.V. (2009). *Solar Energy Mater. Solar Cells* **93**, p. 1540.

Yuranova T. et al. (2006). *J. of Molecular Catalysis A: Chemical* **244**, pp. 160–167.

12

Plasma textile technology status, techno-economics, limitations and industrial usage potential

P.B. Jhala and S.K. Nema

Abstract: Although plasma is being researched since 1920s, its broader industrial exploitation started only in the 1960s and focused on low-pressure plasma technology driven by the need of the microelectronics industry for ultrahigh performance batch processing. Operating at low pressure, however, restricted its move into mainstream industry such as textiles needing in-line roll-to-roll processing. This led to the development of atmospheric pressure plasmas.

The use of atmospheric pressure cold plasma is now well established as a versatile technology for modifying the surfaces of textiles. It produces no more than a surface reaction and does not alter the bulk properties of textiles. The energetic species of cold plasma can break the covalent bonds of the fiber at its surface and etch or functionalize its surface. It is an environment-friendly process and has an edge over chemical processes. It has opened up a host of opportunities for the production of innovative finishes, branding and marketing.

The chapter discusses the technology types, advantages and applications in textiles giving the current status in the world and country. In the world, efforts have been made under various European Union Framework Programs to develop innovative plasma solutions for various textile applications keeping in view the techno-economic aspects. In India too, FCIPT has commercialized Angora wool treatment technology for spinning of 100% Angora fiber for the cottage industry. The technology has also been developed for technical textile in order to enhance the adhesion strength of coating on polyester substrate. The techno-economics for the above processes have been presented.

Key words: Cold plasma, textiles, surface modification, techno-economics

12.1   Introduction

The 21st century is a green and environment protection century. As people's living standard is rising, choice of environmentally sound, healthy and safe clothing has become a characteristic of future consumption. The green process and environment-friendly product development has been the focus of market competitiveness.

The plasma is regarded as a dry, green process compared with traditional textile wet processing; and its potential for exploitation in the textile and clothing industries has been under active study, now for more than three decades, in a number of research and development centers worldwide.

Technologies developed initially were based on the properties of low pressure plasmas, and it cannot be used for production lines operating at room temperature with machines processing wide width fabric at high speed. Despite all the benefits, plasma processing failed to make an impact in the textile sector mainly due to this particular constraint which is incompatible with industrial mass production. This led to the development of atmospheric pressure plasmas.

Plasma treatments at atmospheric pressure enable to combine it with in-line spraying or aqueous aerosol treatments. In contrast, the low-pressure plasma treatments are carried out in batch process. Further, compare to low pressure plasma, equipments used to generate atmospheric plasma are less expensive and their maintenance cost is also low.

12.2   Technology status

In the textile field, significant research work has been going on since the early 1980s in many laboratories across the world dealing with low-pressure plasma treatments of a variety of fibrous materials showing very promising results regarding the improvements in various functional properties in plasma-treated textiles (Hocker, 2002; Shishoo, 2007; Sparavigna, 2008; Buyle, 2009).

Lately, many EU-financed projects within the 4th, 5th and 6th Framework Programme such as Plasmatex (1997–2000), Plasmatech (200–2005), Acteco (2005–2009) have had the objective of developing and demonstrating the feasibility of plasma-based industrial processes to meet the needs of the textile industry and offer tools for product development and innovation. At the end of Plasmatex Program in 2001, three different proto-type atmospheric pressure plasma processing systems (APPS) were built by Plasma Ireland and were supplied to consortium's industrial partners for further research. The first in Sweden was used to improve adhesion of coatings applied on polymer surface, automobile textiles and hygiene paper, the second in Germany was used in elimination of felting in wool and the third in Britain was used in textile printing. Also, as a part of an EU Project Leapfrog CA, an extensive literature survey was carried out in 2005 in the area of plasmas and plasma-induced functionality of textiles. This survey has shown that a very large number of patents and articles were available in the field of plasma treatment of fibres, polymers, fabrics, nonwovens, coated fabrics, filter media, composites, etc. The plasma treatment was mainly used for enhancing their functions and performances. The literature survey clearly points out the potential of plasma treatments for various applications which include:

(i)   Anti-felting and shrink-resistance of woolen fabrics
(ii) Enhancement of hydrophilicity for improving wetting and dyeing

(iii) Enhancement of hydrophilicity for improving adhesive bonding
(iv) Increasing hydrophobicity and oil repellency in textiles
(v) Facilitating the removal of sizing agents
(vi) Scouring of cotton, viscose, polyester and nylon fabrics
(vii) Anti-bacterial fabrics by silver particle deposition using plasma process
(viii) Durable antistatic properties using PU-resin and plasma processing
(ix) Electrical conductivity of textile yarns by plasma-assisted deposition

This shows that plasma technology has a strong potential to offer new functionalities in textiles. In recent years, considerable efforts have been made by many plasma technology suppliers to develop both low-pressure and atmospheric-pressure plasma machinery and processes designed for treatment of textiles and nonwovens to impart a broad range of functionalities at industrial level. The standard and custom-designed plasma systems are being offered by following companies:

(i) Low-pressure plasma systems for batch treatment by Europlasma (Belgium), P_2i (UK) and Mascioni (Italy)
(ii) Atmospheric pressure plasma system for in-line continuous treatment by Dow Corning Plasma Solutions (Ireland), Ahlbrandt (Germany), AcXys (France), APJeT (USA) and Tri-Star (USA), InspirOn Engineering Pvt. Ltd. (India).

For the first time, in an International Textile Machinery Exhibition (ITMA) 2007 at Munich, four manufacturers showcased atmospheric pressure and low-pressure plasma processing systems for commercial applications in textiles namely Arioli (Italy), Grinp (Italy), HTP Unitex (Italy) and Mageba (Germany).

In India too, the textile educational and research institutes have been carrying out research work in plasma processing of textile at the laboratory scale contributing to the knowledge base and technology development. Recently, the indigenous development of an atmospheric pressure plasma system for Angora wool (APPAW) treatment to improve cohesion among wool fibers by Institute for Plasma Research and National Institute of Design has generated a lot of interest amongst the Indian textile technologists and researchers for using plasma based technologies in textiles (Jhala et al., 2009).

12.3   Industrial use

As of today, industrial application of plasma in textiles is still very limited because some hurdles exist at various levels. In spite of that, within the textile world, plasma is already being integrated by first adopters for niche applications and it seems fair to state that wider application is close to breakthrough (Buyle, 2009; Home, 2009; Benisek, 2009).

12.3.1   Stain and water repellent cotton

Avondale Mills, Georgia, USA, was one of the first textile companies that made use of the plasma technology called TexJet developed by APJet Inc., USA in April 2006 to treat cotton and cotton/polyester fabric to provide stain resistance and water repellency (Isaac, 2006).

TexJet incorporating eight electrodes 10 kW each uses Atmospheric Pressure Plasma Jet (APPJ) Technology with recycling of helium gas used in the process. APJeT uses a downstream operation by adding the precursors not into the plasma but into what is called the afterglow. Instead of precursor reacting inside of the plasma, the reaction takes place between atomic and metastable species generated by plasma and the associated chemical precursor gas in the afterglow. As shown in Fig. 12.1, fabric enters the machine through a pre-treatment processor which removes excess moisture and entrapped air. It can treat fabric at speed of 35 m/min. Fabric may be treated on one or both sides. A unique textile fabric has been made as shown in Fig. 12.2 which behaves hydrophobic on its one side and hydrophilic on the other side. Morrison Textile Machinery, USA, is now the sole licensee for manufacturing APJeT plasma equipment.

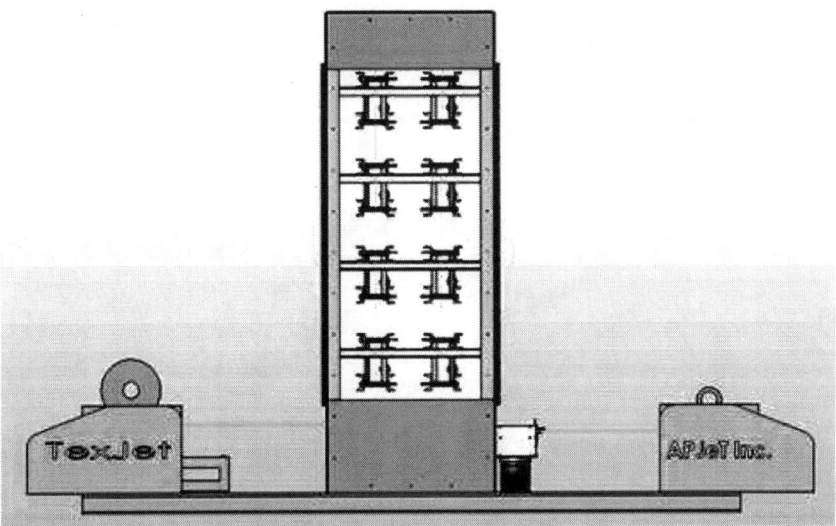

Figure 12.1 TexJet plasma unit for fabric [*Source:* Issac, 2006]

Figure12.2 Dual functionality fabric [*Source:* Issac, 2006]

12.3.2   Functionally coated fabrics

Dow Corning Plasma Solutions (DCPS) licensed its Atmospheric Pressure Plasma Liquid Deposition (APPLD) for commercial coating on its SE-1000 AP4 System to Freudenberg Forschungsdienste KG, Germany in 2007 for pushing in European market. The company provided product development and toll manufacturing using APPLD technology.

DCPS APPLD SE-1000 AP4 System with 4 active legs having helium glow discharge can treat 1.2 m wide fabric at line speed up to 70 m/min as shown in Fig. 12.3. It enables the use of a wide range of liquid precursors not being restricted to gas or high volatility liquid precursors. The precursor is introduced as an atomized liquid directly into the glow discharge by ultrasonic nebulizer. The plasma does not damage the precursor molecules, which are polymerized into a coating with their functionality intact. It allows users to engineer the surface of virtually any substrate by applying thin coatings of various materials that chemically bond to the underlying substrate as shown in scanning electron micrograph in Fig. 12.4. Coatings can have a wide range of characteristics, such as waterproofing, low-friction stickiness, adhesion promotion or antimicrobial properties (Hynes, 2006; Anon, 2007).

Figure 12.3 DCPS APPLD SE-1000 AP4 [*Source:* Hynes, 2006]

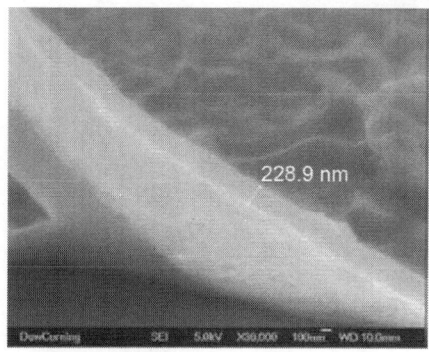

Figure 12.4 APPLD Fluorocarbon Coating [*Source:* Anon, 2007]

12.3.3   Shrink-resist woolens

Softal, Hamburg, Germany built Sortex, the largest atmospheric pressure plasma treatment machine in the world with 100 kW plasma output distributed onto 100 specially designed electrodes, which was installed at Richter F & A Company, Stadtallendorf, Germany in March 2008 for continuous shrink-resist treatment of wool tops shown in Fig. 12.5.

Figure 12.5 Sortex Corona Station [*Source:* Benisek, 2009]

Plasma can replace the chlorination pre-treatment with a dry effluent step followed by a special continuous polymer treatment free of adsorbable halogen compounds (AOX).

Richter F & A had been industrially processing wool tops in Sortex Corona Station (Benisek, 2009) and SEM of untreated and plasma-treated wool fibre is shown in Fig. 12.6(a) and (b), respectively. Richter International, Canada introduced the product into market with the name PlasmaWool, the first ecologically friendly collection of superfine merino wool socks using water and chlorine-free manufacturing.

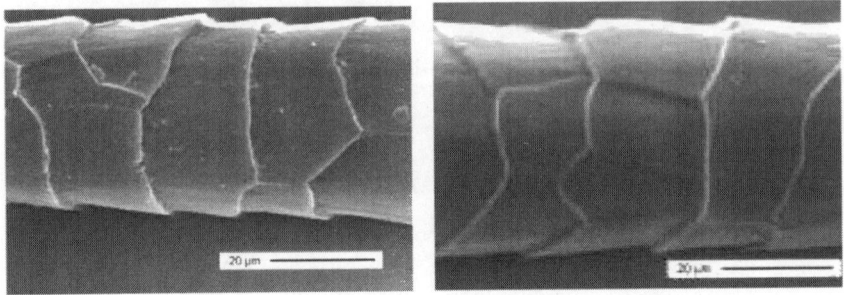

Figure 12.6 SEM of (a) untreated and (b) plasma-treated fiber
[*Source:* Benisek, 2009]

12.3.4   Deposition of coating

AS Coating Star, manufactured and commercialized by Ahlbrandt System GmbH, Germany, combines classical corona treatment with an aerosol precursor application (Tom, 2013). The working mechanism of the AS Coating Star system is shown in Fig. 12.7. The plasma is made in a box with temperature-controlled housing enabling the creation of specific microclimate in the box. The aerosol can be injected between the two corona electrodes. The intimate contact between aerosol and corona discharge will lead to a polymerization/cross-linking of the chemical product and finally a deposition of a coating onto the plasma-activated material surface.

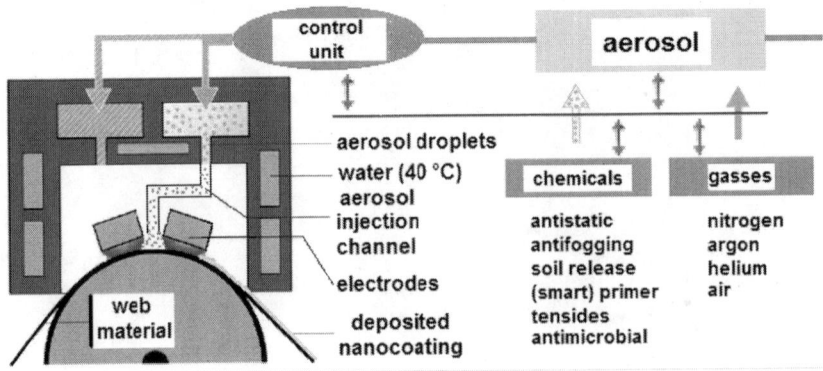

Figure 12.7 AS Coating Star: Working Mechanism [*Source:* Tom, 2013]

AS Coating Star technology can be considered as a supporting step for existing production processes like printing, coating, laminating, gluing, dyeing, metallization.

12.3.5   Fading of denim/jeans

GFK, Spain, introduced the latest technology called G2 as shown in Fig. 12.8 for the textile industry in April 2009 to lessen the environment impact of denim processing. Rather than relying on the traditional combination of water and chemicals to produce various shades of denim, G2 uses a process that relies on air. Atmospheric air is converted into a blend of active oxygen and ozone in plasma environment and pressure of 6 bars is maintained in the machine. The active species (O, O_3) formed in plasma interacts with denim surface and does color fading as shown in Fig. 12.9. It has garment processing capacity of 50–60 kg/cycle (Silla, 2010).

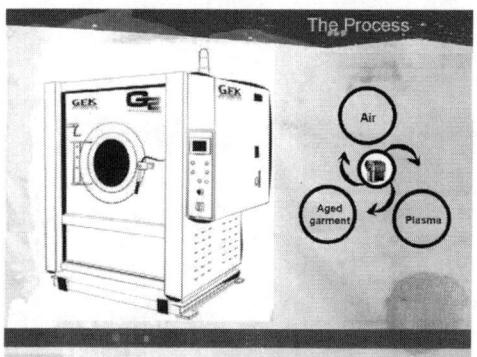

Figure 12.8 GFK G2 Machine [*Source:* Silla, 2013]

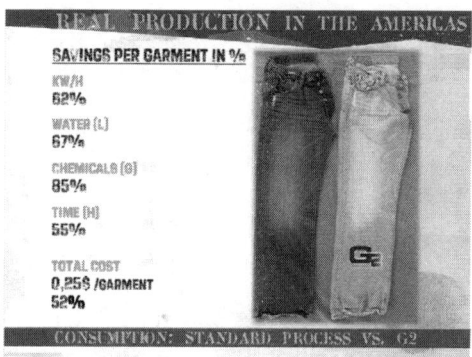

Figure 12.9 G2 Faded Jeans [*Source:* Silla, 2013]

In addition to eliminating the use of water and chemicals, the G2 rids the finishing process of toxic emissions and dumping and reduces overall energy usage. In India, Action Laundry, Ahmedabad, is using G2 Machine for jeans fading.

12.3.6   Shedding-free Angora apparels

FCIPT, Institute for Plasma Research and National Institute of Design carried out a pioneering research work by way of developing an innovative atmospheric pressure plasma processing system for Angora wool (APPAW). It is the first large size atmospheric pressure plasma system developed in the country for textile processing.

APPAW generates plasma at atmospheric pressure using air as plasma-forming gas. Angora fibre surface is modified using dielectric barrier discharge which is shown in SEM image illustrated in Fig. 12.10 (a) and (b). Plasma treatment assists in increasing the coefficient of friction and cohesion among the fibers and assists in spinning of 100% Angora yarn without any difficulties such as static, shedding, fibrosity (Jhala, 2008; Jhala et al., 2009).

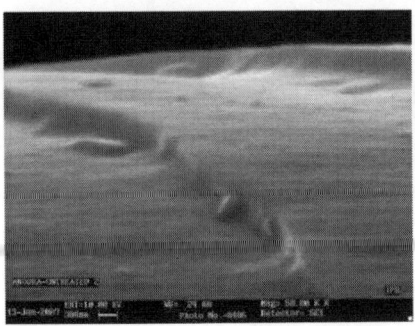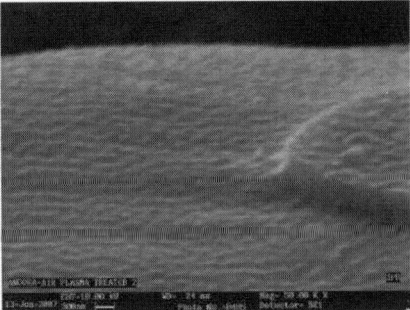

Figure 12.10 SEM image of (a) untreated Angora fibre and (b) plasma-treated Angora fibre [*Source:* Jhala, 2009]

The prototype system was successfully established in the Angora Cottage Industry at Kullu in 2009 for spinning of 100% Angora yarn and making newer products. This activity was financially supported by Department of Science and Technology, New Delhi. The technology has been licensed to a leading textile machinery manufacturer, InspirOn Engineering Private Ltd., Ahmedabad, for commercialization. The company has manufactured and supplied two industrial scale plants for plasma processing of Angora fibres, as shown in Fig. 12.11; one has been commissioned at Ranichauri (Uttarakhand) and the other one at Kullu (Himachal Pradesh) in 2012. The third plant is under fabrication and will be installed at Chungthang (Sikkim) during May 2014.

Figure 12.11 APPAW Industrial Scale Machine commissioned at HIFEED, Ranichauri, India

12.3.7   Adhesion improvement

FCIPT, Institute for Plasma Research in collaboration with Man Made Textiles Research Association (MANTRA), Surat, has also developed atmospheric pressure plasma system for technical textile (APPTT) for pre-treatment of polyester fabric in order to increase the adhesion between substrate and coating/ lamination. APPTT is 0.5 m wide machine as shown in Fig. 12.12. It has been installed at the Centre of Excellence in Technical Textiles, MANTRA, Surat. The plasma-treated samples have shown significant (30–40%) improvement in adhesion strength of coating. This adhesion enhancement is crucial for attaining the highest level of operational performance in the end use.

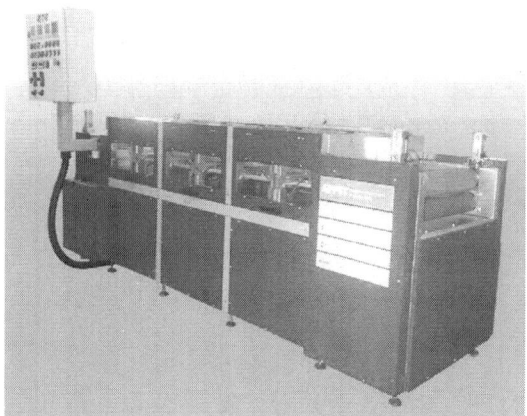

Figure 12.12 APPTT Machine commissioned at MANTRA, Surat, India

12.4   Techno-economics

Until recently, plasmas for industrial processing were only available under reduced pressure. This limited manufacturing because of the high cost of vacuum equipment and batch processing. Plasma processing is now available at atmospheric pressure resulting in the ability to impart broad range of functionality in textiles in a continuous full width process. The techno-economics of some systems for different applications is given below.

12.4.1   World status

At the Institute of Textile Technology, NCSU, Raleigh, USA, David Wade Tyner (2007) carried out research to evaluate water repellant fluoropolymer finish on textiles by Dow Corning Atmospheric Pressure Plasma Liquid Deposition (APPLD) technology and Conventional Pad-Dry-Cure method in terms of its repellency, durability and cost in 2007. The core objective of this project was to determine if the APPLD process could be a viable replacement for the conventional pad-dry-cure method at the state of technology for the cotton, nylon, polyester and polyester/cotton fabrics tested in the research.

It showed that all fabrics treated by the atmospheric plasma process exhibited equal levels of water repellency as a commercial product finished by the conventional pad-dry-cure method. Only the cotton and polyester/cotton fabrics treated by atmospheric plasma showed a decrease in repellency after multiple wash cycles when compared to the conventional finishes under identical washings.

It has been determined that the atmospheric plasma cost associated with the fabric used in the research was US$ 1.13 per square yard as compared to the conventional method cost US$ 0.20. Although there is a large discrepancy, a theoretical cost projection for the APPLD process in a fully engineered industrial scenario was estimated at US$ 0.15 per square yard. It was suggested that APJeT should be investigated as they also have a continuous full width atmospheric plasma machine having ability to run faster due to higher plasma density and can recycle the helium gas used in the process leading to a projected cost per square yard of fabric US$ 0.10.

As a part of Acteco Project, Tony Herbert (2009) of Dow Corning Plasma Solutions carried out a comparative study of Atmospheric Pressure Plasma (APP) Functionalisation, which is given in Table 12.1

Table 12.1 APP technologies comparisons [*Source:* Herbert, 2009]

S. no.	Atmospheric Pressure Plasma (APP) Type	System specifications	Applications	Cost, €	
				Capital	Running
1.	Standard Corona	Width: 1–2 m Speed: 100–1000 m/min	Printing, gluing, laminating, coating	17–110 k	0.0001 per m^2
2.	Controlled Atmospheric Corona	Width: 1–2 m Speed: ≤ 300 m/min	Flexible packaging, plastic Film	300–370 k	0.002 per m^2
3.	Liquid Deposition Corona	Width: 0.75 m Speed: 150 m/min	Printing, gluing, laminating, coating	60 k	0.005 per m^2
4.	APP Liquid Deposition	Width: 2 m Speed: 30–70 m/min	Textiles, plastics, medicals, paper	800 k	0.02 per m^2

The study has concluded that standard and controlled atmospheric corona have been matured with well-defined capability, market and steady growth. While APP liquid deposition is an emerging technology with relatively high cost which must be justified by high added value and unlike traditional market, it must seek new high value markets such as biotechnology, electronics and medical.

12.4.2   Indian status

The atmospheric pressure plasma processing system (APPAW) developed by FCIPT, Institute for Plasma Research, in India for Angora wool uses air as plasma-forming gas to enhance friction and cohesion between fibres to facilitate the spinning and weaving without shedding and fibrosity. APPAW Industrial Scale Plants are already operating at Ranichauri and Kullu in India. It is a cost effective eco-friendly process as the plasma treatment cost is US$ 0.8 per kg of Angora wool which is much less if it is compared with the cost of Angora fibers that is US$ 32 per kg. In the above process electricity and air are the only consumables. Similarly, the system developed for improving adhesion between polyester surface and laminating layer, the plasma treatment cost is around US$ 0.03 per square meter. It has been observed that by minimizing the use of helium or other costly gases to produce plasma in commercial plasma systems, it is possible to make plasma technology economically viable for textile industry.

12.5   Other applications

There is significant potential for profitable applications of plasma technology in textiles. The scope is very wide as it is evident from some of the current research work going on worldwide.

12.5.1   Functional sports and workwear

Switzerland-based knitwear manufacturer Eschler AG and Austria-based textile finisher Grabher Group have partnered to develop new functional sports and work wear fabrics featuring environmentally friendly Plasma Technology by Eschler. The technology uses minimal water and chemicals, which is energy efficient and free of fluorocarbon compounds. Eschler was to offer the first apparel products at retail in summer 2012.

12.5.2   Functional polymers and fabrics

The plasma team of SMITA, Textile Department, IIT, Delhi, has successfully carried out textile functionalisation such as imparting hydrophobic or hydrophilic properties in textiles by atmospheric pressure cold plasma. Also, Bombay Textile Research Association, Mumbai, with GRINP Make Atmospheric Pressure Plasma Laboratory Machine has imparted functional finishes to cotton, polyester and blended fabrics by plasma treatment. The efforts are being made in taking these processes to industrial use.

12.5.3   Itch-free woolen apparels

Wool Research Association, Mumbai, has developed itch-free woolens to be worn next to the skin by improvement of surface properties by plasma/enzyme treatment followed by polymer/softener treatment (Udakhe, 2011). User trial has shown that the treated fabric apparels are totally itch proof, and they can be readily acceptable as next to skin garments. There is a considerable interest from the industry in this eco-friendly technology. It is expected that it will soon be taken to the commercial level.

12.5.4   Banana fibre apparels

Banana fibre being biodegradable and naturally occurring has great potential for use in textile, paper and other industries. Also, due to its excellent intrinsic properties, banana fibre apparels have good demand in international market. Institute for Plasma Research along with textile mill and machinery manufacturer are jointly working on development of technology using plasma and other processes for modification of Banana fibres in environment-friendly and cost-effective way for making it spinnable to finer counts to make apparels for niche market.

12.6   Challenges and opportunities

Plasma industrial application is still very limited because important hurdles exist at various levels such as cleanliness of substrates, investments, off-line treatment and scale up. Key factors are also the three-dimensional structure and the large surface area. These two intrinsic textile properties are posing major challenges for plasma treatment and are, at the moment, still limiting the maximum through puts that can be realized (Buyle, 2009).

The major advantage of plasma treatments is that they offer a dry alternative method of processing textile materials. Aqueous-based processes generate a pollution load in the wastewater from the process, leading to increased costs for effluent treatments and disposal. In addition the removal of water from textile materials is energy intensive. Thus plasma treatments offer economic and ecological advantages.

12.7   Conclusions

As compared to conventional wet processing, plasma has the crucial advantage of reduced usage of chemicals, water and energy. It offers dyers, printers and finishers opportunities for exploring innovative approaches to novel effects; in some cases it is impossible to obtain through traditional process without changing the key textile properties.

This potential has driven the various research laboratories and machinery manufacturers to extensive investigation. Integrating plasma processes at different stages of the production process have been investigated for a whole range of different materials and applications.

12.8   References

Anon (2007). 'Freudenberg Licenses Innovative Plasma Technology from Dow Corning', *Finishing Today*, **83**(11), p. 10.

Lado B. (2009). 'Industrial Use of Atmospheric Plasma', *International Dyer*.

Buyle G. (2009). 'Nanoscale Finishing of Textiles via Plasma Treatment', *Materials Technology*, **24**(1), pp. 46–51.

Tony H. (2005). 'Open Questions in Atmospheric Plasma Functionalisation', EC ACTECO Public Symposium, 1 June 2005, Dow Corning.

Hocker H. (2002). 'Plasma Treatment of Textile Fibres', *Pure and Applied Chemistry,* **74**(3), pp. 423–427.

Ian H. (2009). 'Plasma Processes for Novel Treatments', *International Dyer.*

Alan H. (2006). 'Atmospheric-Pressure Plasma Liquid Deposition: A New Route to High Performance Textiles from Dow Corning Plasma Solutions', Plasma Processing Updates, Gandhinagar, FCIPT, July, 8–12.

Mac I. (2006). 'APPJ Technology', *ATA Journal for Asia on Textile Apparel,* April.

Jhala P.B. (2008). 'Plasma Glow Brightens Angora Cottage Industry Future', Ahmedabad, National Institute of Design.

Jhala P.B., Nema S.K. and Mukherjee S. (2009). 'Innovative Plasma System to Improve Angora Fibre', *The Indian Textile Journal,* **119**(4), pp. 35–39.

Jhala P.B. (2009). 'Emerging Role of Plasma Technology in Textile Treatments', National Seminar on Textiles: Prospects and Growth Beyond 2020, The Institution of Engineers, Surat.

Shishoo R. (2007). Plasma Technologies for Textiles, Cambridge, Woodhead Publishing Ltd.

Enrique S. (2010). 'Future Trends & Technologies in Jeans Treatment', International Conference on Denims – The New Face of Indian Textile & Apparel Industry, The Textile Association (India), Ahmedabad, October 29–30.

Sparavigna A. (2008). Plasma Treatment Advantages for Textiles, Physics.

Tom Van Hove (2013). 'Depositing Micro- and Nano-sized Coating by means of Aerosol assisted Large Area Cold Atmospheric Plasma Technology. Available from: http://www.docstock.com/search/tom-a-dot-van-de-goor [accessed 16 January 2014]

David T. (2007). 'Evaluation of Repellant Finishes Applied by Atmospheric Plasma' M.Sc. Thesis, NCSU, Raleigh.

Jayant U. (2011). 'Development of Itch-free Woolens by Plasma and Enzyme Treatment', National Seminar cum Workshop on Recent R & D Initiatives and Development Schemes of Wool & Woolens, WRA, Mumbai, May 2011.

Index